Vertebrate Locomotion

SYMPOSIA OF THE ZOOLOGICAL SOCIETY OF LONDON
NUMBER 48

Vertebrate Locomotion

*(The Proceedings of a Symposium held at
The Zoological Society of London
on 27 and 28 March 1980)*

Edited by

M. H. DAY

*Department of Anatomy, St Thomas's Hospital Medical School,
London, England*

Published for

THE ZOOLOGICAL SOCIETY OF LONDON

BY

ACADEMIC PRESS

1981

ACADEMIC PRESS INC. (LONDON) LTD
24/28 Oval Road, London NW1 7DX

United States Edition published by
ACADEMIC PRESS INC.
111 Fifth Avenue, New York, New York, 10003

British Library Cataloguing in Publication Data

Vertebrate locomotion. — (Symposia of the
    Zoological Society of London, ISSN 0084-5612;
    no. 48)
    1. Vertebrates — Congresses
    I. Day, M. H. II. Series
    596'.01'8   QL 605.A3
    ISBN 0-12-613348-4

    LCCCN 81-66401

Printed in Great Britain at the Alden Press
Oxford London and Northampton

Contributors

AIELLO, L., *Department of Anthropology, University College London, Gower Street, London WC1E 6BT, England* (p. 331)

ALEXANDER, R. MCNEILL, *Department of Pure and Applied Zoology, University of Leeds, Baines Wing, Leeds LS2 9JT, England* (p. 269)

ARMSTRONG, R. B., *Department of Physiology, Oral Roberts University, 7777 South Lewis, Tulsa, Oklahoma 74171, USA* (p. 289)

BATTY, R. S., *Scottish Marine Biological Association, Dunstaffnage Marine Research Laboratory, P.O. Box 3, Oban, Argyll, Scotland* (p. 53)

BLAKE, R. W., *Department of Zoology, University of British Columbia, 2354–6270 University Boulevard, Vancouver, B.C. V6T 2A9, Canada* (p. 29)

FLEAGLE, J. G., *Department of Anatomical Sciences, Health Sciences Center, State University of New York at Stony Brook, New York 11794, USA* (p. 359)

GOLDSPINK, G., *Department of Zoology, University of Hull, Hull HU6 7RX, England* (p. 219)

JENKINS, F. A., JR, *Museum of Comparative Zoology, Harvard University, Cambridge, Massachusetts 02138, USA* (p. 429)

JOHNSTON, I. A., *Department of Physiology, Bute Medical Buildings, The University, St Andrews, Fife, Scotland* (p. 71)

JUNGERS, W. L., *Department of Anatomical Sciences, Health Sciences Center, State University of New York at Stony Brook, New York 11794, USA* (p. 359)

LANYON, L. E., *Department of Veterinary Anatomy, University of Bristol, Park Row, Bristol BS1 5LS, England and School of Veterinary Medicine, Tufts University, Boston, USA* (p. 305)

MARTIN, R. D., *Department of Anthropology, University College London, Gower Street, London WC1E 6BT, England* (p. 377)

NORBERG, U., *Department of Zoology, University of Göteborg, Box 250 59, 400 31 Göteborg, Sweden* (p. 173)

PYE, J. D., *Department of Zoology, Queen Mary College, Mile End Road, London E1 4NS, England* (p. 199)

RAYNER, J. M. V., *Department of Zoology, University of Bristol, Woodland Road, Bristol BS8 1UG, England* (p. 137)

REWCASTLE, S. C., *39 Christmaspie Avenue, Wanborough, Guildford, Surrey GU3 2EQ, England* (p. 239)

ROBERTS, B. L., *Marine Biological Association of the United Kingdom, The Laboratory, Citadel Hill, Plymouth PL1 2PB, England* (p. 115)

ROLLINSON, J., *Department of Anthropology, University College London, Gower Street, London WC1E 6BT, England* (p. 377)

STERN, J. T., JR., *Department of Anatomical Sciences, Health Sciences Center, State University of New York at Stony Brook, New York 11794, USA* (p. 359)

SUSMAN, R. L., *Department of Anatomical Sciences, Health Sciences Center, State University of New York at Stony Brook, New York 11794, USA* (p. 359)

VANGOR, A. K., *Department of Anatomy, University of Pennsylvania, School of Veterinary Medicine H1, Philadelphia, Pa. 19104, USA* (p. 359)

VIDELER, J. J., *Department of Zoology, Groningen University, P.O. Box 14, 9750 AA Haren, The Netherlands* (p. 1)

WELLS, J. P., *West Virginia School of Osteopathic Medicine, Lewisburg, West Virginia 24901, USA* (p. 359)

Organizer and Chairmen of Sessions

ORGANIZER

M. H. DAY, on behalf of the Zoological Society of London

CHAIRMEN OF SESSIONS

R. MCNEILL ALEXANDER, *Department of Pure and Applied Zoology, University of Leeds, Baines Wing, Leeds LS2 9JT, England*

M. H. DAY, *Department of Anatomy, St Thomas's Hospital Medical School, London SE1 7EH, England*

JAMES LIGHTHILL, FRS, *Provost, University College London, Gower Street, London WC1E 6BT, England*

C. S. WARDLE, *Department of Agriculture and Fisheries for Scotland, Marine Laboratories, P.O. Box 101, Victoria Road, Torry, Aberdeen AB9 8DB, Scotland*

D E
MOTV ANIMALIVM
IO. ALPHONSI BORELLI
NEAPOLITANI

MATHESEOS PROFESSORIS

Opus Poſthumum.

PARS PRIMA.

R O M AE,
Ex Typographia Angeli Bernabò. M. DC. LXXX.
SVPERIORVM PERMISSV.

Preface

The symposium that gave rise to this volume was held in London on 27 and 28 March 1980 under the joint sponsorship of The Zoological Society of London and The Anatomical Society of Great Britain and Ireland. It brought together an international group of scientists all of whom have made outstanding contributions to their special fields. The choice of title was made by agreement between the two societies, since it seemed to us that the time was ripe for a review of vertebrate locomotion, an aspect of whole animal biology of immense evolutionary significance. The ability to move in an effective way extends the options and evolutionary strategies open to a creature to such an extent that it may well be of prime importance in terms of its survival and evolutionary success.

Research into the ways in which animals move has a long history; indeed the symposium coincided, not entirely by chance, with the tercentenary of the first edition of Giovanni Borelli's book *De Motu Animalium*, perhaps the first book to consider animal locomotion from a quantitative viewpoint. Dr. Nigel James writes:

Essentially we know of Borelli only through his writings. He made the first serious and systematic attempt to formulate biological principles in mathematical terms. His approach to the mechanics of locomotion, both in animals and in man, was highly original. He dealt with items which would not seem out of place today. For example, his studies on mechanics and locomotion relied heavily upon geometry and even the use of calculus. Borelli clearly appreciated the nature and variability of muscles in the body and certainly understood clearly the difference between multipennate and unipennate muscles. He appreciated that the principles of physics regarding levers were applicable to the body when studying the mechanics of locomotion. His analyses were often quantitative as well as descriptive, particularly in his studies based on statics. He was concerned with the position of the centre of gravity of the human body and its relation to movement during locomotion. His work has, sadly, been neglected, yet it contains numerous other items of interest. His book was published prior to Newton's *Principia*, which contains the first published account of what we now call calculus, yet Borelli was using the methods of differential calculus in studying locomotion. Some of his differential equations, if modified to take account of the nomenclature he used, would not seem markedly out of place even in a modern journal. Borelli's book, apart from its contributions to the locomotor system, could also throw important light on whether or not as a result of his geographical connections, he learnt about calculus from Liebnitz, a mathematician who many people regard as the discoverer of calculus.

In approaching as broad a topic as locomotion, it seemed sensible to limit our contributors as little as possible within the Phylum Vertebrata and then to consider the media of locomotion as the key to bringing together papers into sessions that would appeal to zoologists and anatomists tackling similar problems, as well as to students of biology in general.

Water, air, the ground surface and the trees cover most of the vertebrate opportunities for movement, other than fossorial forms, and this lead to sessions on swimming, flight, walking and running, and primate arborealism. The success of this approach became apparent at the meeting and so it was decided that this volume should be arranged in the same way.

Symposium organizers are always heavily dependent on session chairmen and here I have been most fortunate. Dr C. S. Wardle (Swimming), Professor Sir James Lighthill (Flight) and Professor R. McNeill Alexander (Walking and Running) all agreed readily to take on the task. Each provided an array of talent in his field, and chaired his session with distinction. The chairmen subsequently read and, in effect, refereed, the papers in their sessions and the high standard of the contributions reflects their diligence. As a chairman myself of the final session (Primate Arborealism) I know at first hand how much work they put in, and I am indebted to them all.

I must also thank the participants for their contributions, for the promptness of the delivery of their manuscripts and for making my task so much easier than it might have been.

The smoothness of the arrangements for the symposium at The Zoological Society Meeting House was due to the skill and experience of Dr Gwynne Vevers, and the editorial assistance of Miss Unity McDonnell has been invaluable. I am also indebted to Mr Dave Cummings who has seen the volume through the press with patience, expertise and tact.

Finally, it is my hope that those who came to listen, and indeed filled the hall at each session, will find in this volume an accurate and useful account of the researches so ably presented by those who took part.

London, May 1981   M. H. DAY

Contents

Swimming Movements, Body Structure and Propulsion in Cod *Gadus morhua*

J. J. VIDELER

Mechanics of Drag-based Mechanisms of Propulsion in Aquatic Vertebrates

R. W. BLAKE

Locomotion of Plaice Larvae

R. S. BATTY

Structure and Function of Fish Muscles

IAN A. JOHNSTON

The Organization of the Nervous System of Fishes in Relation to Locomotion

B. L. ROBERTS

Flight Adaptations in Vertebrates

J. M. V. RAYNER

Flight, Morphology and the Ecological Niche in some Birds and Bats

ULLA M. NORBERG

Echolocation for Flight Guidance and a Radar Technique Applicable to Flight Analysis
J. D. PYE

The Use of Muscles During Flying, Swimming, and Running from the Point of View of Energy Saving
G. GOLDSPINK

Stance and Gait in Tetrapods: An Evolutionary Scenario
STEPHEN C. REWCASTLE

The Gaits of Tetrapods: Adaptations for Stability and Economy

R. McNEILL ALEXANDER

Recruitment of Muscles and Fibres within Muscles in Running Animals

R. B. ARMSTRONG

Locomotor Loading and Functional Adaptation in Limb Bones

L. E. LANYON

The Allometry of Primate Body Proportions
LESLIE C. AIELLO

Climbing: A Biomechanical Link with Brachiation and with Bipedalism
JOHN G. FLEAGLE, JACK T. STERN, JR, WILLIAM L. JUNGERS, RANDALL L. SUSMAN, ANDREA K. VANGOR and JAMES P. WELLS

Comparative Aspects of Primate Locomotion, with Special Reference to Arboreal Cercopithecines
J. ROLLINSON and R. D. MARTIN

Wrist Rotation in Primates: A Critical Adaptation for Brachiators

FARISH A. JENKINS, JR

Symp. zool. Soc. Lond. (1981) No. 48, 1—27

Swimming Movements, Body Structure and Propulsion in Cod *Gadus morhua*

J. J. VIDELER

Department of Zoology, Groningen University, P.O. Box 14,
9750 AA Haren, The Netherlands

SYNOPSIS

Cod were trained to swim up and down a 14 m long tank. High speed cine film recordings at 100 or 200 frames s^{-1} of straight forward swimming were used to analyse the kinematics of various swimming styles. Cod (ranging in size between 0.25 and 0.50 m) frequently use steady swimming at uniform velocities between $1.4 \, L \, s^{-1}$ and $2 \, L \, s^{-1}$ (where L = body length). Swimming at average velocities slower than $1.4 \, L \, s^{-1}$ usually shows slight acceleration or deceleration. Steady swimming faster than $2 \, L \, s^{-1}$ occasionally occurs during short periods. For prolonged swimming at high velocities cod use a kick-and-glide swimming style. Kinematics of gliding and braking are described. The shape of the body and fins of cod is described and compared with mackerel and eel. Anatomical structures and some of their apparent functions in locomotion are discussed.

Morphological and kinematical arguments are used to choose the right hydrodynamic model for propulsive force calculations in cod. Lighthill's large-amplitude elongated-body theory is used to calculate cost of transport as work per metre distance covered for steady swimming and kick-and-glide swimming at various speeds. At average swimming velocities faster than $2.5 \, L \, s^{-1}$ kick-and-glide swimming is cheaper than steady swimming. Results show that drag on a steady swimming fish is 3.3 times greater than the drag on a gliding fish at the same average speed and that drag on a gliding fish is twice the turbulent drag on a rigid streamlined body.

INTRODUCTION

Locomotion of a fish like cod is the result of complex interaction between varying movement patterns and forces resulting from the medium, water. Movements in fish are accomplished by two muscle—bone systems:

(1) The main power source is the metameric lateral musculature on either side of the vertical septum which is supported by the vertebral column. Blocks of short muscle fibres (myotomes) are connected to one another by sheets of collagenous fibres (the

myosepts). Myotomes are connected to the skull, the vertical septum, the skin and the tail fin (Wardle & Videler, 1980). Trains of muscle contractions run down the myotomes, alternately on the right and left side of the body, causing waves of curvature from head to tail (Gray, 1933).

(2) Paired and unpaired fins consist of fin rays articulating with supporting skeletal elements, and moved by small intrinsic muscles (Videler, 1975, 1977; Geerlink & Videler, 1974; Geerlink, 1979). This system determines to a great extent velocity and direction of a fish.

A variety of swimming styles can be observed even if the observations are confined to movements in one direction of specimens of one size of a single species. A 0.30 m cod performs sustained steady swimming at speeds up to two body lengths per second (Wardle, 1977). Above this speed cod can swim steadily for a few minutes but then usually start to use a cyclic kick-and-glide swimming style. During the kick phase of one cycle the fish accelerates using two tail-strokes followed by deceleration during the glide phase. The fish can accelerate gradually to reach higher steady speed levels or can change to kick-and-glide swimming. During fast swimming and gliding the paired fins are kept pressed against the body and the unpaired dorsal and ventral fins are withdrawn. These fins are actively used for stabilization during slow forward swimming and for generating drag during braking.

Understanding of the relations between movements and body structure requires knowledge of the nature of the interaction between fish and water. Precise knowledge of this kind is still lacking but estimates can be based on a choice of three basically different hydrodynamic models. Two models take the interaction between the whole body and water into consideration. The first is called "resistive", because it describes how every part of the moving body is counteracted by a resistant force. The magnitude of this force depends on the surface area and the velocity squared of each part of the body, on the density of the water and on a coefficient of resistance. The value of this coefficient depends on the shape of the body, the Reynolds number of each part of the body and on the conditions of the flow (G. Taylor, 1952; Gray, 1968). The second model, called "reactive", is based on reactive forces from an inert mass of water around the fish. While swimming, the head of the fish enters motionless water, then movements of the body accelerate a so-called virtual mass of water; the trailing edge of the tail blade leaves this accelerated water behind in a direction dictated by the angle between fin and mean path of motion in a horizontal plane. The magnitude of the propulsive force depends on the rate of change of velocity of the virtual mass of water and on the direction in which this water is left behind (Lighthill, 1960, 1971).

In a third model the tail fin of a fish is treated as an oscillating hydrofoil. The velocity and the direction of the flow over the hydrofoil and the hydrodynamic properties of its profile determine the magnitude of the propulsive forces. Each of the hydrodynamic models has its optimum requirements, in terms of movements and body structures. These requirements will be considered in a comparison between morphological and kinematical data of eel, cod and mackerel, to fit the right model to the right species. High speed cine pictures taken at 100 to 200 frames per second of an undisturbed free swimming 0.30 m-long cod are used to analyse the kinematics of steady straight forward swimming, kick-and-glide swimming, gliding and braking.

The suitable hydrodynamic model for cod is used to calculate the amount of work per unit distance travelled of steady straight forward swimming and of kick-and-glide swimming. These calculations and estimates of deceleration rates during gliding and braking give some preliminary information about the reaction between the hydrodynamic drag force and the velocity of free swimming fish during steady swimming, kick-and-glide swimming, gliding and braking.

TECHNIQUES

Cinematographic Techniques

Cod were trained to swim back and forth between two feeding points by association of underwater flashing lights with the appearance of food (Wardle & Kanwisher, 1974) in a 14 m-long tank at the Marine Laboratory in Aberdeen. The water depth in the tank was 0.8 m and the width of the swimming channel was 1.2 m. A high speed cine-camera (Red Lake, Locam) was placed 2.8 m above the tank floor. A watertight perspex raft floated on the surface below the camera to eliminate distortion of the images by waves. A background illumination lighting system was used which allowed frame rates up to $200 \, s^{-1}$ without disturbing the fish at all. Two fibre optic light sources were fitted vertically as close as possible around the lens. The light was almost fully reflected by reflex traffic light material. The background, as seen by the camera, was very bright and the fish appeared as dark silhouettes with sharp outlines. In a similar way shots of side views of swimming fish were taken. The overall light level in the tank was kept very low to leave fish undisturbed.

Film Analysis

Films were analysed on a Vanguard motion analyser. A few sequences of one 0.30 m cod were carefully selected: five showing continuous swimming, at steady speeds; three of the same fish using the kick-and-glide swimming mode at different speeds; and some examples showing gliding and braking. In the selected shots the fish was swimming at a constant level in mid water and the pectoral fins were kept close to the body, except during braking. Studies of similar movements of the fish on side views show that we may expect that the pelvic fins are kept close to the body in all the selected shots apart from the ones that show braking.

From the top views on cine film the displacement in space of the foremost point of the head and the tip of the tail was traced every 0.01 s. The mean path of motion was found and defined to be the x-axis in a frame of reference, where z is the perpendicular axis in the horizontal plane. The time derivatives of the displacements in the x- and z-direction were calculated using the five points differentiation formula of Lagrange:

$$\text{At } t = n: \; dx/dt = f\{1/12x_{(n-2)} - 2/3x_{(n-1)} + 2/3x_{(n+1)}$$
$$- 1/12x_{(n+2)}\} \tag{1.1}$$

where f is the number of frames per second.

The velocity in the x-direction was estimated to be the mean value of $5 \, dx/dt$ values around $t = n$ (Videler & Wardle, 1978). Values of dx/dt, dz/dt of the tail tip and the angle θ between the end of the tailblade and the x-axis as well as the velocity of the head were plotted against time.

Acceleration and deceleration rates were assumed to be the slopes of linear regression lines through velocity—time curves.

Anatomical Techniques

The structure of the locomotion apparatus was studied using standard morphological and histological techniques. Skeleton preparations were made using W. R. Taylor's (1967) method of clearing with enzyme and staining with alizarine red s.

SWIMMING MOVEMENTS

Key Kinematic Parameters

Gray (1933) showed most clearly how waves of curvature run down the body of a fish swimming straight forward and how these can be understood as a combination of two wave-like phenomena:

(1) Cyclic muscle contractions cause a lateral wave on the body, running in a caudal direction at velocity v with a wave length λ_b and a wave period T, where $v = \lambda_b/T$.

(2) Every single point of the body describes, in consequence of the body wave, a sinusoidal track in a horizontal plane with forward velocity u, wavelength λ_s, period T and amplitude A, where $u = \lambda_s/T$.

The basic relationships between the parameters of the two wave systems have been described by Videler & Wardle (1978) and Wardle & Videler (1980). They used simple models such as the ones shown in Fig. 1a–f to clarify these relationships. In these models a hypothetical fish is represented in a horizontal plane by a single line of a given length. Every line represents an instant of swimming motion

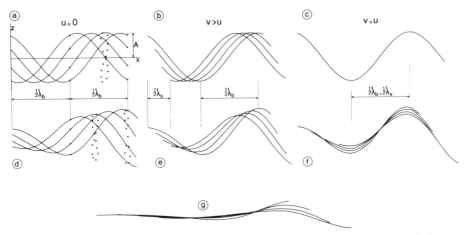

FIG. 1. Simple models to clarify the relationships between the two wave systems existing simultaneously on the body of a swimming fish. The models are explained in the text.

with an equal time separation between adjacent lines. It is assumed that forward motion of the line is resisted by the water and that appropriate lateral movements of the line generate force in interaction with the water. Forward motion is uniform and thrust equals drag. The direction of the body wave is to the right and swimming direction to the left. Figure 1a–c illustrate the case where the amplitude A (the excursion in the z-direction) is equal for all parts of the line. In Fig. 1a forward speed $u = 0$; wave crests move to the right and show the velocity of the body wave v and the wavelength λ_b. The wave period T equals λ_b/v. The foremost point of the line oscillates along a straight line in the z-direction with the same period

T; λ_s and u are both equal to 0. Points of the body at distances $\frac{1}{2}\lambda_b$ and λ_b from the left side of Fig. 1a move also in a straight line of displacement in the z-direction. All other points describe a figure of eight as indicated by a series of dots.

In Fig. 1b, $u > 0$. The foremost point of the line proceeds to the left along a sinusoidal path with wavelength λ_s and speed u. The velocity of the displacement of the wave crests, v, to the right is greater than u. Figure 1c shows the extreme case in which the forward speed u is equal to the speed of the body wave v. λ_s equals λ_b and therefore the line is following one single path.

Figure 1a and c represents purely hypothetical cases; Fig. 1b is slightly more realistic but an equal division of amplitude along the body does not occur in fish. Figure 1d—f is drawn with a linear amplitude increase. Note in Fig. 1d how the foremost point and the point at $\frac{1}{2}\lambda_b$ from the left side of the figure follow a nearly straight track in the z-direction. These two points will therefore proceed at a steady forward velocity in Fig. 1e. The forward velocity of all other points in Fig. 1e will oscillate slightly during uniform motion, owing to the figure of eight effect. Figure 1f shows that with increasing amplitude it is no longer possible for all points of the line to follow the same path.

Figure 1g shows the movements of a 0.42 m cod swimming at a uniform velocity of $0.9 \, \mathrm{m \, s^{-1}}$. Movements of this fish have been recorded from the dorsal side at a rate of 100 frames $\mathrm{s^{-1}}$. Each line is made to divide the silhouette of the fish into a right and a left half. Time separation between the lines is 0.05 s. Computer analyses of the curvature of these midlines show that this fish swims by simple harmonic motion of the body and that the amplitude of the movements increases towards the rear according to a quadratic function (Videler, Reid & Wardle, in preparation). Investigation of the relations between the wave parameters of cod (Videler & Wardle, 1978) showed that swimming at higher speeds, as well as acceleration, is always attended with an increase of v as a result of the decrease of the period T. The wavelength λ_b does not change with T, whereas λ_s slightly decreases with an increase of speed.

Kinematic parameters of the two wave systems are useful tools to describe the fundamental movements of the body during active swimming in cod. A complete description of the movements of fish requires mention of the use of paired and unpaired fins in combination with body movements.

Diversity of Swimming Movements in Forward Direction

Observations of swimming behaviour of 0.25—0.50 m cod in large tanks show different swimming styles used during swimming in one direction.

Steady swimming

At velocities lower than about 1.4 body lengths per second $(1.4\,\mathrm{L\,s^{-1}})$ steady swimming at uniform velocity hardly occurs. Most swimming is slightly irregular because fish are always accelerating or decelera-ating at a low rate. Pectoral fins are usually extended and dorsal and ventral fins are partly erected. The caudal fin is usually scooping, with the dorsal and ventral edge leading in the direction of the stroke.

At velocities higher than $1.4\,\mathrm{L\,s^{-1}}$ cod sometimes swim at uniform

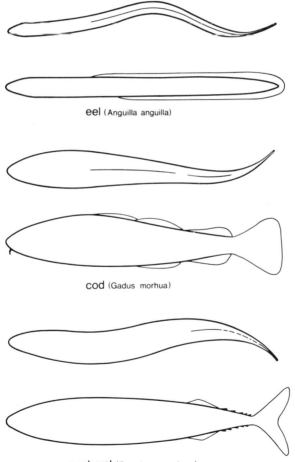

eel (Anguilla anguilla)

cod (Gadus morhua)

mackerel (Scomber scombrus)

FIG. 2. The body shape of eel, cod and mackerel during steady swimming as seen from the dorsal and lateral side.

velocities. The pelvic and pectoral fins are usually pressed against the body. The first and second dorsal fin and the first ventral fin are kept in a collapsed position as shown on the side view of a swimming cod in Fig. 2. The amount of scooping of the tailblade gradually decreases when the velocities become higher. In the lower range of speeds under $2 \, L \, s^{-1}$ the height of the end of the tailblade is about 0.19 L. The tailblade is kept flat during swimming at speeds over $2 \, L \, s^{-1}$ and the height of the end is about 0.21 L. Steady swimming at velocities between $1.4 \, L \, s^{-1}$ and $2 \, L \, s^{-1}$ can be observed most frequently. This is in good accordance with Wardle's (1977) prediction of a maximum sustained swimming speed of $2 \, L \, s^{-1}$. Steady swimming at speeds higher than $2 \, L \, s^{-1}$ could be observed occasionally and only up to $3 \, L \, s^{-1}$. The maximum velocity of a 0.30 m-long cod at 14°C, as estimated by Wardle (1975), of $10 \, L \, s^{-1}$ was never reached by the undisturbed fish in this experiment. Five examples illustrate steady swimming of a 0.30 m cod (0.302 kg). In the first example the velocity decreases from $0.41–0.36 \, m \, s^{-1}$ in 1.0 s, and the velocity increases in the second example from $0.30–0.50 \, m \, s^{-1}$ in 1.13 s. Three examples of steady swimming at uniform velocity were analysed: two near the maximum cruising speed of $2 \, L \, s^{-1}$ and one well above that speed at $2.8 \, L \, s^{-1}$. Mean values of kinematic parameters of these examples are shown in Table I.

TABLE I

Mean values of kinematic parameters of straight forward swimming of a 0.30 m cod

Forward speed, u (m s^{-1})	Wave period, T (s)	Tailbeats (s^{-1})	λ_s (m)	$2A$ max (m)
0.38(0.41–0.36)	0.45	2.2	0.17	0.045
0.44(0.3–0.5)	0.38	2.6	0.17	0.055
0.59	0.30	3.3	0.18	0.05
0.60	0.30	3.3	0.18	0.05
0.85	0.20	5.0	0.18	0.06

No attempt was made to determine the wavelength (λ_b) and velocity (v) of the body wave. Videler & Wardle (1978: table II) show that λ_b is fairly constant around 0.85 L in cod and that v (and consequently u) changes with period T. (Values of v and λ_b were not required for the calculation of propulsive force.)

Detailed analyses of steady swimming at $0.60 \, m \, s^{-1}$ and $0.85 \, m \, s^{-1}$ are shown in Figs 3 and 4. The velocity of the foremost point of the head is plotted against time in the lower graphs of these figures. The

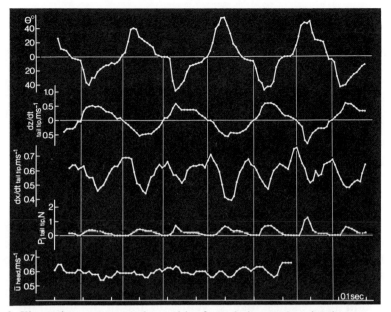

FIG. 3. Kinematic parameters and propulsive force during steady swimming at an average velocity of 0.60 m s^{-1} of a 0.30 m cod.

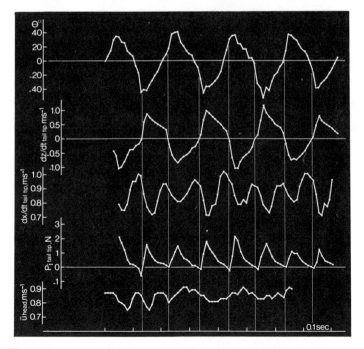

FIG. 4. Kinematic parameters and propulsive force during steady swimming at an average velocity of 0.85 m s^{-1} of a 0.30 m cod.

head velocity is constant compared to the forward velocity (dx/dt, Figs 3 and 4) of the tailtip. dx/dt of the tailtip changes rhythmically due to the figure of eight effect. The velocity curve of the head in Fig. 4 suggests that the same effect is shown by the head at high velocities. Comparison of dx/dt of the tailtip with the velocity curve of lateral displacement of the tailtip (dz/dt) indicates that the forward velocity of the tailtip is high just before the lateral velocity is zero near the most lateral point of displacement and that the forward velocity is low in the middle of the tailstroke.

The graph of dz/dt at 0.85 m s^{-1} (Fig. 4) shows how the lateral velocity suddenly increases just after the tailtip passes the most lateral point of the stroke and how the highest velocity is reached in the beginning of the stroke followed by a more gradual decrease towards the end of the stroke. Bainbridge (1963) described a similar asymmetric movement in goldfish swimming at 2.8 L s^{-1}. In steady swimming of cod the asymmetry becomes less prominent at lower speeds.

The angle between the end of the tailblade and the mean path of motion (θ) reaches highest values in the beginning of the tailstroke and lowest values just before the end. The highest lateral velocities of the tailtip coincide with the highest values of angle θ.

Kick-and-glide swimming

Cod use a cyclic kick-and-glide swimming style for prolonged swimming at speeds greater than the maximum sustained swimming speed of 2 L s^{-1}. Data from two examples of this type of swimming (0.3 m cod 0.302 kg) are used to draw Figs 5 and 6. The mean velocity of the kick-and-glide of Fig. 6 was 1.08 m s^{-1} (3.6 L s^{-1}), and 0.76 m s^{-1} (2.5 L s^{-1}) for Fig. 5, which was the lowest speed observed in this style. The movements of the fish during the kick phase are rather standard and can be best explained using the fast example. At the end of the glide, the kick phase starts with a muscle contraction running down the body on the left side. The tailblade performs a stroke to the left. Before the tailtip reaches the ultimate lateral position on the left, a fast muscle contraction train runs down the body on the right, the tailtip flicks back like a whiplash and accelerates fast towards the right. Before the muscle contraction train on the right hand side reaches the caudal peduncle, a new contraction starts on the left and causes a fast tailstroke to the left. The whole sequence ends with a less powerful stroke to the right after which the fish restores the straight posture shown throughout the gliding phase.

During the first fast stroke to the right the tailfin was kept like a flat plate (tail height 0.21 L). The fin was scooping during the next strokes to the left and right.

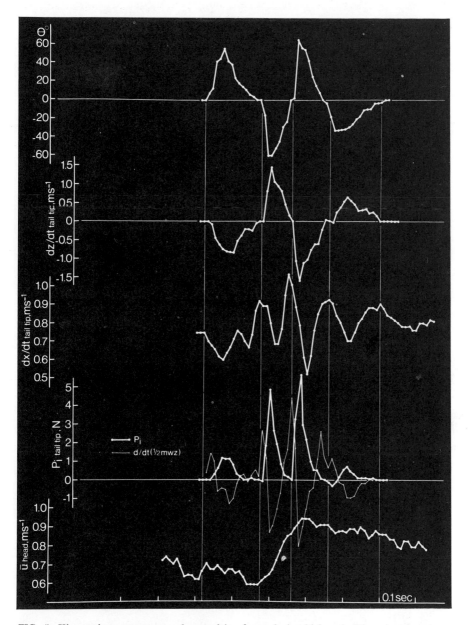

FIG. 5. Kinematic parameters and propulsive force during kick-and-glide swimming at an average velocity of $0.76\,\mathrm{m\,s^{-1}}$ of a 0.30 m cod.

12

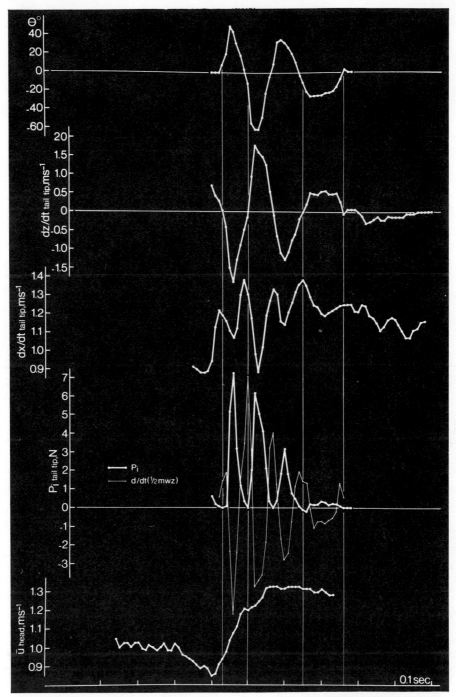

FIG. 6. Kinematic parameters and propulsive force during kick-and-glide swimming at an average velocity of $1.08\ \mathrm{m\,s^{-1}}$ of a 0.30 m cod.

The unpaired fins were usually pressed against the body during both the kick and the glide phase. In the kick phase cod kept dorsal and ventral fins as during steady swimming; during the glide all the unpaired fins were completely collapsed.

The velocity of the fish increases rapidly during the kick phase and decreases gradually during the glide. The rate of change of the velocities during both phases, as calculated with the five points differentiation formula of Lagrange, was close to uniform and was determined by calculating the slope of the regression line through the velocity against time curve. The usual acceleration rate was around $3 \, \mathrm{m \, s^{-2}}$. (The acceleration rate of the examples of Figs 5 and 6 was $3.1 \, \mathrm{m \, s^{-2}}$ in each case.) Occasionally faster kicks were observed with acceleration rates up to $6.2 \, \mathrm{m \, s^{-2}}$. None of the filmed shots of kick-and-glide swimming contained a complete cycle. Shots were selected for analysis on which the complete kick phase and part of the glide phase could be studied. The total distance covered during the glide phase (D_g) had to be calculated. At the end of the acceleration phase the fish had initial velocity u_i, and at the end of the glide phase a final velocity u_f, which is the starting velocity of the kick phase. The deceleration rate (A_d) was used to calculate the duration of the glide (t_g):

And now
$$t_g = (u_i - u_f)/A_d \tag{1.2}$$

$$D_g = u_i \cdot t_g + \tfrac{1}{2} A_d \cdot t_g{}^2 \tag{1.3}$$

Table II gives the parameters of three examples of kick-and-glide swimming. The example with a mean velocity $(\bar{u})$ of $0.76 \, \mathrm{m \, s^{-1}}$ is illustrated in Fig. 5, the one with $\bar{u} = 1.08 \, \mathrm{m \, s^{-1}}$ in Fig. 6. The third example has the highest acceleration rate (A_a) shown by the $0.3 \, \mathrm{m}$ cod on the present cine film recordings.

TABLE II

Parameters of kick-and-glide swimming of a 0.30 m cod

$\bar{u}$ (m s^{-1})	u_i (m s^{-1})	u_f (m s^{-1})	A_a (m s^{-2})	A_d (m s^{-2})	D_k (m)	D_g (m)	t_k (s)	t_g (s)
1.08	1.33	0.85	3.1	−0.6	0.16	0.87	0.15	0.8
0.76	0.94	0.59	3.1	−0.41	0.084	0.65	0.11	0.85
0.95	1.23	0.67	6.2	−0.46	0.085	1.16	0.09	1.22

$\bar{u}$ = mean velocity; u_i = initial velocity at the beginning of the glide phase; u_f = final velocity at the end of the glide phase; A_a = kick acceleration rate; A_d = glide acceleration rate; D_k = distance covered during kick; D_g = distance covered during glide; t_k = kick duration; t_g = glide duration.

J. J. Videler

Gliding and braking

Gliding does not necessarily occur in combination with kick swimming. Bouts of steady swimming can be terminated by gliding to a
lower velocity or to a halt. The unpaired fins are usually erected and
pectoral fins extended at low speeds. Various deceleration rates can
be observed, mainly depending on the use of paired and unpaired
fins.

Braking is different from gliding. The body is not straight but
frozen in an S shape. Figure 7 was drawn from a photograph of a
braking cod. The tailblade at the end of the S-shaped body is bent

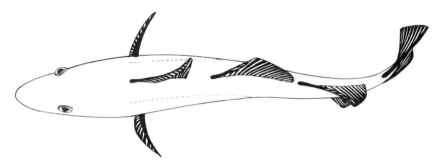

FIG. 7. Drawing from a photograph of a braking cod

sideways. Note how the fin rays in the centre of the fin are more
extremely bent than the ones near the edges. The fin rays are fully
spread. The dorsal and ventral fins are erected and form an S-shaped
path along which the water is forced to flow. Pelvic fins are fully
spread and lowered. The fin rays of the pectoral fins are extended
and the flat surfaces of the fins face the direction of movement. A
deceleration rate as low as $2.3\,\mathrm{m\,s^{-2}}$ was measured for a braking
$0.42\,\mathrm{m}$ cod with a body weight of $0.676\,\mathrm{kg}$. Braking started at an
initial velocity of $1.26\,\mathrm{m\,s^{-1}}$.

STRUCTURE AND FUNCTION OF THE LOCOMOTION APPARATUS

External Characteristics

The external shape during swimming is determined by the shape of
the body as well as by the number, shape and posture of the fins.
Movement in the water requires a streamlined body for minimum
resistance. Ideally this is a tear-drop shaped body, typically rounded
at the front end and tapered at the rear end. The maximum diameter
should be situated neither near the front nor close to the rear. Such

a body has minimum resistance when the ratio between the length
and the maximum diameter (the fineness ratio) is about 4.5. Figure 2
shows the body of a cod during steady swimming as seen from the
dorsal and lateral side. For comparison the bodies of an eel and a
mackerel are drawn in the same way. The fineness ratio of the lateral
and dorsal views of these fishes is shown in Table III.

TABLE III

*Fineness ratios (length/maximum diameter) of three different
fishes*

Fineness Ratio	Cod	Mackerel	Eel
Lateral view	5.0	6.0	15.0
Dorsal view	7.5	8.8	21.0

This Table shows that the bodies of cod and mackerel are streamlined
and that the body of an eel is not.

A fish makes lateral movements with its body for the generation
of propulsive force. Lateral surface areas in interaction with the
water will generate thrust. The rear end of the cod is laterally flattened; the last dorsal and ventral fins, the tailfin and the shape of the
caudal peduncle contribute to that effect. The tailfin of cod provides
a large lateral surface area with a straight rear end. For comparison,
the body of the eel shows lateral flattening to the rear which is
enhanced by long dorsal and ventral fins. There is no separate tailfin
and no straight rear end. The body of the mackerel is rounded on
transection. The caudal peduncle is narrow and carries a high caudal
fin. The centre of the fin does not extend as far backwards as the
dorsal and ventral lobes.

Features of the Main Body

As stated in the Introduction, the body of a fish like cod contains a
muscle—bone system which acts as the main power source for locomotion. Three parts can be distinguished:

(1) The vertical median septum supported by a flexible but incompressible vertebral column.

(2) The lateral musculature with metamerically arranged myotomes and myosepts.

(3) The tendinous layers of the skin.

Cod have about 53 vertebrae (for comparison: European eel, 114;
mackerel, 31) varying in length between 0.22 L and 0.06 L, and in
diameter from 0.24 L near the head to 0.08 L in the caudal peduncle.

(For a full description of the connection between vertebrae see Symmons, 1979.) The vertebrae show dorso-ventral symmetry: the neural arch and spine on the dorsal side is very similar to the haemal arch and spine on the ventral side. This symmetry is disturbed between head and anus by the presence of the abdominal cavity. The angle between the spines and the vertebral axis in the plane of the septum varies in cod between a mean value of 50° for the neural spine in the abdominal region and a mean value of 30° for neural and haemal spines in the caudal region. The dorsal and ventral spines are connected by the collagenous fibres of the median septum. This septum is attached to the skin on the dorsal and ventral side and divides the fish into two lateral halves.

The complicated shape of cod myosepts and myotomes has been described by Wardle & Videler (1980). The lateral musculature is divided by the horizontal septum into symmetric dorsal and ventral portions. Muscle fibres of the myotomes are attached to the myosepts and their direction is approximately in line with the longitudinal body axis. Wardle & Videler (1980) recall how the outermost layer of red muscle fibres in the myotomes of cod is used aerobically during prolonged slow swimming and how the vast amount of white fibres can be used during short periods of fast swimming. The same paper shows how myosepts are attached to the vertical septum (Wardle & Videler, 1980: fig. 6) and to the skin (fig. 7). It gives a plausible explanation of how muscle contractions cause lateral bending of the vertical septum and simultaneously pull the skin forward.

The dermal layer of the skin of teleosts like cod consists of a well vasculated stratum laxum in which the scales have their basis and a mechanically important stratum compactum, a thick layer of mainly collagenous fibres. This part of the skin has a plywood structure: layers of fibres running in one direction alternate with layers containing fibres in a different direction. This construction makes a strong tendinous sheet. Videler (1975) showed how such a cross-fibre structure can be used either to transmit forces in a longitudinal direction or to stretch according to the bulging of the muscles below.

The Unpaired Fins

The last dorsal and ventral fins and the tail fin are involved in steady straight forward locomotion (see Fig. 2). This part of the body has the largest amplitude and so the largest lateral velocity.

The muscle—bone system of this part of the body will exchange propulsive forces with the water. The structures exert pressure on to the water and the water reacts with propulsive force on to the

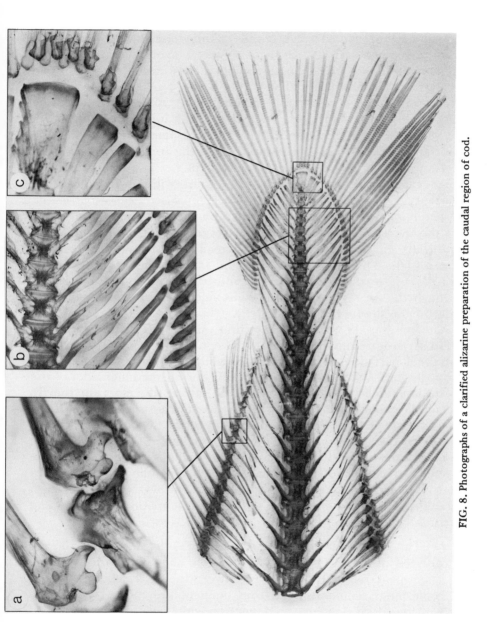

FIG. 8. Photographs of a clarified alizarine preparation of the caudal region of cod.

structures. This propulsive force has to be transferred to the main incompressible body axis without losses due to elasticity.

Figure 8 is a photograph of a cleared alizarine preparation of the caudal part of cod. The vertebral column ends straight, and the whole structure shows a strong dorso-ventral symmetry. The fin rays of the tailfin are in connection with a large number of modified neural and haemal spines. The insets b and c in Fig. 8, show how these spines are reinforced compared with spines in the caudal peduncle. The structure and function of the joint between the fin rays and the laterally flattened ends of the spines was described by Videler (1975, 1977). He showed that this type of joint allows freedom for various movements of the fin ray and that propulsive force can be transferred across this joint without losses due to elasticity.

The dorso-ventral symmetry shown by the skeleton is also shown by the overlying musculature. The last myotome inserts on the fin ray heads and lies on top of a rim of small intrinsic muscles (Hinderssson, 1910). The fin ray heads of the tailfin show a strong lateral process on the forward side. A photograph of a frontal section through the fin ray head on the right side, shows two thick tendons inserting on the fin ray head (Fig. 9).

The tendon-like structure inserting on the lateral process is the stratum compactum of the skin. The other tendon on the most proximal part of the fin ray is a combination of two tendons, one from the last myotome and one (the fibres on the inner side) of the intrinsic muscles. The position of the lateral processes on the fin rays indicates that pulling of the skin would cause abduction of the fin rays as well as active lateral bending of the rays towards the side on which the skin is pulling. (For the mechanism behind this action see McCutchen, 1970; Videler, 1977.) Abduction will increase the lateral surface area of the fin and the active bending force counteracts the bending of the fin rays due to the water pressure. Two anatomical features emphasize the active role of the last dorsal and ventral fin in swimming.

(1) The basal elements of the supporting skeleton of these fins are situated in the vertical septum. Figure 8 shows that the vertebral spines penetrate into the spaces between the basal elements. Strong collagenous tendinous structures connect the ends of the spines with the basal elements, a connection which would allow forces to be transferred from the fin to the vertebral column without losses due to elasticity, because collagenous fibres are non-stretchable structures. The supporting elements of the first and second dorsal fins and the first ventral fin do not reach the ends of the neural

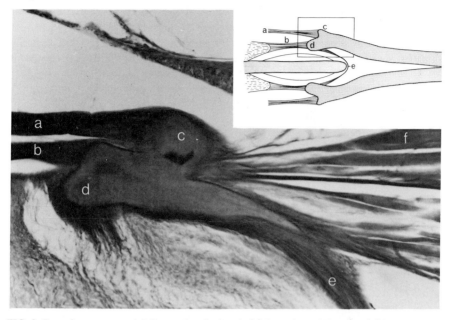

FIG. 9. Insertion on a caudal fin ray head of cod. (a) Insertion of the skin; (b) insertion of last myotomes and intrinsic muscles; (c) lateral process on fin ray head; (d) fin ray head; (e) capital ligament (Videler, 1975); (f) interradial muscles. The body of the fish is to the left and the tail to the right.

and haemal spines and consequently there is no firm connection between these fins and the vertebral column.

(2) The other feature is the joint between the fin rays of the unpaired fins and their supporting elements. Geerlink & Videler (1974) described the structure and movements of such a joint. This description shows that the large lateral processes shown in Fig. 8 (inset a) indicate strong activity of the intrinsic muscles, and that this activity will serve to trim the trailing edge of the dorsal and ventral fins during lateral motion.

HYDRODYNAMIC APPROACH

The Choice of a Hydrodynamic Model

Requirements of the three types of hydrodynamic models for fish propulsion are compared with morphological and kinematical data of cod, eel and mackerel in order to make the right choice for cod.

The propulsive force in resistive models depends on the instantaneous velocity relative to the water (Lighthill, 1975). A fish with a large amplitude over a major part of its body would make good use of this dependence.

Figure 10 shows the double amplitude (see A, Fig. 1a) relative to the body length as a function of a (a is the position on the body

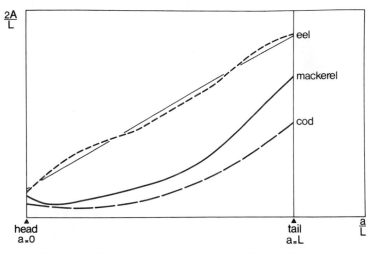

FIG. 10. Division of amplitude over the body from head to tail measured during steady swimming at about 2 L s^{-1} for eel, mackerel and cod.

between head ($a = 0$) and tailtip ($a = L$) of the three species). Swimming velocity was about 2 L s^{-1} for each species. The amplitude of the eel is similar to the amplitude of the model in Fig. 1e and fits the requirements of resistive force theory because it is large over a large part of the body. The amplitude of the head has to be small for obvious reasons.

The amplitude and consequently the lateral velocity increase linearly to the rear. The propulsive force will increase similarly if resistive forces play an important role. The amplitude curves of both mackerel and cod do not meet the requirements of the resistive models.

A resistive model demands a body to be rounded on transection to avoid reactive forces. The eel is laterally flattened towards the rear but the body height hardly changes from head to tail (see Fig. 2). The caudal peduncle and fins of cod make the rear part of the cod extremely laterally flattened. The body of the mackerel is rounded apart from the tailblade and the very small dorsal and ventral fins. The rounded part of the body with the highest amplitude (the caudal peduncle) is extremely narrow which avoids resistance during lateral movements.

The third type of hydrodynamic theory mentioned in the Introduction, required a fin with the shape of a hydrofoil. This fin has to move as fast as possible through preferably undisturbed water. Morphological features and the division of amplitude along the body of mackerel match this type of interaction with the water.

Reactive forces depend on the rate of change of velocity of the body with respect to the water and on the dimensions of the virtual mass. This rate of change would be uniform in the case of the eel because the amplitude increases linearly. The amplitude of cod increases quadratically and so does the lateral velocity of parts of the body towards the tailtip. Mackerel has an even greater increase of amplitude but the lateral surface area decreases strongly towards the rear and so does the virtual mass. In reactive theory the accelerated mass of water has to be shed by the trailing edge, preferably in the best possible direction. A fish with a more or less pointed trailing edge like the eel, cannot control this shedding process as well as a fish with a straight tailfin end like cod. Figure 2 shows a sudden decrease in body height just behind the last dorsal and ventral fin in cod. This could have a dramatic effect on the virtual mass, but Lighthill (1969) expects a vortex sheet to fill these gaps sufficiently to avoid this effect.

This comparison between three species and three types of models suggests that cod uses reactive forces for its propulsion.

Lighthill's (1971) "large-amplitude elongated-body theory for fish locomotion" was used here to calculate the propulsive force of cod during steady and kick-and-glide swimming. Propulsive force can be calculated from the equation:

$$P = \left[mw\frac{dz}{dt} - \frac{1}{2}mw^2 \cos\theta \right]_{\text{tailtip}} + \frac{d}{dt}\int_0^L mw\frac{dz}{da}\,da \tag{1.4}$$

where m is the virtual mass of water per unit length ($m = \pi\rho\,(s/2)^2$; ρ is the density of water and s the height of the tailblade). θ is the angle between the tailblade and the mean path of motion in a horizontal plane, dz/dt is the velocity in the z-direction, perpendicular to the mean path of motion, and w is the velocity vector at the tailtip:

$$w = \frac{dz}{dt}\cos\theta - \frac{dx}{dt}\sin\theta \tag{1.5}$$

where dx/dt is the velocity along the mean path of motion of the tailtip.

The rate of change of the integral term will be zero over a symmetrical tail cycle. So for steady swimming instantaneous propulsive force could be calculated from:

$$P_i = \left[mw \frac{dz}{dt} - \frac{1}{2} mw^2 \cos \theta \right]_{\text{tailtip}} \qquad (1.6)$$

In case of kick swimming the rate of change of the integral term was estimated using Lighthill's (1971) crude approximation:

$$\frac{d}{dt} \left[\tfrac{1}{2} mwz \right]_{\text{tailtip}} \qquad (1.7)$$

where z is the displacement of the tailtip in the z-direction. See, for further explanation of the method, Wardle & Reid (1977).

Thrust and Drag Calculations for Different Swimming Styles

Examples of the results of propulsive force calculations using Lighthill's model are included in Figs 3–6.

Figures 3 and 4 show graphs of the propulsive force in Newtons against time during steady swimming. High thrust values coincide with high values of dz/dt and θ, just after the tailtip reaches the most lateral point of its excursion in the z-direction. (This becomes more obvious during fast swimming as shown in Fig. 4.) In the kick phase (Figs 5 and 6) high values of propulsive force (P_i) occur during the second and third tail stroke. The estimates of the rate of change of the integral term ($d/dt[\tfrac{1}{2} mwz]_{\text{tailtip}}$) show positive and negative peak values during the same period of the kick phase. The sum of the positive and negative thrust values from these estimates during the complete kick phase was zero in the three examples studied (see Table II and Figs 5 and 6). The amount of work per metre distance travelled was calculated using the equation:

$$\frac{1}{D} \cdot \sum_{t=0}^{t=n} P_i \cdot U_t \cdot \Delta t \qquad (1.8)$$

where D is the total distance covered between $t = 0$ and $t = $ n, U_t is the instantaneous velocity of the head and Δt is the time interval between frames. The time between $t = 0$ and $t = $ n contains a kick and a complete glide when the equation is used for kick-and-glide swimming and it contains a number of complete tailstroke cycles when work done during steady swimming is calculated. The results are used to compare the cost of transport of different swimming styles at different speeds, as shown in Fig. 11. The solid circles in Fig. 11 show the values for steady swimming. (Kinematic data of the shots used in Fig. 11 are summarized in Tables I and II.)

The open circle of the example at 0.39 m s^{-1} indicates the amount of work m^{-1} done by the fish during deceleration from 0.41—

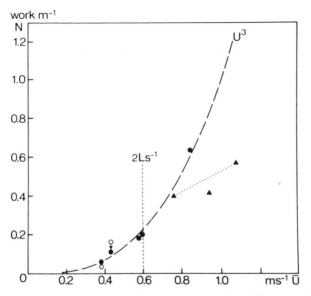

FIG. 11. The amount of work per metre distance covered during different speeds of steady swimming (solid circles) and kick-and-glide swimming (triangles) of cod. See text for further explanation.

$0.36 \, \mathrm{m \, s^{-1}}$ $(= 0.03 \, \mathrm{N})$. The amount of work $\mathrm{m^{-1}}$ gained from the loss of kinetic energy $(0.015 \, \mathrm{N})$ has to be added to find the work $\mathrm{m^{-1}}$ to overcome drag at this velocity, shown as a solid circle.

The open circle at $0.44 \, \mathrm{m \, s^{-1}}$ is the total work $\mathrm{m^{-1}}$ of a shot in which the fish accelerated from $0.3 \, \mathrm{m \, s^{-1}}$ to $0.5 \, \mathrm{m \, s^{-1}}$ (see Table I). The amount of thrust to overcome drag was calculated by subtracting the force needed to accelerate the $0.302 \, \mathrm{kg}$ fish. The solid circle below the open circle shows the work done per metre to overcome drag. The other solid circles are calculated from swimming at uniform velocities. The solid circles indicate that the work $\mathrm{m^{-1}}$ during steady swimming increases with U^3.

The solid triangles in Fig. 11 are the values for kick-and-glide swimming. The one at $0.76 \, \mathrm{m \, s^{-1}}$ represents the lowest mean speed at which kick-and-glide swimming has been observed in this fish. This point nearly coincides with the line U^3, indicating that at this speed kick-and-glide swimming and steady swimming require almost equal amounts of work $\mathrm{m^{-1}}$. The other triangles in Fig. 11 show that above this speed, kick-and-glide swimming is much cheaper than steady swimming. Theoretical studies of Weihs (1974) suggest that bursts of swimming activities followed by gliding at average velocities below maximum cruising speed could reduce the amount of work $\mathrm{m^{-1}}$

drastically. Under the circumstances described here, cod did not perform intermittent swimming at velocities below maximum cruising speed. Extrapolation from the slowest and fastest kick-and-glide swimming examples of Fig. 11 suggests that kick-and-glide swimming requires more work m^{-1} than steady swimming at velocities below 0.76 m s^{-1}.

Electromyographic evidence (i.e. Bone, Kiceniuk & Jones, 1978) shows that fish use red muscle fibres at velocities below 2 L s^{-1}. These muscles work aerobically and are virtually unexhaustible at low speeds. Above 2 L s^{-1} cod will use the bulk of fast white anaerobic musculature which can be exhausted within a few minutes when contracting at full rate. It seems that there is no need for a different cheap swimming style at low velocities and that for prolonged swimming at high speeds such a style is necessary. The present calculations suggest that cod use the kick-and-glide swimming mode for cheaper cruising at high velocities.

The question remains, why is kick-and-glide swimming observed to be cheaper than steady swimming?

The equation of motion of an accelerating fish during the kick phase is:

$$F = M \cdot a + \tfrac{1}{2}\rho A_w u^2 C_a \qquad (1.9)$$

where F is the propulsive force; the first term on the right is mass (M) times acceleration rate (a) and the second term is the drag of the accelerating body: ρ is the density of the water, A_w the wetted surface area, u the average velocity and C_a the drag coefficient during acceleration. In the gliding phase $F = 0$ and so the equation of motion is

$$M \cdot a = \tfrac{1}{2}\rho A_w u^2 C_d \qquad (1.10)$$

where C_d is the drag coefficient when gliding. (Note that deceleration requires a negative acceleration rate so the term on the left is positive.)

During steady swimming at uniform velocity, the propulsive force equals the drag and so:

$$F = \tfrac{1}{2}\rho A_w u^2 C_n \qquad (1.11)$$

where C_n is the drag coefficient during steady swimming.

The present data allow calculations of F and $M \cdot a$. From the equations of motion, drag and drag coefficients were calculated as presented in Table IV.

Kick-and-glide swimming is cheaper than steady swimming if the average of very high drag during the short kick period and very low drag during the long glide phase is lower than the constant drag during steady swimming at the same average velocity. Interpolation

TABLE IV

Comparison of drag during steady swimming and kick-and-glide swimming of a 0.30 m cod

	u (m s^{-1})	Drag (N)	Acceleration rate (m s^{-2})	Drag Coefficient
Steady swimming	0.39	0.04	(−0.05)	0.019
	0.44	0.13	(+0.18)	0.049
	0.59	0.18	−	0.037
	0.60	0.20	−	0.040
	(0.76	0.41	−	0.051)
	0.85	0.64	−	0.064
Kick phase	0.76	2.51	3.10	0.310
	0.95	4.50	6.20	0.360
	1.08	2.81	3.10	0.170
Glide phase	0.76	0.12	−0.41	0.015
	0.95	0.14	−0.46	0.011
	1.08	0.18	−0.60	0.011

along the line U^3 in Fig. 11 gives a propulsive force equal to the drag force of 0.41 N for steady swimming at 0.76 m s^{-1} (between brackets in Table IV). Kick-and-glide swimming at 0.76 m s^{-1} with an acceleration rate of 3.1 m s^{-2} and a deceleration rate of 0.42 m s^{-2} will have the same average drag of 0.41 N as for steady swimming if the duration of the glide is 7.3 times the duration of the kick. Kick-and-glide swimming at this velocity showed a glide phase which was 7.7 times longer than the kick phase and so the average drag will be lower than the drag for steady swimming even at this velocity. The same calculations give a ratio between steady swimming drag and gliding drag of 3.3. This figure is higher than figures found by Lighthill (1971) and Webb (1971). These authors compared calculated swimming drag values for dace and trout with drag values for streamlined bodies calculated with equations for drag coefficients used by Hoerner (1965):

$$C_d = 0.074 \, R_e^{-0.2} \left[1 + 1.5 \left(\frac{d}{L}\right)^{3/2} + 7 \left(\frac{d}{L}\right)^3 \right] \qquad (1.12)$$

This equation gives the drag coefficient when the flow around the body is turbulent, R_e is the Reynolds number, d is the maximum depth and L is the length of the body.

Figures for the drag coefficient of a gliding cod calculated with this equation are about half as high as the values shown in Table IV. Drag coefficients during gliding according to Hoerner's equation are 0.0072 at 0.76 m s^{-1}, 0.0069 at 0.95 m s^{-1} and 0.0068 at 1.08 m s^{-1}.

The drag coefficients found for steady swimming of cod are about six times greater than the estimates based on Hoerner's equation. This is twice as much as the discrepancy found by Webb (1971), and 1.5 times Lighthill's estimate of the difference between the drag coefficient of a swimming fish and a streamlined body.

The highest drag coefficient measured from deceleration during braking (as shown in Fig. 7) was 0.09 which is about nine times higher than during gliding. The drag coefficients during the kick phase are between two and four times as high as the highest values found during braking. This indicates that, due to the unsteady motion of the kick phase, propulsive force values from Lighthill's model are too high. The energy required for kick-and-glide swimming is therefore even less than the figures in this paper suggest (D. Weihs, personal communication).

Results shown in this paper suggest that drag on a steady swimming fish is 3.3 times greater than drag on fish gliding at the same average speed, and that the drag on a gliding fish is twice the turbulent drag on a rigid streamlined body of similar dimensions and moving at the same velocity.

ACKNOWLEDGEMENTS

I am grateful for a Royal Society Fellowship in the European Science Exchange Programme, which enabled me to use the excellent facilities for fish swimming studies at the Marine Laboratory in Aberdeen during a whole year. I thoroughly enjoy the stimulating cooperation with C. S. Wardle and I thank him for the discussions on various topics related to this paper and for his help with this foreign language.

REFERENCES

Bainbridge, R. (1963). Caudal fin and body movement in the propulsion of some fish. *J. exp. Biol.* **40**: 23–56.

Bone, Q., Kiceniuk, J. & Jones, D. R. (1978). On the role of the different fibre types in fish myotomes at intermediate swimming speeds. *Fish. Bull. US.* **76**: 691–699.

Geerlink, P. J. (1979). The anatomy of the pectoral fin in *Sarotherodon niloticus* Trewavas (Cichlidae). *Neth. J. Zool.* **29**: 9–32.

Geerlink, P. J. & Videler, J. J. (1974). Joints and muscles of the dorsal fin of *Tilapia nilotica* L. (Fam. Cichlidae). *Neth. J. Zool.* **24**: 279–290.

Gray, J. (1933). Studies in animal locomotion. I. The movement of fish with special reference to the eel. *J. exp. Biol.* **10**: 88–104.

Gray, J. (1968). *Animal locomotion*. London: Weidenfeld & Nicholson.

Hindersson, H. A. (1910). Über die Schwanzflossenmuskulatur der Teleostier. *Anat. Anz.* **36**: 465—471.

Hoerner, S. F. (1965). *Fluid-dynamic drag*. Midland Park, N.J.: S. F. Hoerner.

Lighthill, M. J. (1960). Note on the swimming of slender fish. *J. Fluid Mech.* **9**: 305—317.

Lighthill, M. J. (1969). Hydromechanics of aquatic animal propulsion. *A. Rev. Fluid Mech.* **44**: 265—301.

Lighthill, M. J. (1971). Large-amplitude elongated-body theory of fish locomotion. *Proc. R. Soc. Lond.* (B) **179**: 125—138.

Lighthill, M. J. (1975). *Mathematical biofluid dynamics*. Philadelphia, Pennsylvania: Society for Industrial and Applied Mathematics.

McCutchen, C. W. (1970). The trout tail fin: a self-cambering hydrofoil. *J. Biomech.* **3**: 271—281.

Symmons, S. (1979). Notochordal and elastic components of the axial skeleton of fishes and their functions in locomotion. *J. Zool., Lond.* **189**: 157—206.

Taylor, G. (1952). Analysis of the swimming of long narrow animals. *Proc. R. Soc.* (A) **214**: 158—183.

Taylor, W. R. (1967). An enzyme method of clearing and staining small vertebrates. *Proc. U.S. natn. Mus.* **122**(3596): 1—17.

Videler, J. J. (1975). On the interrelationships between morphology and movement in the tail of the cichlid fish *Tilapia nilotica* (L.). *Neth. J. Zool.* **25**: 143—194.

Videler, J. J. (1977). Mechanical properties of fish tail joints. *Fortschr. Zool.* **24**: 183—194.

Videler, J. J. & Wardle, C. S. (1978). New kinematic data from high speed cine film recordings of swimming cod (*Gadus morhua*). *Neth. J. Zool.* **28**: 465—484.

Videler, J. J., Reid, A. & Wardle, C. S. (In preparation). *Kinematic ananlysis of fish swimming: High velocities of Mackerel, Saithe, Mullet, Cod and Salmon.*

Wardle, C. S. (1975). Limit of fish swimming speed. *Nature, Lond.* **255**: 725—727.

Wardle, C. S. (1977). Effects of size on swimming speeds of fish. In *Scale effects in animal locomotion*: 229—313. Pedley, T. J. (Ed.). New York and London: Academic Press.

Wardle, C. S. & Kanwisher, J. W. (1974). The significance of heart rate in free swimming Cod, *Gadus morhua*: some observations with ultrasonic tags. *Mar. Behav. Physiol.* **3**: 311—324.

Wardle, C. S. & Reid, A. (1977). The application of large amplitude elongated body theory to measure swimming power in fish. In *Fisheries mathematics*: 171—191. Steele, J. H. (Ed.). London and New York: Academic Press.

Wardle, C. S. & J. J. Videler, (1980). Fish swimming. *Soc. exp. Biol. Sem. Ser.* **5**: 125—150. Elder, H. T. & Trueman, E. R. (Eds). Cambridge: University Press.

Webb, P. W. (1971). The swimming energetics of trout. I. Thrust and power output at cruising speeds. *J. exp. Biol.* **55**: 489—520.

Weihs, D. (1974). Energetic advantages of burst swimming of fish. *J. theor. Biol.* **48**: 215—229.

Symp. zool. Soc. Lond. (1981) No. 48, 29—52

Mechanics of Drag-based Mechanisms of Propulsion in Aquatic Vertebrates

R. W. BLAKE

Department of Zoology, University of British Columbia, 2354—6270 University Boulevard, Vancouver, B.C. V6T 2A9, Canada

SYNOPSIS

A general hydromechanical model of the paddling propulsor is developed based on a blade-element approach. It is applied to an analysis of labriform (pectoral fin) swimming in the angelfish (*Pterophyllum eimekei*) and employed to calculate the thrust force produced and the work done over a pectoral fin-beat cycle.

During the power stroke the distal half of the fin produces over 90% of the total thrust (only small amounts of reversed thrust are produced at the base of the fin at the beginning and end of the stroke) and work produced. In the recovery stroke the fin is "feathered" and the impulse of a drag force acting in the direction of the animal's body is only 1/20th of that associated with the thrust force produced during the power stroke. The fin-beat cycle propulsive efficiency is of the order of 0.2.

Through the introduction of simple numerical constants (e.g. shape factors) the model can be employed to assess the influence of paddle shape and kinematics on thrust production. The wedge-shaped pectoral fins of the angelfish investigated have a shape factor of 0.43, which is 14% less than the value for a triangular planform (0.5).

The model's underlying assumptions, its scope for further application, experimental data on the influence of paddle shape on rowing and paddle bending mechanics are considered in a general discussion.

INTRODUCTION

References in the literature to the kinematics and hydrodynamics of the paddling propulsor in aquatic animals are few and largely anecdotal. In view of the fact that members of many major systematic groups employ drag-based mechanisms of propulsion (e.g. Polychaeta, Dytiscidae, Gyrinidae, Hydrophilidae, Heteroptera, Tricoptera, Conchostraca, Cladocera, Brachyura, Actinopterygii, Anura, Testudinata, Gaviiformes, Podicipitiformes, Monotremata, Sirenia), this is surprising.

Biologists and physical scientists have concentrated their efforts on the kinematics, hydrodynamics and energetics of undulatory (see Lighthill, 1975; Webb, 1975 — for reviews) and jet propulsion (e.g. Siekmann, 1963; Johnson, Soden & Trueman, 1972; Weihs, 1977), which are now relatively well understood.

Here, a general model of the mechanics of paddling is developed. In addition to providing a basis for the analysis of the power and recovery stroke phases of any rowing appendage's beat-cycle, the model considers the influence of the gross morphological and kinematic factors (e.g. paddle shape and stroke pattern) on thrust production. The model is applied to labriform (pectoral fin) locomotion in the angelfish (*Pterophyllum eimekei*) and is discussed in relation to paddling in other aquatic and semi-aquatic vertebrates.

A GENERAL MODEL OF THE MECHANICS OF THE PADDLING PROPULSOR

The model which follows is general, in the sense that it is applicable to any rowing animal. We assume:

(1) That the rowing appendage is free to rotate at its point of attachment to the body which it propels.

(2) That the beat-cycle is divisible into power stroke and recovery stroke phases, over which the positional angle of the appendage varies as a sinusoidal function of time.

(3) That the body is propelled steadily forward at a constant velocity and level.

(4) That during the power stroke the appendage presents a geometrical angle of attack (angle between the appendage and its projection on to the horizontal plane) of $90°$, throughout its length, for the entire power stroke duration.

(5) That at the end of the power stroke the appendage rotates about its base and presents a low geometrical angle of attack (i.e. the structure is "feathered") throughout its length and for the entire recovery stroke duration.

For the purpose of analysis the appendage is considered to be composed of a number of arbitrarily defined blade-elements. The characteristics of the entire appendage are determined by integrating over the length of the structure and time.

The Power Stroke

Velocities and the hydrodynamical angle of attack

The appendage moves from its starting position at a positional angle

γ_1 (γ, the angle between the projection of the structure on to the horizontal plane and the median axis of the body being propelled) to γ_2, its position at the end of the stroke in a time t_p (the duration of the power stroke). Assuming the positional angle to vary as a harmonic function of time, we have:

$$\gamma = \left(\frac{\gamma_1 + \gamma_2}{2}\right) - \left(\frac{\gamma_2 - \gamma_1}{2}\right) \cos\left(\frac{\pi}{t_p}\right)t \qquad (2.1)$$

and the angular velocity (Ω, the angular velocity of the projection of the appendage on to a horizontal plane) is:

$$\Omega = \left(\frac{\gamma_2 - \gamma_1}{2}\right)\left(\frac{\pi}{t_p}\right) \sin\left(\frac{\pi}{t_p}\right)t \qquad (2.2)$$

The normal (v_n) and spanwise (v_s) velocity components are:

$$v_n = \Omega r - V \sin \gamma$$
$$= \left[\left(\frac{\gamma_2 - \gamma_1}{2}\right)\left(\frac{\pi}{t_p}\right) \sin\left(\frac{\pi}{t_p}\right)t\right] r - V \sin \gamma \qquad (2.3)$$

where V is the forward velocity of the body and r is the distance out from the base of the appendage which is of total length R.

$$v_s = V \cos \gamma \qquad (2.4)$$

The resultant relative velocity (v_r) is therefore:

$$v_r = [(\Omega r - V \sin \gamma)^2 + (V \cos \gamma)^2]^{1/2}$$
$$= \left\{ \left(\left[\left(\frac{\gamma_2 - \gamma_1}{2}\right)\left(\frac{\pi}{t_p}\right) \sin\left(\frac{\pi}{t_p}\right)t\right] r - V \sin \gamma)^2 + (V \cos \gamma)^2 \right\}^{1/2} \qquad (2.5)$$

The hydrodynamical angle of attack (α, the angle between the projection of the longditudinal axis of the appendage on to the direction of the relative fluid velocity) is given by:

$$\tan \alpha = (\Omega r - V \sin \gamma)/V \cos \gamma$$
$$= \frac{\left[\left(\frac{\gamma_2 - \gamma_1}{2}\right)\left(\frac{\pi}{t_p}\right) \sin\left(\frac{\pi}{t_p}\right)t\right] r - V \sin \gamma}{V \cos \gamma} \qquad (2.6)$$

Figures 1 and 2 give a schematic impression of the beat-cycle and diagrammatic definitions of some of the power stroke parameters.

Force, normal force coefficients and impulse
The normal force (dF_n) acting on a blade-element is given by:

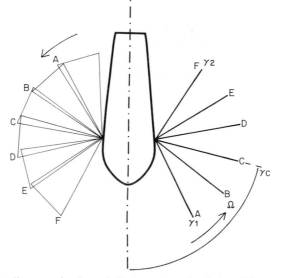

FIG. 1. Schematic diagram of a hypothetical rowing animal viewed from above, showing appendage positions during the power and recovery stroke. All notation is defined in the text.

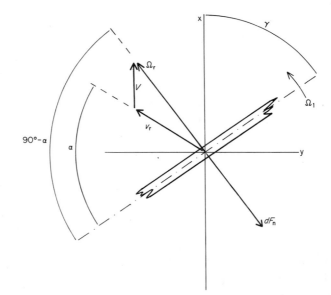

FIG. 2. A blade-element during the power stroke. All notation is defined in the text.

$$dF_n = \frac{1}{2}\rho dA C_n v_r^2$$

$$= \frac{1}{2}\rho c\, dr\, C_n \left[(\Omega r - V \sin \gamma)^2 + (V \cos \gamma)^2 \right]$$

$$= \frac{1}{2}\rho c\, dr\, C_n \left\{ \left(\left[\left(\frac{\gamma_2 - \gamma_1}{2} \right) \left(\frac{\pi}{t_p} \right) \sin \left(\frac{\pi}{t_p} \right) t \right] \right.$$

$$\left. \times\ r - V \sin \gamma)^2 + (V \cos \gamma)^2 \right\} \qquad (2.7)$$

where ρ is the water density, dA is the frontally projected area of an element, c is the chord and C_n is a normal force coefficient.

The value of C_n depends on:

(1) Reynolds Number (R_e = velocity × length/kinematic viscosity).
(2) Hydrodynamical angle of attack.
(3) Geometry.

C_n is plotted against R_e in Fig. 3 for a circular cylinder and a square plate over a R_e range $(10–10^3)$ where large changes in C_n occur. For values of $R_e > 10^3$ the value of C_n for a square plate is constant at about 1.1; the value for a circular cylinder, however, dips down to a minimum of about 0.9 at $R_e \simeq 2 \times 10^3$ and then rises again to about $C_n = 1.2$ at $R_e = 3 \times 10^4$.

Values of C_n for flat plates of various breadth (b) to chord ratios are plotted against α in Fig. 4, which shows that:

(1) An initial phase of increase in C_n with increasing α (which can be described by a simple equation of the form $C_n = h \sin \alpha$, where h is a constant) is followed by an abrupt fall in C_n, due to the flow separating from the edges of the plate. We shall refer to the angle of attack at which separation occurs as α_{crit}. After α_{crit} the value of C_n is relatively insensitive to changes in α.

(2) $dC_n/d\alpha$ increases with increasing b/c. Values of C_n at $\alpha > \alpha_{crit}$ increase with increasing b/c.

(3) α_{crit} and the value of C_n at which separation occurs increase as b/c decreases.

Figure 5 illustrates the variation of C_n with fineness ratio (defined as b/c for plates and l/d (length/diameter) for circular cylinders). The values for square and rectangular sections and circular cylinders placed normal to the incident flow will be of most interest to biologists studying paddling locomotion. The value of C_n for rectangular plates $\rightarrow 1.86$ as $b/c \rightarrow \infty$.

The thrust force (dT) produced by the rowing appendage is given by:

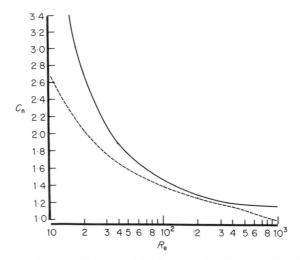

FIG. 3. The normal force coefficient (C_n) is plotted against Reynolds Number (R_e) for the case of a square flat plate (solid line) and a circular cylinder (broken line) of finite span, placed at right angles to the incident flow. (Data from: Fage & Johansen (1927); Knight & Wenzinger (1928); Lindsey (1938); Wick (1954); Lilley (1970); and Wiles (1970).)

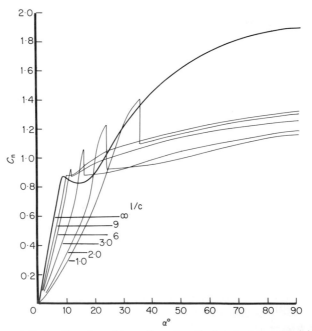

FIG. 4. Values of C_n for flat plates (three-dimensional) of various breadth (b) to chord (c) ratio ($f = b/c$) are plotted against α. (Data from: Fage & Johansen (1927); Knight & Wenzinger (1928); Lindsey (1938); Wick (1954); Lilley (1970); and Wiles (1970).)

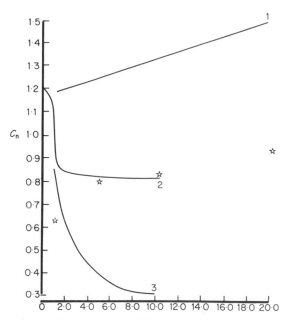

FIG. 5. Variation of C_n with fineness ratio (f) for square and rectangular plates set at right angles to the incident flow (1), blunt circular cylinder parallel to the flow (2), half-streamlined circular cylinder parallel to the flow (3), and circular cylinder set at right angles to the flow (☆). (Data from: Hoerner (1958); Lilley (1970); and Wiles (1970).)

$$dT = dF_n \sin \gamma$$

$$= dF_n \left[\frac{(\Omega r - V \sin \gamma)}{v_r} \right]$$

$$= \frac{1}{2}\rho c dr C_n \left(\left[\left(\frac{\gamma_2 - \gamma_1}{2} \right) \left(\frac{\pi}{t_p} \right) \sin \left(\frac{\pi}{t_p} \right) t \right] r - V \sin \gamma \right)^2 \quad (2.8)$$

Integrating over the length of the appendage (R), the total thrust (T) is:

$$T = \int_{r=0}^{R} \frac{1}{2}\rho v_n{}^2 C_n c\, dr \quad (2.9)$$

the mean thrust at any given instant ($\bar{T}_i$) is:

$$\bar{T}_i = \frac{1}{R} \int_{r=0}^{R} \frac{1}{2}\rho v_n{}^2 C_n c\, dr dr \quad (2.10)$$

Integrating over the length of the appendage and the stroke duration, the mean thrust ($\bar{T}$) can be written as:

$$\bar{T} = \frac{1}{t_p} \int_{t=0}^{t_p} \int_{r=0}^{R} \frac{1}{2}\rho v_n{}^2 C_n c\, dr dt \quad (2.11)$$

The impluse of the thrust force produced over the stroke (P) is given by

$$P = \int_{r=0}^{R} \int_{t=0}^{t_p} dT\, dt$$

$$= \int_{r=0}^{R} \int_{t=0}^{t_p} \frac{1}{2}\rho v_n{}^2 C_n c\, dr\, dt$$

$$= \int_{r=0}^{R} \int_{t=0}^{t_p} \frac{1}{2}\rho C_n c\, dr\, \left\{ \left[\left(\frac{\gamma_2 - \gamma_1}{2} \right)\left(\frac{\pi}{t_p} \right) \right.\right.$$

$$\times \left. \sin\left(\frac{\pi}{t_p} \right)t \right] r - V \sin\gamma \right\}^2 dt \qquad (2.12)$$

The impluse of the drag force acting on the body being propelled (P_b) is given by:

$$P_b = \frac{1}{2}\rho V^2 S_w C_D t_0 \qquad (2.13)$$

where C_D is the drag coefficient of the body, t_0 is the total append-age-beat cycle time (i.e. including the recovery stroke duration time) and S_w is the total wetted surface area of the body. From equations 2.12 and 2.13 we can write:

$$\frac{1}{2}\rho V^2 S_w C_D t_0 = n \int_{r=0}^{R} \int_{t=0}^{t_p} \frac{1}{2}\rho v_n{}^2 C_n c\, dr\, dt \qquad (2.14)$$

where n refers to the number of paddles: for the case of $n = 2$, we can write:

$$\frac{1}{2}\rho V^2 S_w C_D t_0 = 2 \int_{r=0}^{R} \int_{t=0}^{t_p} \frac{1}{2}\rho C_n c\, dr\, \left\{ \left[\left(\frac{\gamma_2 - \gamma_1}{2} \right)\left(\frac{\pi}{t_p} \right) \right.\right.$$

$$\times \left. \sin\left(\frac{\pi}{t_p} \right)t \right] r - V \sin\gamma \right\}^2 dt \qquad (2.15)$$

Equation 2.15 can be employed to infer a value of C_D, which can be compared with experimentally determined values.

Torque, power and propulsive efficiency
The contribution of a blade-element to the torque (dQ) is given by:

$$dQ = dF_n r = \frac{1}{2}\rho dA C_n v_r{}^2 dr$$

$$= \frac{1}{2}\rho c\, dr\, C_n \left[(\Omega r - V \sin\gamma)^2 + (V \cos\gamma)^2 \right] dr$$

$$= \frac{1}{2}\rho c\, dr\, C_n \left\{ \left(\left[\left(\frac{\gamma_2 - \gamma_1}{2} \right)\left(\frac{\pi}{t_p} \right) \sin\left(\frac{\pi}{t_p} \right)t \right] r - V \sin\gamma \right)^2 \right.$$

$$+ (V \cos\gamma)^2 \right\} dr \qquad (2.16)$$

Denoting the torque, mean instantaneous torque and mean torque operative over the stroke as Q, $\bar{Q}_i$ and $\bar{Q}$ respectively, we have:

$$Q = \int_{r=0}^{R} \frac{1}{2}\rho v_r^2 C_n c\, dr\, dr \qquad (2.17)$$

$$\bar{Q}_i = \frac{1}{R} \int_{r=0}^{R} \frac{1}{2}\rho v_r^2 C_n c\, dr\, dr\, dr \qquad (2.18)$$

$$\bar{Q} = \frac{1}{t_p} \int_{r=0}^{R} \int_{t=0}^{t_p} \frac{1}{2}\rho v_r^2 C_n c\, dr\, dr\, dt \qquad (2.19)$$

The rate of working (dW) is given by

$$
\begin{aligned}
dW &= dF_n \Omega r \\[4pt]
&= \frac{1}{2}\rho dA C_n v_r^2 \Omega r \\[4pt]
&= \frac{1}{2}\rho c\, dr\, C_n \left[(\Omega r - V\sin\gamma)^2 + (V\cos\gamma)^2 \right] dr \\[4pt]
&= \frac{1}{2}\rho c\, dr\, C_n \left\{ \left(\left[\left(\frac{\gamma_2 - \gamma_1}{2} \right) \left(\frac{\pi}{t_p} \right) \sin\left(\frac{\pi}{t_p} \right) t \right] r - V\sin\gamma \right)^2 \right. \\[4pt]
&\qquad \left. + (V\cos\gamma)^2 \right\} dr \qquad (2.20)
\end{aligned}
$$

So, the mean power output over the stroke $(\bar{W})$ is given by:

$$\bar{W} = \frac{1}{t_p} \int_{t=0}^{t_p} \int_{r=0}^{R} \frac{1}{2}\rho v_r^2 C_n c\, dr\, dr\, dt \qquad (2.21)$$

From equations 2.8 and 2.16, the instantaneous propulsive efficiency (η^*) can be written as:

$$\eta^* = V dT / \Omega dQ$$

$$
\frac{V\frac{1}{2}\rho c\, dr\, C_n \left\{ \left(\left[\left(\frac{\gamma_2 - \gamma_1}{2} \right)\left(\frac{\pi}{t_p} \right)\sin\left(\frac{\pi}{t_p} \right)t \right] r - V\sin\gamma \right)^2 \right\}}{\Omega\frac{1}{2}\rho c\, dr\, C_n \left\{ \left(\left[\left(\frac{\gamma_2 - \gamma_1}{2} \right)\left(\frac{\pi}{t_p} \right)\sin\left(\frac{\pi}{t_p} \right)t \right] r - V\sin\gamma \right)^2 + (V\cos\gamma)^2 \right\} dr}
$$

$$(2.22)$$

and a mean stroke propulsive efficiency $(\bar{\eta})$ given by:

$$\bar{\eta} = V / \overline{\Omega r} \qquad (2.23)$$

Added mass

A paddling appendage in unsteady motion has to be given kinetic energy to accelerate its own mass plus that of the fluid it entrains owing to its motion. The work done in accelerating the structure is therefore greater than if only the appendage itself had to be moved. One way of explaining this result is to propose that the actual mass of the appendage itself increases by an amount equal to the mass of the fluid which it entrains. The additional component of mass is termed the added or induced mass. Added mass depends on the size, volume, shape, mode of motion and the fluid density. A simple approach to the influence of added mass on a rowing appendage (which we will assume can be likened to a plate placed normal to the incident flow) follows.

The added mass (dm_a) is given by:

$$dm_a = \pi \left(\frac{c}{2}\right)^2 \rho \, dr \tag{2.24}$$

the added mass force (dF_a) and the impulse of the added mass force (P_a) can be calculated from:

$$dF_a = m_a \left(\frac{dv_r}{dt}\right)$$

$$= \pi \left(\frac{c}{2}\right)^2 \rho \, dr \left\{ \frac{d[(\Omega r - V \sin \gamma)^2 + (V \cos \gamma)^2]^{1/2}}{dt} \right\} \tag{2.25}$$

$$P_a = \int_{r=0}^{R} \int_{t=0}^{t_p} dF_a \, dt$$

$$= \int_{r=0}^{R} \int_{t=0}^{t_p} \pi \left(\frac{c}{2}\right)^2 \rho \, dr \left\{ \frac{d[(\Omega r - V \sin \gamma)^2 + (V \cos \gamma)^2]^{1/2}}{dt} \right\} dt \tag{2.26}$$

The power required is given by:

$$dW_a = m_a \left(\frac{dv_r}{dt}\right) v_r$$

$$= \pi \left(\frac{c}{2}\right)^2 dr\rho \left\{ \frac{d[(\Omega r - V \sin \gamma)^2 + (V \cos \gamma)^2]^{1/2}}{dt} \right\}$$

$$\times [(\Omega r - V \sin \gamma)^2 + (V \cos \gamma)^2]^{1/2} \tag{2.27}$$

An alternative, more rigorous, approach would involve calculating the added mass matrix for the appendage under consideration, from which the added mass coefficients appropriate to the particular motion could be derived and the kinetic energy required to accelerate the appendage and its associated added mass calculated.

Most of the experimental and theoretical literature on added mass relevant to the understanding of the mechanics of the paddling pro- pulsor, however, relates to elliptical and rectangular flat plates and aerofoils in simple translation. However, Scrutton (1941) gives experimental data on elliptical and rectangular plates of various aspect ratio (Aspect Ratio = S/c, where S refers to the span of the plate) oscillated about a fixed point. Scrutton (1941) calculates the added mass for flat plates from:

$$m_a = k\rho\pi c^2 s/2 \qquad (2.28)$$

where k is a coefficient of additional mass and s is the semi-span of the plate. Scrutton's experimentally determined values of k are in good agreement with the corresponding theoretical predictions (Jones, 1941). Values of k are plotted against aspect ratio for flat plates in Fig. 6.

As the velocity of the appendage is assumed to be periodic in time the mean value of the added mass force over the stroke is zero (see

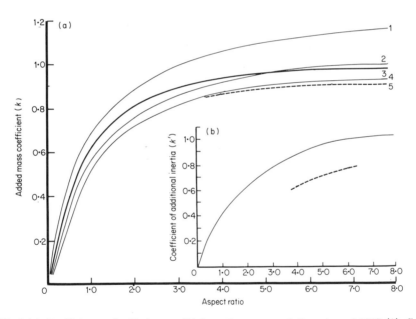

FIG. 6.(a) Coefficients of added mass (k) for a German aerofoil section of 1937 (1), flat plates (2), elliptical plates (3, theoretical values), N.A.C.A. aerofoil of 1943 (4) and theor- etical values for flat plates (5) are plotted against aspect ratio (A). Curves (1), (2), (3) and (4) are drawn from data in Scrutton (1941). Values for curve (5) are taken from Jones (1941). (b) Coefficients of additional inertia (k') are plotted against aspect ratio for flat plates (solid line: experimental values from Scrutton (1941); broken line: theoretical values from Jones (1941)).

Batchelor, 1967: 408) and therefore it does not influence the force balance (given by equation 2.14). However, it can affect the power required (see pp. 42–43).

The Recovery Stroke

Velocities and forces

The normal (v_n') and spanwise (v_s') velocity components are given by:

$$v_n' = \Omega'r + V \sin \gamma \qquad (2.29)$$

$$v_s' = V \cos \gamma \qquad (2.30)$$

Denoting the spanwise, chordwise and normal components of force as dF_s, dF_c and dF_n' respectively, we have:

$$dF_s = \frac{1}{2}\rho v_s'^2 A_{(t)} C_s$$

$$= \frac{1}{2}\rho(V \cos \gamma)^2 A_{(t)} 1.33 R_s^{-1/2}$$

$$= \frac{1}{2}\rho(V \cos \gamma)^2 A_{(t)} 1.33 \left(\frac{RV \cos \gamma}{v}\right)^{-1/2} \qquad (2.31)$$

$$dF_c = \frac{1}{2}\rho(v_n' \cos \beta)^2 A_{(t)} 1.33 \left[\frac{c(v_n' \cos \beta)}{v}\right]^{-1/2} \qquad (2.32)$$

$$dF_n' = \frac{1}{2}\rho v_n' A_{(t)} C_n \qquad (2.33)$$

where $A_{(t)}$ is the total wetted area of the appendage, C_s is a frictional drag coefficient, R_s is a Reynolds Number (based on appendage length and the spanwise velocity component) and β is the geometrical angle of attack of the appendage.

A drag force acting in the direction of the body (dF_b) is given by:

$$dF_b = dF_s \cos \gamma + dF_c \cos \beta \sin \gamma + dF_n \sin \beta \sin \gamma \qquad (2.34)$$

The impulse of this force (P') is given by:

$$P' = \int_0^R \int_0^{t_r} dF_d \, dt$$

$$= \int_0^{t_r} \int_0^R (dF_s \cos \gamma + dF_c \cos \beta \sin \gamma + dF_n' \sin \beta \sin \gamma) dt \qquad (2.35)$$

where t_r is the recovery stroke duration.

Torque and power

The torque due to the drag force acting in the direction of the body (Q') and the mean torque operative over the stroke ($\bar{Q}'$) are given by:

$$Q' = \int_{r=0}^{R} (dF_s \cos \gamma + dF_c \cos \beta \sin \gamma + dF'_n \sin \beta \sin \gamma)\, dr \tag{2.36}$$

$$\bar{Q}' = \frac{1}{t_r} \int_{t=0}^{t_r} \int_{r=0}^{R} (dF_s \cos \gamma + dF_c \cos \beta \sin \gamma$$

$$+ dF'_n \sin \beta \sin \gamma)\, dr\, dt \tag{2.37}$$

The mean power required ($\bar{W}'$) is given by:

$$\bar{W} = \frac{1}{t_r} \int_0^{t_r} (dF_s v'_s \cos \gamma + dF_c v'_n \cos \beta \sin \gamma + dF'_n v'_n \sin \gamma \sin \beta)\, dt \tag{2.38}$$

The energy required to produce the recovery and power stroke can be summed in order to derive an estimate of the appendage-beat cycle propulsive efficiency (e.g. Blake, 1980). In those cases where the appendage is "feathered" on the recovery stroke (e.g. labriform locomotion in teleosts), the energy required to return it will be relatively small. However, if the propulsive surface of the appendage is not "feathered", but merely folded (e.g. the paddling appendages of semi-aquatic birds and mammals), a substantial pressure drag term may arise and the energy required for the recovery stroke may be a significant proportion of the total energy required to perform the appendage-beat cycle.

Influence of Appendage Shape and Kinematics on Thrust

Confining our consideration to the power stroke and neglecting the forward velocity of the body being propelled, we can write:

$$dF''_n = \tfrac{1}{2}\rho(\Omega r)^2\, c\, dr\, C_n$$

$$dT'' = \tfrac{1}{2}\rho(\Omega r)^2\, c\, dr\, C_n \sin \gamma \tag{2.39}$$

The impulse of the stroke can be written as:

$$P'' = \int_{r=0}^{R} \int_{t=0}^{t_p} dT''\, dt$$

$$= \int_0^{R} \int_0^{t_p} \frac{1}{2}\rho(\Omega r)^2\, c\, dr\, C_n \sin \gamma\, dt$$

$$= \frac{1}{2}\rho C_n \int_0^{R} r^2 c\, dr \int_0^{t_p} \Omega^2 \sin \gamma\, dt \tag{2.40}$$

Calling:

$$\int_0^R r^2 \, c \, dr = I_1 R^2 A \tag{2.41}$$

and

$$\int_0^{t_p} \Omega^2 \sin \gamma \, dt = \int_{\gamma_2}^{\gamma_2} \left(\frac{d\gamma}{dt}\right) \sin \gamma \, d\gamma$$

$$= \frac{I_2(\gamma_2 - \gamma_1)^2}{t_p} \tag{2.42}$$

where I_1 and I_2 are numerical constants (of the order of 1.0) which depend on the paddle geometry and kinematics respectively. Recalling that $P_b = 2P$, we can write:

$$\frac{1}{2}\rho V^2 S_w C_D t_0 = \frac{C_n I_1 R^2 A I_2 (\gamma_2 - \gamma_1)^2}{t_p} \tag{2.43}$$

Values of I_1 can be derived graphically (see Blake, 1981) or analytically. $I_1 = 0.33$ for squares and rectangles and 0.5 for triangular planforms.

LABRIFORM LOCOMOTION IN THE ANGELFISH (*PTEROPHYLLUM EIMEKEI*): A CASE STUDY

The model of the mechanics of paddling developed in the previous sections of this chapter has been applied to labriform (pectoral fin) locomotion in the angelfish (*Pterophyllum eimekei*). The analysis of the power stroke and recovery stroke is based on a single individual (length = 8.05 cm s^{-1}) swimming steadily forward at a constant velocity ($V = 4.06$ cm s^{-1}) and level in the water. Information on experimental methods and details concerning the fin-beat cycle kinematics are given in Blake (1979a, b, 1980, 1981).

The fin is divided into four blade-elements (e1–e4), the boundaries of which are the same for the power and recovery phases of the fin-beat cycle. Information on the positional angle of the fin (γ), its angular velocity (Ω), normal velocity (v_n) and hydrodynamical angle of attack (α) during the power stroke ($t_p = 0.1$ s) are given in Fig. 7, which shows that:

(1) Flow reversal and negative hydrodynamical angles of attack occur at the base of the fin at the beginning and end of the stroke.
(2) Values of v_n and α at any given instant are higher at the distal end of the fin.

Figure 8 illustrates the variation of the hydrodynamic thrust force (a value of $C_n = 1.1$ was selected for values of $\alpha > 40°$ and a value calculated from: $C_n = 2.5 \sin \alpha$ for $\alpha < 40°$), added mass force and total force operative throughout the power stroke. Small amounts of reversed thrust are generated at the base of the fin (not shown in

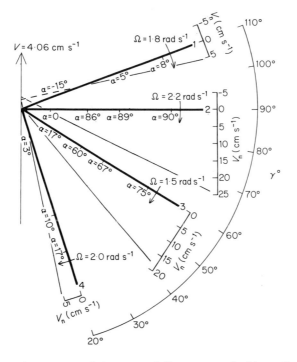

FIG. 7. Diagrammatic summary of the pectoral fin power stroke kinematics of a rowing stroke of an angelfish, swimming in the drag-based labriform mode. Data from Blake (1979a, b). All notation is defined in the text.

Fig. 8). The outermost 40% of the fin produces over 80% of the total hydrodynamic thrust force produced during the stroke. Substituting in values into equation 2.14 ($t_0 = 0.2$ s), a value of $C_D \simeq 0.10$ is inferred.

The power required to produce the various force components is plotted against time in Fig. 9. A propulsive efficiency for the power stroke of 0.26 is calculated. Taking into account the energy required to accelerate the mass of the fin and its added mass, the value of the propulsive efficiency is reduced to about 0.18. When the energy needed to produce the recovery stroke is taken into account the propulsive efficiency is further reduced to 0.16. The impulse of the recovery stroke is only about 1/20th of that of the power stroke.

DISCUSSION

The Model: Features and Assumptions

In applying a blade-element theory to the analysis of the paddling propulsor, the following assumptions are made:

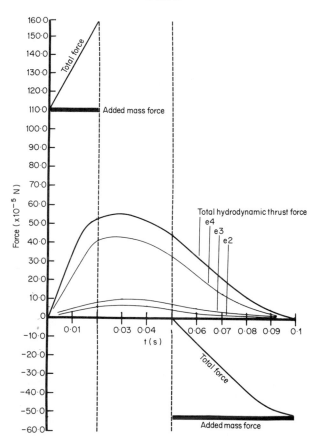

FIG. 8. The contributions to the total hydrodynamic thrust force from three arbitrarily defined pectoral fin blade-elements, the midpoints of which are located at $0.38R$, $0.54R$ and $0.81R$ for elements e2, e3 and e4 respectively ($R = 1.30$ cm), the total thrust force, the added mass force (F_a) and the total force are plotted against time. Values of F_a could not be accurately calculated between $t = 0.02$ s and $t = 0.05$ s, due to the small accelerations. Data from Blake (1979a, b).

(1) In selecting a value of C_n we can assume that:

(i) The drag force acting on the appendage is due solely to pressure drag.

(ii) The appendage can be likened to a series of three-dimensional flat plates set at high angles of incidence to the flow.

(iii) The flow is steady.

(2) That the exact nature of the induced velocity field can be neglected.

(3) That no hydrodynamic interactions occur between the paddling appendage and the body that is being propelled.

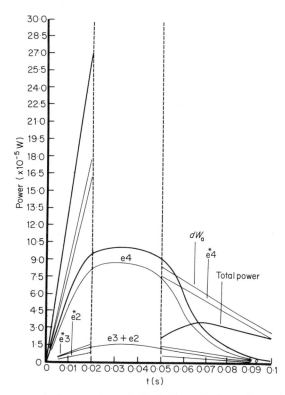

FIG. 9. The power required to produce the hydrodynamic thrust force, the contributions to this total of the blade-elements (e2 + e3) and e4, the power needed to accelerate and decelerate the added mass of the fin, the contributions to this total of the blade elements (e*2, e*3 and e*4) and the total power required are plotted against time. Data from Blake (1979a, b).

The value of α at which the flow separates from the rear surface of a flat plate in a steady flow (α_{crit}) is dependent on the fineness ratio of the plate (see Fig. 4). For $\alpha > \alpha_{crit}$ (for any given case) it is reasonable to assume that the flow has separated. Fage & Johansen (1927) recorded a frictional drag coefficient of about 0.004 for a two-dimensional flat plate (the corrections for the three-dimensional case are small; see Elder, 1960) at zero incidence (for $R_e > 10^3$) and state that corrections for frictional drag need not be made for flat plates at high angles of attack.

The value of C_n can be allowed to vary along the length of the appendage and in time. It could be argued however, that the arbitrarily selected elements are not equivalent to three-dimensional flat plates as they are bounded by the borders of their neighbours.

Unfortunately, little work has been done on the variation of C_n

during the unsteady motion of flat plates. Theoretical treatments exist for the inviscid flow around an oscillating, two-dimensional plate (Anton, 1956; McNown, 1957). Keulegan & Carpenter (1956) give experimental data on the continuously oscillating, two-dimensional plate. Gadd (1964) points out that the flow pattern for a continuously oscillating plate will be very different from that established by a single pass. Unsteady effects have been shown to be significant in the hovering flight of insects (e.g. Weis-Fogh, 1973) and birds (e.g. Norberg, 1975) and it is likely that they are also important for small, slow swimming animals which are propelled by paddles.

The influence that the body of the animal being propelled has on the induced velocity field is not usually known and for practical purposes such complications are generally ignored.

Influence of Appendage Shape

The impulse of the thrust force produced during the power stroke by a rowing appendage is directly proportional to both the shape factor and the kinematic factor (I_1 and I_2 respectively; see equation 2.43). Therefore, we can expect the propulsive appendages of rowing animals to be characterized by planforms of high values of I_1 and I_2. For structures of a given area, $I_1 = 0.33, 0.39, 0.5, 0.6, 0.67$ for square and rectangular (irrespective of the ratio of major to minor axis), semi-elliptical, triangular (irrespective of height to base ratio), parabolic and cubic planforms respectively. The pectoral fins of teleosts which swim in the drag-based labriform mode are commonly triangular or wedge-shaped, with I_1 in the range of 0.4–0.5 ($I_1 = 0.43$ for the pectoral fins of *P. eimekei*).

Experimental work on model angelfish and pectoral fin shape combinations (Blake, 1979a, 1981) shows that appendages of a triangular planform of a given area create less interference drag over the body to which they are attached (see Hoerner, 1958; von Mises, 1959 – for a definition and explanation of interference drag) than square or rectangular ones.

Bending of a Paddling Appendage

Deflections of a paddling appendage at any instant during the power stroke can be predicted from cantilever beam theory. Assuming that the resultant of the forces operating act at a point, the deflection (δ) is given by:

$$\delta = \frac{FR^3}{3EJ} \tag{2.44}$$

where F, R, E and J are the load, length, Youngs Modulus and second moment of area of the appendage respectively. For efficient rowing δ should not exceed $0.2R$ at the distal tip of the paddle.

Vertebrate paddles are composed of bone and cartilage. In Table I the Youngs Modulus and density of these materials is compared with that of materials which are commonly employed in the construction of rowing boat oars.

TABLE I

Youngs Modulus and density of materials from which oars are made

Material	Youngs Modulus $(E : Nm^{-2})$	Density $(\rho : kg\, m^{-3})$
Mild steel	4.3×10^9	7,840
Aluminium	1.4×10^9	2,720
Fibre glass	2.8×10^8	1,184
Wood (ash)	2.4×10^8	704
Bone	2.0×10^{10}	2,000
Cartilage	8.0×10^6	1,500

In teleosts which swim in the drag-based labriform mode deflections of the pectoral fins are brought about by bending at the fin-ray joints (Blake, 1979a). The fin-rays of labriform swimmers are commonly unjointed proximally.

Incidence of Drag-based Mechanisms of Propulsion in Aquatic Vertebrates

Labriform locomotion occurs in many major teleost families (e.g. Embiotocidae, Chaetodontidae, Cichlidae, Serranidae, Scaridae, Labridae, Pomacentridae). Most labriform swimmers are inhabitants of coral reefs or grassy undersea meadows and are designed for high manoeuvrability rather than speed. Many possess a laterally compressed, inflexible body (e.g. Chaetodontidae, Scatophagidae, Platacidae) and are therefore poorly designed for undulatory swimming in the anguilliform or sub-carangiform modes.

It is likely that not all teleosts which swim in the labriform mode employ a drag-based mechanism of propulsion. Lift-based mechanisms probably occur among forms characterized by high aspect ratio pectoral fins (e.g. many Chaetodontidae). Others (e.g. some Serranidae) seem to employ "jet-effects", produced by "clapping" their pectoral fins synchronously against the sides of their body. A comparative study of the swimming mechanics and design of teleosts which are propelled by pectoral fins would clearly be rewarding.

Walker (1971) has made observations on the swimming of representatives of four genera of sea turtle (*Chelonia mydas*, *Eretmochelys
imbricata*, *Lepidochelys kempii* and *Caretta caretta*) and found no
differences in the kinematics of the basic pectoral flipper beat-cycle.
The flipper moves up and down along an axis inclined from 40—70°
to the horizontal, with the leading edge inclined antero-ventrally on
the downstroke and antero-dorsally on the upstroke.

The distal tip of the flipper describes a figure of eight over a
complete beat-cycle. On the basis of this Carr (1952) and Walker
(1971) have likened the motion and mechanics of the flipper to that
of a bird's wing in flight. This is probably not the case, as for most of
the downstroke the flipper presents high geometrical angles of attack
(60—80°) which is consistent with a drag-based mechanism. On the
upstroke however, the geometrical angle of attack of the flipper
rarely exceeds 30°.

It has been suggested (on the basis of anatomical evidence) that
the extinct Placodontia and Sauropterygia (pleisiosaurs and notho-
saurs) employed drag-based mechanisms of propulsion (Aleyev,
1977). It is possible that the well developed hindlimbs of the meso-
saurs (characterized by spreading digits which were probably webbed)
were employed as paddles. The thaunatosaurs of the Jurassic also
have paddle-like limbs (Romer, 1966).

Among the birds, the extinct Hesperornithes (e.g. *Hesperornis*)
were probably propelled by the paddling action of webbed feet,
perhaps in a manner similar to that observed in living Gaviidae and
Podicipitiformes (see Dabelow, 1925).

Fish (1980) has studied paddling locomotion in the muskrat
(*Ondatra zibethecus*), which is propelled by the alternate strokes of
its hindlimbs. Fish recorded a relatively constant limb-beat frequency
(about 2.4 Hz) over a range of swimming speeds ($20—75 \text{ cm s}^{-1}$). In
the angelfish swimming in the drag-based labriform mode, increases
in forward velocity are facilitated by an increase in the fin-beat frequency (Blake, 1979a). By actively controlling the angle between
the femur and tibia (at the knee joint) and that between the tibia
and foot (at the ankle joint), the normal force component produced by the hindlimbs of the muskrat can be usefully directed over
most of the power stroke. This kinematic strategy also reduces the
net pitching moments operating over the stroke (similar adaptations
occur in the Gaviidae and Podicipitiformes).

Many pinnipeds (e.g. Otariidae) are propelled by the paddling
action of their flipper-like anterior limbs. In the Monotremata, the
platypus (*Ornithorhynchus*) swims using its webbed fore- and hindlimbs as paddles. The Sirenia (dugongs and manatees) possess forelimbs

which are modified as paddles. As in the platypus, muskrat and beaver the paddling locomotion of forms such as *Manatus* and *Dugong* is supplemented by undulations of a dorso-ventrally compressed tail.

SUMMARY OF NOTATION

Unless otherwise stated, all notation refers to a single appendage.

General Mode of the Paddling Propulsor

Power stroke

r	—	distance from the base of an appendage
c	—	the chord
R	—	total length
dA	—	projected area of a blade-element
γ	—	positional angle
γ_1	—	positional angle at the beginning of a stroke
γ_2	—	positional angle at the end of a stroke
Ω	—	angular velocity
V	—	forward velocity of body being propelled
v_n	—	normal velocity component
v_s	—	spanwise velocity component
v_r	—	resolved, relative velocity
t	—	time
t_p	—	time of power stroke duration
α	—	hydrodynamical angle of attack
M_a	—	added mass
dF_n	—	normal force component
dF_a	—	added mass force
dT	—	thrust
$\bar{T}_i$	—	mean thrust force at any instant
$\bar{T}$	—	mean thrust force acting over the stroke
P	—	impulse of thrust force
P_b	—	impulse of the drag force acting on the body being propelled
P_a	—	impulse of the added mass force
C_D	—	drag coefficient of the body
n	—	number of paddling appendages
C_n	—	normal force coefficient
h	—	a constant
dQ	—	torque

$\bar{Q}_i$ — mean torque at any instant
$\bar{Q}$ — mean torque over the stroke
dW — rate of working
$\bar{W}$ — mean rate of working over the stroke
dW_a — rate of working associated with the added mass force
η^* — instantaneous propulsive efficiency
$\bar{\eta}$ — mean propulsive efficiency

Recovery stroke

v'_n — normal velocity component
v'_s — spanwise velocity component
Ω' — angular velocity
$A_{(t)}$ — total area of appendage
t_r — time of recovery stroke duration
C_s — spanwise force coefficient
R_s — Reynolds Number based on a spanwise velocity component
β — geometrical angle of attack
dF_s — spanwise force component
dF_c — chordwise force component
dF'_n — normal force component
dF_b — force acting in the direction of the body being propelled
P' — impulse of force acting in the direction of the body being propelled
Q' — torque
$\bar{Q}'$ — mean torque produced over a stroke
$\bar{W}'$ — mean rate of working over a stroke

Other Notation

dF''_n — normal force component in simplified power stroke model
dT'' — thrust force in simplified power stroke model
P'' — impulse in simplified power stroke model
I_1 — shape factor
I_2 — kinematic factor
δ — deflection
F — load
E — Youngs Modulus
J — second moment of area
b — breadth of a flat plate
d — diameter of a circular cylinder
l — length of a flat plate

α_{crit} —   angle of attack for a flat plate at which flow separation occurs

k —   added mass coefficient

k' —   additional inertia coefficient

S —   span

s —   semi-span

v —   kinematic viscosity of the fluid

R_e —   Reynolds Number

ρ —   fluid density

REFERENCES

Aleyev, Yu. G. (1977). *Nekton*. The Hague: W. Junk.

Anton, L. (1956). Formation of a vortex at the edge of a plate. *Mem. Nat. Advis. Comm. Aero. Tech.* No. 1938.

Batchelor, G. K. (1967). *An introduction to fluid dynamics*. Cambridge: Cambridge University Press.

Blake, R. W. (1979a). *The mechanics of labriform locomotion*. Ph. D. Thesis: Cambridge University.

Blake, R. W. (1979b). The mechanics of labriform locomotion. I. Labriform locomotion in the Angelfish (*Peterophyllum eimekei*): an analysis of the power stroke. *J. exp. Biol.* 82: 255–271.

Blake, R. W. (1980). The mechanics of labriform locomotion. II. An analysis of the recovery stroke and the overall fin-beat cycle propulsive efficiency in the Angelfish, *J. exp. Biol.* 85: 337–342.

Blake, R. W. (1981). Influence of pectoral fin shape on thrust and drag in labriform locomotion. *J. Zool., Lond.* 194: 53–66.

Carr, A. (1952). *Handbook of turtles of the United States, Canada and Baja California*. New York: Comstock Press.

Dabelow, A. (1925). Die Schwimmanpassung der Vögel. Ein Beitrag zur biologischen Anatomie der Fortbewegung. *Gegenbaurs Morph. Jb.* 54: 288–321.

Elder, J. W. (1960). The flow past a plate of finite width. *J. Fluid Mech.* 9: 133–153.

Fage, A. & Johansen, F. C. (1927). On the flow of air behind an inclined flat plate of infinite span. *Rep. Br. Aeronaut. Res. Counc. R & M* No. 1104.

Fish, F. (1980). Energetics and mechanics of aquatic locomotion in the muskrat. *Am. Zool.* 19: 898.

Gadd, G. E. (1964). Bilge keels and bilge vanes. *Rep. Nat. Phys. Lab. Ship* No. 64.

Hoerner, S. F. (1958). *Fluid dynamic drag*. Published by the author.

Johnson, W., Soden, P. D. & Trueman, E. R. (1972). A study in jet propulsion: an analysis of the motion of the squid, *Loligo vulgaris*. *J. exp. Biol.* 56: 155–165.

Jones, W. P. (1941). The virtual inertias of a tapered wing in still air. *Rep. Br. Aeronaut. Res. Counc. R & M* No. 1946.

Keulegan, G. H. & Carpenter, L. H. (1956). Forces on cylinders and plates in an oscillating flow. *Rep. Nat. Bur. Stds* No. 4821.

Knight, M. & Wenzinger, C. J. (1928). Wind tunnel tests on a series of wing models through a large angle of attack range. Part 1. Force tests, *N.A.C.A. Rep.* No. 317.

Lighthill, M. J. (1975). *Mathematical biofluiddynamics.* Philadephia: S.I.A.M.

Lilley, G. M. (1970). Fluid forces acting on circular cylinders for application in general engineering. Part II: Finite-length cylinders. *Engineering Sciences Data Unit. Item* No. 70014.

Lindsey, W. F. (1938). Drag of cylinders of simple shapes. *N.A.C.A. Rep.* No. 619.

McNown, J. S. (1957). Drag in unsteady flow. *Int. Congr. App. Mech.* 9(3): 124–134.

Norberg, U. M. (1975). Hovering flight in the Pied flycatcher (*Ficedula hypoleuca*). In *Swimming and flying in nature* 2: 869–881. Wu, Y. T., Brokaw, C. J. & Brennen, C. (Eds). New York: Plenum Press.

Romer, A. S. (1966). *Vertebrate palaeontology.* Chicago: University of Chicago Press.

Scrutton, C. (1941). Some experimental determinations of the apparent additional mass effect for an aerofoil and for flat plates. *Rep. Br. Aeronaut. Res. Counc. R & M* No. 1931.

Siekmann, J. (1963). Theoretical studies of sea animal locomotion. Part 2. *Ing. Arch.* 32: 40–50.

von Mises, R. (1959). *Theory of flight.* New York: Dover Publications.

Walker, W. (1971). Swimming in sea turtles of the family Cheloniidae. *Copeia* 1971: 229–233.

Webb, P. W. (1975). Hydrodynamics and energetics of fish propulsion. *Bull. Fish. Res. Bd Can.* 190: 1–159.

Weihs, D. (1977). Periodic jet propulsion of aquatic creatures. In *The physiology of movement: biomechanics.* Nachtigall, W. (Ed.). Stuttgart: Gustav Fischer Verlag.

Weis-Fogh, T. (1973). Quick estimates of flight fitness in hovering animals, including novel mechanisms for lift production. *J. exp. Biol.* 56: 79–104.

Wick, B. H. (1954). Study of the subsonic forces and moments on an inclined plate of infinite span. *N.A.C.A. Tech. note* No. 3221.

Wiles, W. F. (1970). Fluid forces and moments on flat plates. *Engineering Sciences Data Unit. Item* No. 70015.

Symp. zool. Soc. Lond. (1981) No. 48, 53—69

Locomotion of Plaice Larvae

R. S. BATTY

Scottish Marine Biological Association, Dunstaffnage Marine Research Laboratory, P.O. Box 3, Oban, Argyll, Scotland

SYNOPSIS

The locomotion of larval plaice is demonstrated using a new technique — silhouette cinematography. This method, which uses no lenses, gives a very high contrast which is necessary to photograph the almost transparent fins of fish larvae.

Kinematic analyses of cruising speed swimming from the films show many differences from adult swimming. The pectoral fins are used together with body waves at cruising speeds, but pectoral fins are not important in producing propulsive thrust. Their purpose is to counteract recoil caused by lateral tail movements and therefore prevent yaw of the head. This is achieved by synchronized tail and fin movements with a $180°$ phase difference between the strokes of the left and right pectoral fins. Body wave speed v and swimming speed u were measured and show low values of u/v (0.2 to 0.4) which are positively correlated with swimming speed. Other parameters of the body waves (wavelength, amplitude, frequency) were also measured.

Recordings of burst speeds showed very high tail beat frequencies and a change in swimming style. The pectoral fins are no longer used at these speeds, at which the wavelength of the body wave appears to increase.

INTRODUCTION

The kinematics and hydrodynamics of adult fish swimming have been extensively studied by biologists and physical scientists. This effort, which has been concentrated on undulatory propulsion, has been reviewed by Gray (1968) and Webb (1975). "Paddling" has received much less attention, which is surprising since its use for slow swimming speeds and manoeuvering is so widespread, but it is the subject of another paper in this volume (Blake, this volume).

Little is known, however, about the kinematics of larval fish swimming. This is not so surprising when it is realized that, owing to the transparency of the larvae, special techniques have to be employed to photograph their movements (Hunter, 1972; Arnold & Nuttall-Smith, 1974). Most attention has been given to clupeoid larvae, either anchovy (*Engraulis mordax*) by Hunter (1972) and Vlymen (1974), or herring (*Clupea harengus*) by Rosenthal (1968) and Rosenthal & Hempel (1969). Clupeoid larvae are typical of one group of larvae that have relatively long slender bodies. Other larvae, with shorter, deeper bodies, have not been studied at all. The plaice larva (*Pleuronectes platessa* L.) is typical of this shorter, deeper body shape. This species is bilaterally symmetrical on hatching at a length of 5 mm and during the larval stages until it metamorphoses at a length of approximately 12 mm to lie on the left side.

Preliminary work with plaice larvae had shown that they swim with body undulations and by moving their pectoral fins, a type of swimming that turbot and lemon sole larvae also employ, which indicates that this may be typical of all flatfish larvae. This chapter describes a new photographic technique called silhouette cinematography which has been used to make a kinematic analysis of the swimming of plaice larvae.

METHODS

Source of Animals

Eggs were stripped from a brood stock kept in the laboratory and fertilized artificially. The eggs were incubated and larvae reared to metamorphosis in 20-litre round black plastic tanks. Temperature was not controlled and the ambient sea water temperature in the aquarium rose from 6 to 12°C during the season. Larvae were fed on a mixture of *Artemia salina* nauplii and natural plankton.

Post yolksac larvae of 7–11 mm body length were used for these experiments.

Silhouette Cinematography

Preliminary work, using closed circuit television and a "Scotchlite" background illumination technique similar to that used by Wardle & Reid (1977), showed that the pectoral fins were used simultaneously with body movements. These fins are nearly transparent and were not always clearly visible using this technique.

Edgerton (1977) had shown that high speed silhouette photography, a simple technique using no lenses, could give high resolution photographs of small rapidly moving subjects. This method would, in theory, give the highest possible contrast and was thought to be the ideal method to demonstrate larval plaice swimming.

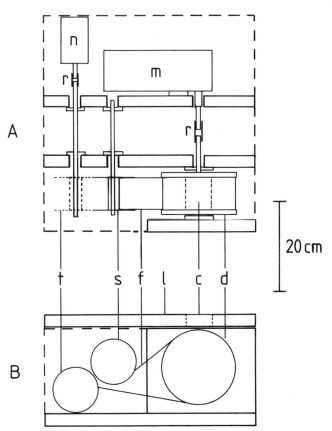

FIG. 1. The silhouette "camera". (A) plan view, (B) side view. c, container for subjects; d, drum; f, 70 mm film; l, lid; m, drum motor; n, take up spool motor; r, rubber coupling; s, supply spool; t, take-up spool.

Edgerton's technique was modified so that the film moved beneath the subject while a strobe flashlight, 2.2 m vertically above the subject in a dark room, gave a series of separate exposures along the length of the film.

The field of view is determined by the area of film exposed and is limited by the film format used. Therefore, to allow a large 57 mm × 57 mm field of view, 70 mm (Kodak TriXpan) film was used. A diagram of the apparatus is shown in Fig. 1. The supply spool contains the film to be exposed preceded by a 10 m leader of scrap film. With the 10 μs, 0.15 J strobe running at 70 Hz and the subjects in place, both motors are started together. The take-up spool motor tends to tension the film against the drum and the drum motor determines the maximum speed of film movement of 4.9 m s^{-1} at which the

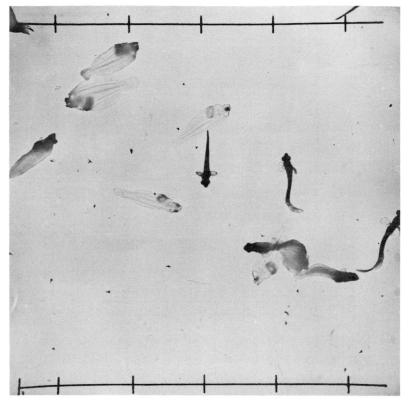

FIG. 2. A print from one frame of a silhouette film showing plaice larvae at various stages. The reference scales have 1 cm divisions and are 5 cm long.

70 Hz strobe gave separate images on the film. After exposure the film was developed in Kodak D 76 developer for 9 min. A print from one frame of a resulting negative is shown in Fig. 2.

Analytical Techniques

The films were analysed by a computer-aided method. Individual frames were projected and traced onto sheets of paper which were then fixed to a digitizing table linked to a Hewlett Packard 9825 desktop computer. Whole outlines of the body and pectoral fins were traced with the digitizer's cursor (Fig. 3), recorded by the computer and stored on magnetic tape for subsequent analysis.

Body waves
The first step was to reduce the data to a number of points centred

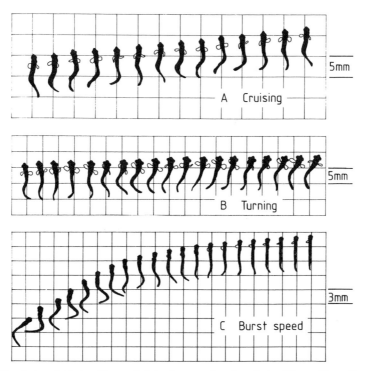

FIG. 3. Computer plotted outlines of plaice larvae swimming, derived from silhouette photographs. (A) an example of cruising swimming; (B) turning; (C) burst speed swimming. There are 14 ms between frames.

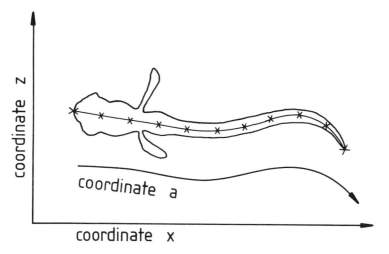

FIG. 4. A diagram showing the co-ordinate system and the 11 points on the co-ordinate a at fixed 0.1 a increments from the head ($a = 0$) to the tail ($a = 1$).

on the notochord of fixed position along a length co-ordinate a (see Fig. 4). To achieve this a series of points making up a line midway between the two sides of the larva was found. This was done by taking pairs of points, one from each side of the body at a similar distance from the head and then finding the co-ordinates of the point midway between these two points. The a co-ordinate of each of these points was found. The x and z co-ordinates for 11 points at 0.1 a increments, along a from head = 0 to tail = 1 were found by interpolation. It was then possible to follow similar methods to those used by Videler & Wardle (1978) to find the various parameters of the body waves, which are summarized by Fig. 5. These authors estimated wave speed v from wave crest da/dt, i.e. speed along the co-ordinate a. This value was found by plotting z against time for each a increment and so finding the time at which a wave crest (maximum or minimum) passes through each point. These values were then plotted as graphs of a against time (Fig. 6). Videler & Wardle (1978) multiplied the value for da/dt, found by fitting regression lines, by a factor to correct for the effect of body bending on the distance occupied by the body along the x axis.

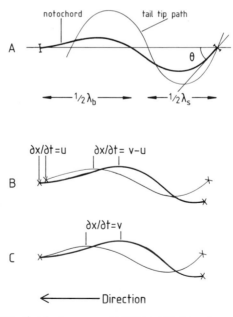

FIG. 5. Diagrams demonstrating body wave parameters. (A) shows λ_b, λ_s, and θ (the angle between any part of the body and the direction of motion. (B) and (C) define swimming speed u and wave speed v from the passage of wave crests down the body. In (B) the two consecutive notochord lines are in their real positions and in (C) they are moved so that the heads are superimposed.

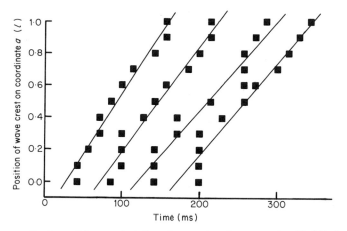

FIG. 6. *a* co-ordinate positions against time for a series of wave crests with fitted regression lines.

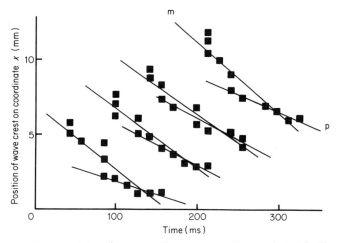

FIG. 7. *x* co-ordinate positions (corresponding to *a* co-ordinates plotted in Fig. 6) against time. The slopes of fitted regression lines (m) to all points and (p) to the last 6 points give mean and posterior values for $v - u$.

   In the present case this shortening was increased by the larger body angles θ, and therefore it was considered better to measure actual wave crest dx/dt and then find the corresponding x co-ordinate for each a co-ordinate found by Videler & Wardle's (1978) method. Figure 6 shows that plots of a against time are straight lines and contrast with Fig. 7, showing plots of x against time (for calculation of dx/dt on the same sequence). This value of dx/dt is equal to $v - u$. Figure 7 shows that $v - u$, and therefore v, can decrease along the length of the larva and demonstrates the value of using this method for calculating v.

Two values for v were calculated from each wave, using either all the points on a or the last six; in other words a mean or posterior v (v_m, v_p).

Pectoral fin movements

A simple analysis was made in order to describe pectoral fin movements. It was not possible to distinguish and follow individual fin rays through a fin stroke cycle as Blake (1979) did. Left and right fin tip co-ordinates were defined for each frame as the point on the fin outline furthest from the base of the fin (the first pair of co-ordinates recorded for each fin). Having found the fin tip, the fin angle was defined as the angle between the tip of the larva's head, the fin base and the fin tip. The phase and frequency of fin movements and body movements could now be compared. Since it was not possible to find the positions of individual fin rays the three-dimensional shapes of the fins were not known, and for this reason a measurement of angle of attack to the water flow, could not be made. An estimation of this angle of attack was taken as the projected area of the fin seen from above. When this projected fin area was low, angle of attack to the water flow was high and vice versa.

RESULTS

Cruising Speeds

Cruising swimming of plaice larvae is characterized by the simultaneous use of pectoral fins and body waves. On the only recorded occasion when pectoral fins were used alone no forward progress was made. Body waves without the use of pectoral fins were only used at burst speeds.

Body waves

Wave characteristics were analysed for two sizes of larvae, 7 mm and 11 mm. An example of amplitude of z displacement along the body is shown in Fig. 8, together with angle θ for the intervals between each a increment. These amplitude curves are very similar to those for adult fish swimming in the subcarangiform mode (Grillner & Kashin, 1976). There is a major difference between larval and adult fish in the body angle θ curve. A much larger tail θ occurs at the moderate cruising speed employed here than has been found in

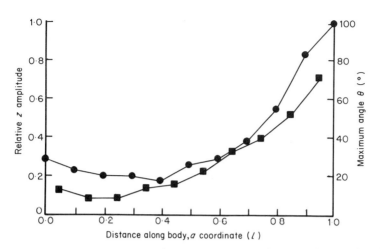

FIG. 8. Maximum relative amplitudes for each length co-ordinate a point ●, and maximum angle θ of each interval between the points ■, for a 10 mm larva.

similar analyses of adult fish cruising. Indeed, in larvae, tail angles of greater than 90° have been observed at higher cruising speeds and during acceleration. It is notable that the measured wave speeds v show no positive correlation with forward swimming speed u (Fig. 9). In adult fish u/v is relatively constant (Videler & Wardle, 1978). The technique used to find v has some bearing on the result since v varies with position on the body (Fig. 7), decreasing from head to tail, when u was above $30 \, \text{mm s}^{-1}$. For this reason both v_m and v_p are plotted in Fig. 7 showing the changing swimming style at higher cruising speeds. Figure 10, a plot of u/v against speed u, shows a clear increase in u/v with swimming speed and the change in style.

If v does vary along the body then so must wavelength λ_b, otherwise the larvae would be ripped apart if frequency changed. Since frequency is equal to v/λ_b, when speed v decreases the wavelength λ_b must get shorter.

Tail tip amplitude varies little with speed, therefore maximum tail θ must increase with swimming speed as tailwards decrease in v becomes more pronounced. This may be the means of increasing thrust force with speed. Unfortunately, maximum measured values for θ at the tail are not very close to the true maximum value since the technique did not provide sufficient frames per second. Films at a higher frame rate would be necessary to confirm this conclusion.

Mean values of λ_b/L (body wavelength in units of body length), the number of waves included on the body and amplitude (A), are

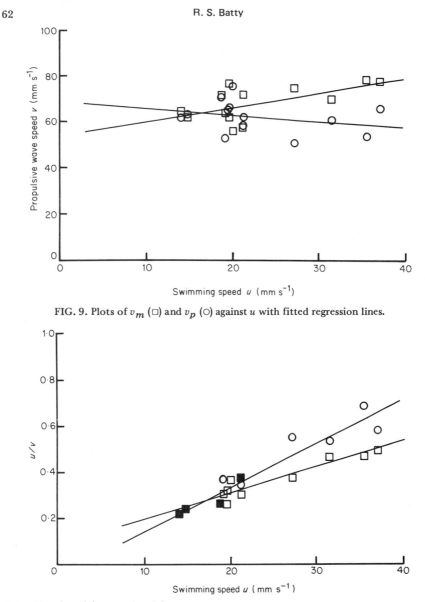

FIG. 9. Plots of v_m (□) and v_p (○) against u with fitted regression lines.

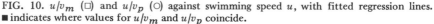

FIG. 10. u/v_m (□) and u/v_p (○) against swimming speed u, with fitted regression lines. ■ indicates where values for u/v_m and u/v_p coincide.

shown in Table I, for two body lengths, 7.2 mm and 10.0 mm. There is little difference between the two sizes of larva and both include slightly more than one wave on their body when swimming at cruising speeds.

TABLE I

Mean values of body wave parameters from two plaice larvae swimming at cruising speeds

Length on a (mm)	Length on x (mm)	λ_b/L (L)	Waves on body	A/L (L)
7.2	6.7	0.84	1.1	0.32
10.0	9.2	0.82	1.1	0.35

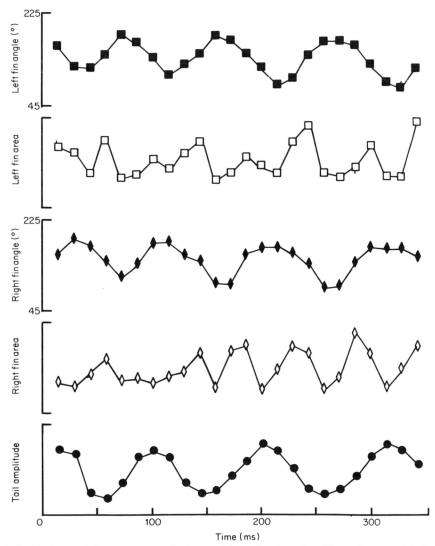

FIG. 11. Pectoral fin movements during straight swimming of a 10 mm larva. ■, left fin angle; □, left fin area; ◆, right fin angle; ◇, right fin area; ●, tail amplitude.

Pectoral fin movements

An example of an analysis of pectoral fin movements during steady forward swimming is shown in Fig. 11. This diagram clearly shows the synchronization of pectoral fins with body waves. Their movements are of the same frequency but there is a 180° phase difference between the left and the right pectoral fin. Projected fin areas are plotted on the same axis as fin angle and give an indication of angle of attack. The projected fin area (an indication of angle of attack) varies at twice the frequency of fin beating, showing that lift or negative lift may be produced on both strokes.

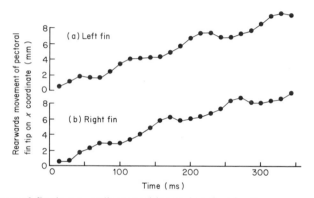

FIG. 12. Pectoral fin tip x co-ordinate position against time for the same sequence shown in Fig. 11.

Rearward strokes of the pectoral fins may have little propulsive effect. Figure 12, a graph of fin tip x co-ordinate against time demonstrates the low or zero effect on the water of rearward strokes; but the synchronization of pectoral fin strokes with propulsive waves will tend to counteract head yaw by counteracting the recoil effect produced by the tail fin strokes. The pectoral fin strokes are perfectly timed to do this and improve efficiency by reducing drag induced by swimming movements.

Turning

Pectoral fin movements change during turning. In the example shown in Fig. 3, the larva is turning to the right. The right fin moves between a point perpendicular to the body and a point 180° from the head as in straight swimming, but the left fin moves between 0 and

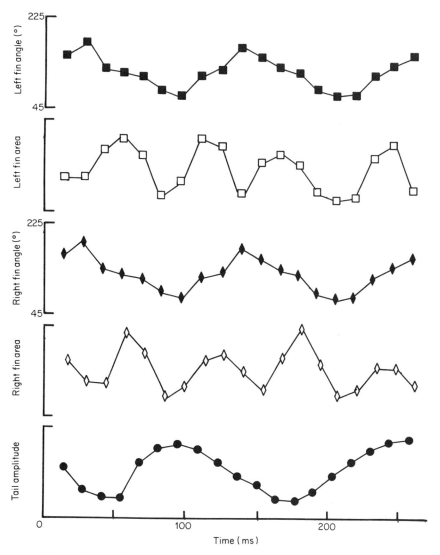

FIG. 13. Pectoral fin movements during a right turn. Labelled as in Fig. 11.

90° from the head; this situation is reversed in left turns. An analysis of this turn (Fig. 13) shows a movement similar to straight swimming except for this change in the stroke of the left fin to move in the range 0 to 90° from the head.

The part played by the pectoral fins during turning is not clear, especially since the turning couple on the head would tend to rotate the head against the desired direction.

Burst Speeds

At burst speeds only body waves are used (Fig. 3c) and are apparently of a different form to cruising speed swimming. Unfortunately the photographic method used has limited the number of frames per tail beat to two, at this speed, so that a proper analysis of the propulsive waves cannot be made. An examination of the outlines of a 7 mm larva shown in Fig. 4 indicates that wavelength λ_b is longer than one body length (L) and that amplitude is much greater than during cruising. These two changes cause less than one complete wave to be included on the body. It seems that the larva has "changed gear" in order to swim at this higher speed of 140 mm s^{-1} (20L s^{-1}) whilst using a very high tail beat frequency of 35 Hz. A considerable yaw of the head is seen in this sequence, compared with cruising swimming (Fig. 4a) when pectoral fin movements are used. Measurements of maximum tail beat frequencies of the plaice larvae are plotted in Fig. 14 together with similar data obtained by Bainbridge (1958) for adult fish and by Hunter (1972) for anchovy larvae.

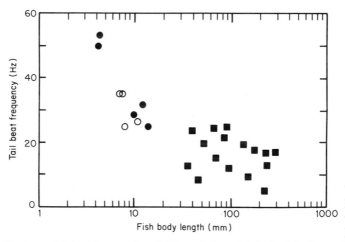

FIG. 14. Maximum tail beat frequencies of different sizes of fish. ○, plaice larvae; ●, anchovy larvae (Hunter, 1972); ■, adult fish (Bainbridge, 1958).

The graph demonstrates that maximum tailbeat frequency is strongly dependent on size, increasing as size decreases. This observation is confirmed by the findings of Wardle (1975) on the contraction time of adult muscle which shows the same type of size effect.

DISCUSSION

The results indicate that the pectoral fins are not important for producing thrust to propel the larva forward. It seems that their role is to produce lift, which is important since the larvae are negatively buoyant (Ehrlich, 1972), and at the same time, during cruising, to reduce head yaw.

Head yaw is caused by angular recoil forces which are generated by the lateral movements of the tail (Lighthill, 1977). The opposite forces generated by the pectoral fins of the larvae form a couple tending to turn the head against the recoil caused by the tail. Any reduction in head yaw should lead to a decrease in drag caused by swimming movements (Lighthill, 1977), but in this case the improvement in swimming movements is due in part to an increase in drag caused by the forward movement of one of the fins.

If alternating fin movements were not used lift would be produced by a pair of fixed fins, resulting in some drag. The plaice larva's alternating pectoral fin movements increase swimming efficiency by using energy spent in producing lift to reduce the amount of drag that is induced by swimming movements.

The one disadvantage is that roll might be induced, but this was too small to be observed in the films made in this study and would be expected to be very small in a deep-bodied fish larva which will have a large resistance to forces producing roll.

If the body waves of larvae are compared with those of adult fish, they are most like adult fish swimming in the subcarangiform mode (Breder, 1926). But there are many differences between larvae and adults. Speed of the propulsive wave v and tail beat frequency are not firmly linked to swimming speed at cruising speeds as in adults. This means that u/v, found to decrease only slightly with increasing speed for an adult fish, actually increases in the larval plaice: u/v is much lower (0.2 to 0.4) than in adult fish where it is in the range 0.6 to 0.8.

Values of u/v may be used to calculate propeller efficiency η_p (Lighthill, 1960) provided that amplitude increases from near zero at the head to a maximum at the tail tip.

$$\eta_p = \tfrac{1}{2}(1 + u/v) \qquad (3.1)$$

Using this formula, η_p would be in the range 0.6 to 0.7 for plaice larvae or 0.8 to 0.9 for most adult fish. Webb (1977) used results obtained by C. C. Lindsey to make estimates of propeller efficiency for a wide range of sizes. He showed a size dependence of propeller efficiency of similar magnitude to the results presented here.

ACKNOWLEDGEMENTS

The work described here was carried out while I held a NERC research studentship at the Scottish Marine Biological Association. I thank Dr J. H. S. Blaxter, my supervisor, for his encouragement and advice. I also acknowledge the many fruitful discussions I have had with my colleagues, in particular D. Booth. The cooperation of D. A. Neave and A. J. Geffen in rearing the plaice larvae was invaluable.

REFERENCES

Arnold, G. P. & Nuttall-Smith P. B. N. (1974). Shadow cinematography of fish larvae. *Mar. Biol.* **28**: 51–53.
Bainbridge, R. (1958). The speed of swimming of fish as related to size and to the frequency and amplitude of the tail beat. *J. exp. Biol.* **35**: 109–133.
Blake, R. W. (1979). The mechanics of labriform locomotion. I. Labriform locomotion in the Angelfish (*Pherophyllum eimekei*): an analysis of the power stroke. *J. exp. Biol.* **82**: 255–271.
Blake, R. W. (1981). Mechanics of drag-based mechanisms of propulsion in aquatic vertebrates. In *Vertebrate locomotion*: 29–52. Day, M. H. (Ed.). London and New York: Academic Press.
Breder, C. M. (1926). The locomotion of fishes. *Zoologica, N.Y.* **4**: 159–256.
Edgerton, H. E. (1977). Silhouette photography of small active subjects. *J. Microsc.* **110**: 79–81.
Ehrlich, K. F. (1972). *Morphometrical, behavioural and chemical changes during growth and starvation of herring and plaice larvae*. Ph.D. Thesis: Stirling University.
Gray, J. (1968). *Animal locomotion*. London: Weidenfeld & Nicolson.
Grillner, S. & Kashin, S. (1976). On the generation and performance of swimming in fish. In *Neural control of locomotion*: 181–201. Herman, R. M., Grillner, S., Stein, P. S. G. & Stuart, D. G. (Eds). New York: Plenum Press.
Hunter, J. R. (1972). Swimming and feeding behavior of larval anchovy *Engraulis mordax*. *Fish. Bull., U.S.* **70**: 821–838.
Lighthill, M. J. (1960). Note on the swimming of slender fish. *J. Fluid Mech.* **9**: 305–317.
Lighthill, M. J. (1977). Mathematical theories of fish swimming. In *Fisheries mathematics*: 131–144. Steele, J. H. (Ed.). London & New York: Academic Press.
Rosenthal, H. (1968). Schwimmverhalten und Schwimmgeschwindigkeit bei den Larven des Herings *Clupea harengus*. *Helgoländer wiss. Meeresunters.* **18**: 453–486.
Rosenthal, H. & Hempel, G. (1969). Experimental studies in feeding and food requirements of herring larvae. In *Symp. Marine Food Chains, Univ. of Aarhus, Denmark 1968*: 344–364. Steele, J. H. (Ed.). Edinburgh: Oliver and Boyd.
Videler, J. J. & Wardle, C. S. (1978). New kinematic data from high speed cine film recordings of swimming cod (*Gadus morhua*). *Neth. J. Zool.* **28**: 465–484.

Vlymen, W. J. (1974). Swimming energetics of the larval anchovy, *Engraulis mordax*. *Fish. Bull., U.S.* 74: 36—51.

Wardle, C. S. (1975). Limit of fish swimming speed. *Nature, Lond.* 225: 725—727.

Wardle, C. S. & Reid, A. (1977). The application of large amplitude elongated body theory to measure swimming power in fish. In *Fisheries mathematics*: 171—191. Steele, J. H. (Ed.). London & New York: Academic Press.

Webb, P. W. (1975). Hydrodynamics and energetics of fish propulsion. *Bull. Fish. Res. Bd Can.* 190: 1—159.

Webb, P. W. (1977). Effects of size on performance and energetics of fish. In *Scale effects in animal locomotion*: 315—331. Pedley, T. J. (Ed.). London & New York: Academic Press.

Symp. zool. Soc. Lond. (1981) No. 48, 71—113

Structure and Function of Fish Muscles

IAN A. JOHNSTON

Department of Physiology, University of St Andrews,
St Andrews, Fife, Scotland

SYNOPSIS

Fish swim using a combination of paired and unpaired fins and undulations of the segmental myotomal muscles. Although there is a simple anatomical separation of fibre types within the myotome, the myoseptal organization and orientation of fibres is complex. The number of distinct fibre types described varies from two to five depending on species. Slow red fibres form either a thin superficial or an internalized strip which constitutes between 0.5% and 29% of the total muscle. Red fibres are multiterminally innervated, being activated by local junction potentials. Characteristically, the fraction of red fibre volume occupied by mitochondria (25—38%) is comparable to mammalian heart muscle. The bulk of locomotory muscle consists of larger diameter fast fibres which have a highly developed glycogenolytic capacity. In elasmobranchs, holosteans, chondrosteans and some primitive teleosts fast fibres are innervated by a single basket-like end-plate formation at one myoseptal end. The available electromyographical evidence suggests that in such cases red fibres alone support sustained activity and the fast muscle is reserved exclusively for short periods of burst swimming. In contrast, most teleosts recruit fast fibres for higher sustainable as well as burst swimming speeds. Fast muscles in such fish are unusual among vertebrates in having extensive polyneuronal innervation. Isolated polyneuronally innervated fast fibres require much higher stimulation frequencies (200—300 Hz) to elicit maximum tensions than fibres with single end-plates (15—20 Hz). Full activation of polyneuronally innervated muscles probably requires simultaneous and perhaps asynchronous activity of a number of different motor neurones. This may give additional flexibility to polyneuronally innervated fast muscles allowing their recruitment at a wider range of swimming speeds than fibres with single end-plates. Finally, special features of the regulation of contractility and energy metabolism in fish muscles are discussed in relation to locomotion.

INTRODUCTION

Adaptations for aquatic locomotion account for many of the specialized features of fish muscle. The energy expenditure of locomotion in water is not linearly related to speed as it is for many terrestrial

vertebrates (see Bennett, 1978). In general, the power requirements
of swimming are thought to rise as a function of body size and velocity[3]. This reflects the way drag-forces on the body increase with
speed (see Webb, 1975). Thus burst speeds require proportionally
more effort than sustained swimming. This explains the predominance
of fibre types that can develop power rapidly and essentially independently of the circulation. Typically, 90% of fish skeletal muscle
is composed of anaerobic white fibres giving the flesh both its characteristic white colour and culinary importance. The possession of a
large muscle mass reserved for burst activity does not constitute a
serious weight penalty as it would in a terrestrial animal since most
fish preserve neutral buoyancy.

In the older fish groups (elasmobranchs, dipnoans, primitive teleosts) the different energetic requirements of sustained and burst
swimming have led to a complete anatomical and functional division
between fast and slow motor systems (Bone, 1966). Indeed the
physical separation of red and white fibres in the swimming muscles
of *Torpedo* led to one of the earliest descriptions of fibre types in
vertebrates (Lorenzini, 1678). Thus, in dogfish, sustained swimming
is entirely supported by slow red fibres which account for only 5–
10% of the myotomal musculature. In contrast, the fast motor system
which consists of phasically active white fibres is reserved for burst
activity (Bone, 1966). The red and white muscles of elasmobranchs
have somewhat similar innervation and physiological properties to,
respectively, the slow tonic and fast twitch muscles of other vertebrate groups (Bone, 1964, 1966, 1978a).

However, in most teleosts there is not a simple division of labour
between red and white muscles and both fibre types are recruited at
sustainable swimming speeds (Hudson, 1973; Johnston, Davison &
Goldspink, 1977; Bone, Kicznuik & Jones, 1978). The fast muscles of
advanced teleost groups are unusual among vertebrates in having
multiple innervation. For example, each fast fibre in the scorpaeni-
form fish, *Myoxocephalus scorpius*, is innervated by two to five
separate axons from each of four spinal nerves (Hudson, 1969).
Characteristically isolated polyneuronally innervated fast muscles
require much higher stimulation frequencies to elicit maximum tensions than fibres with single end-plates (Flitney & Johnston, 1979;
Johnston, 1981). It seems likely that full activation of polyneuronally
innervated muscles requires simultaneous and perhaps asynchronous
activity of a number of different motor neurones. This may well give
additional flexibility to power development by the musculature over
and above that offered by the hierarchical recruitment of motor units
of different sizes. The control of contractility and the regulation of

metabolism in multiply innervated fast muscles remains one of the outstanding problems for physiologists and biochemists interested in fish muscle.

LOCOMOTORY MUSCLES

A wide variety of body forms and modes of swimming have been described among the fishes (see Webb, 1975; Lindsey, 1978). Forward swimming is usually achieved by lateral undulations of the segmental myotomal muscles passing backwards along the trunk. However, a large number of fish rely to some degree on the paired and unpaired fins to provide forward motion (e.g. rays, trigger fish, etc.). In many cases, for example the surf perch *Cymatogaster aggregata*, as swimming speed increases the main propulsive thrust switches from the enlarged pectoral fins to oscillations of the trunk and caudal fin (Webb, 1973). The boxfish *Ostracion* provides an extreme example of the use of fins for locomotion in that the trunk is totally inflexible and the fish swims by fan-like oscillations of the caudal fin.

The overwhelming majority of papers published on fish muscle have dealt exclusively with the segmental myotomal musculature. In teleosts and elasmobranchs the myotomes have a complex W-shape and form a series of overlapping cones. The muscle fibres in each myotome insert into a connective tissue sheath or myocommata. Functional aspects of the shape and arrangement of the myotomes have recently been reviewed both in different groups of fishes and in development (Bone, 1978a). The orientation of fibres within the myotome is complex and varies both along the body and with distance from the vertebral column (see Alexander, 1969). In general, superficial fibres run parallel to the surface whereas deeper fibres make angles of up to $40°$ with the long axis of the body (Alexander, 1969). The significance of this arrangement is thought to be to allow similar degrees of sarcomere shortening at different body flexures. Thus optimal overlap between thick and thin filaments and hence maximum tension generation is achieved at all depths within the myotome.

FIBRE TYPES

Nomenclature

Although classifications based on muscle colour have lost favour with workers on other vertebrate groups, they are still widely applied to

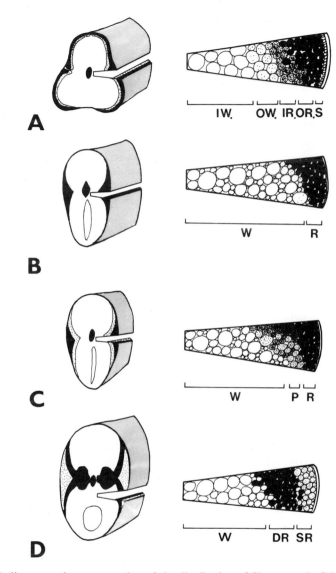

FIG. 1. A diagrammatic representation of the distribution of fibre types in fish segmental myotomal muscle. (A) Dogfish (*Scyliorhinus canicula*) after Bone & Chubb (1978). S, superficial; OR, outer red; IR, inner red; OW, outer white; and IW, inner white muscle fibres. (B) Brook trout (*Salvelinus fontinalis* Mitchill) after Johnston & Moon (1980b). R, red; and W, white muscle fibres. (C) Common carp (*Cyprinus carpio* L.) after Johnston, Davison *et al.* (1977). R, red; P, pink; and W, white muscle fibres. (D) Skipjack tuna (*Katsuwonus pelamis* L.) after Rayner & Keenan (1967) and Bone (1978b). SR, superficial red; DR, deep red; and W, white muscle fibres. The intensity of shading represents the relative histochemical staining reactions for succinic dehydrogenase, a mitochondrial marker enzyme.

fish muscle. In other vertebrates, locomotory muscles contain a heterogeneous mixture of different fibre types. However, in fish, fibre types are largely present in different positions within the myotome (Fig. 1). Thus problems of classification which arise due to the presence of both fast and slow red fibres are not as serious in fish as in other animals. In addition, the complex orientation of myotomal fibres has made it technically difficult to obtain information about relative contraction speeds. The similar innervation of fast and slow muscles in most teleosts also makes this an inappropriate way of distinguishing between fibre types. Since most studies are based entirely on histochemical or ultrastructural criteria it is often safer to retain the rather unsatisfactory nomenclature based on colour than to adopt unsubstantiated physiological criteria such as fast or slow, tonic or phasic. Thus in this review the terms red and white have been used except where data exist to allow classification according to physiological properties.

Histology and Histochemistry

Most investigations have concerned the segmental myotomal musculature, although there have been a few studies of the paired and unpaired fins (Bergman, 1964; Kryvi & Totland, 1978; Nishihara, 1967; Walesby & Johnston, 1980a). In a survey of 84 species of marine fish Greer-Walker & Pull (1975) reported that red fibres constituted between 0.5 and 29% of the myotomal muscle mass. The proportion of red fibres is highest in active pelagic families such as Scombridae and Clupeidae and lowest in bottom dwelling predators, deep-sea fishes and those species which use their fins as a primary means of locomotion.

Cartilaginous fishes which have been studied include *Chimaera monstrosa* (Kryvi & Totland, 1978), dogfish *Scyliorhinus canicula* L. (Bone, 1966, 1978a) and the sharks *Etmopterus spinax* and *Galeus melastomus* (Kryvi, 1977). Bone (1978a) has described five fibre types in the dogfish on the basis of differences in innervation, ultrastructure and histochemical staining characteristics (Fig. 1A). In dogfish, the outer border of the myotome consists of a single interrupted layer of large diameter superficial fibres which show negative histochemical staining for succinic dehydrogenase (SDHase) and Ca^{2+}-activated myofibrillar ATPase activity (Bone & Chubb, 1978). Superficial fibres are characterized by an intense staining for glucose-phosphate-isomerase. The functional significance of these fibres is unclear. They are unlikely to represent growth stages of the underlying red fibres since they appear relatively late during post-embryonic development (Bone, 1978a). Two types of red fibre can be distinguished on the basis of size, aerobic capacity and myofibrillar ATPase activity (Bone & Chubb, 1978). Outer red fibres have a higher SDHase and lower myofibrillar ATPase activity than the somewhat larger diameter red fibres adjacent to the white muscle. The outer few layers of white fibres are differentiated from deep white fibres by having a higher proportion of mitochondria, more abundant capillary supply, and a somewhat different myofibrillar ATPase activity (Fig. 1A) (Bone, 1978a; Bone & Johnston, in press). Biochemical measurements of myofibrillar ATPase of the five fibre types of dogfish are presented in Table I and correlated with the histochemical observations.

TABLE I

Mg^{2+}- Ca^{2+}-stimulated myofibrillar ATPase activities of fish skeletal muscle fibre types. Activities expressed as μmol $P'Pi$ released, mg myofibrillar protein^{-1} min^{-1}

Fibre types	Dogfish[a] (15°C)	Trout[b] (15°C)	Carp[c] (25°C)	Tuna[d] (25°C)
Superficial	0.03	—	—	—
Outer red	0.06 ⎫	0.25 ⎫	0.25	0.51
Inner red	0.18 ⎭	⎭		0.50
Pink	—	—	0.62	—
Outer white	0.55 ⎫	0.73 ⎫	1.09 ⎫	0.99
Inner white	0.40 ⎭	⎭	⎭	

Data from Bone & Johnston (in press)[a], Johnston & Moon (1980b)[b], Johnston, Davison & Goldspink (1977)[c] and Johnston & Tota (1974)[d]. Numbers of fish used and standard errors of the mean are given in the original publications. Assay temperatures are shown in brackets.

The arrangement of fibre types of some representative teleosts is shown in Fig. 1B–D. The simplest case is exemplified by species such as brook trout (*Salvelinus fontinalis* Mitchill) (Johnston & Moon, 1980b: fig. 2) and the Atlantic mackerel (*Scomber scombrus*) (Bone, 1978b). In these fish the myotome is differentiated into two distinct fibre types on the basis of histochemical staining for aerobic enzymes and myofibrillar ATPase (Fig. 1B; Johnston & Moon, 1980b). Biochemically the myofibrillar ATPase activity of the white fibres is around three times that of the red (Johnston, Frearson & Goldspink, 1972: table 2). White fibres are SDHase negative and form a heterogeneous group with respect to fibre size. In the case of rainbow trout there is a continuous distribution of white muscle fibre diameter between 15 and 90 μm, overlapping that of red fibres, 5—40 μm (Johnston, Ward & Goldspink, 1975). Most workers have considered the range of fibre size in teleost white muscle to represent different stages in growth rather than distinct fibre types (Johnston, Ward *et al.*, 1975; Korneliussen, Dahl & Paulsen, 1978; Johnston & Moon, 1980b). Greer-Walker (1970) has shown that fibre number continues to increase throughout life in many species of fish. This contrasts with the situation found in most other vertebrates where fibre number is fixed at or shortly after birth.

There have been relatively few studies of capillary supply and blood flow to fish red and white muscle fibres (Boddeke, Slijper & van der Stelt, 1959; Bone, 1978b; Stevens, 1968; Mosse, 1978). Capillary density is related to the aerobic capacity of the muscle. In a study of three teleost species Mosse (1978) found a mean capillary fibre ratio of 0.2—0.9 for white and 1.9—2.5 for red muscle. In *Platycephalus bassensis* 49% of white fibres have no peripheral capillaries (Fig. 2). Interestingly, the capillarization of white muscles in certain pelagic species is more highly developed. The white fibres of yellow tail scad (*Trachurus maccullochi*), pilchard (*Sardinops neopilchardus*) and mackerel (*Scomber australasicus*) have fewer fibres with no peripheral capillaries (17.8%, 16.5% and 23% respectively) (Mosse, 1979).

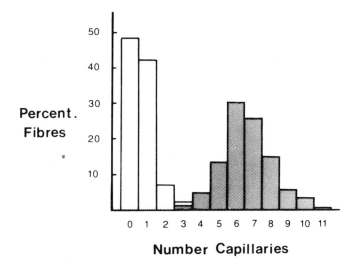

Percent. Fibres

Number Capillaries

FIG. 2. Percentage of red and white muscle fibres surrounded by a given number of capillaries. The data is for the flathead (*Platycephalus bassensis*) and counts were made by direct visualization of capillaries in semi-thin (1 μm) durcupan sections. From Mosse (1978: table 3).

In some species another type of fibre occurs in an intermediate position between the red and white muscle layers (Bokdawala & George, 1967; Johnston, Patterson, Ward & Goldspink, 1974; Mosse & Hudson, 1977). These so-called pink fibres can be distinguished histochemically from other fibre types on the basis of their stability to alkaline (pH 10.3–10.4) preincubation prior to staining for myofibrillar ATPase activity (Johnston, Patterson *et al.*, 1974). In carp (*Cyprinus carpio*), pink fibres constitute a larger proportion of the myotome than red fibres ($\sim 10\%$) (Fig. 1C) (Davison, Johnston & Goldspink, 1976). Biochemical studies of the fibre types in carp have shown pink fibres to have intermediate aerobic enzyme and Mg^{2+} Ca^{2+} myofibrillar ATPase activities between red and white fibres (Johnston, Davison *et al.*, 1977). The light chain compositions of vertebrate myosins have been shown to be characteristic of fibre phenotype (Lowey & Risby, 1971). Carp red fibres have two, and pink and white fibres, three species of light chain, characteristic of slow and fast muscle myosins respectively (Focant, Huriaux & Johnston, 1976; Johnston, Davison *et al.*, 1977).

Some fishes of the family Scombridae have internalized red muscle masses (Fig. 1D). This feature is associated with a counter-current vascular heat-exchanger which enables elevated red muscle and brain temperatures to be maintained over a wide range of ambient temperatures. The extent to which capacities for regulating body temperature have developed within the tunas varies considerably between species (see Sharp & Pirages, 1978). In the skipjack tuna (*Katsuwonus pelamis*) the superficial wedge of red muscle is differentiated from the internalized red muscle by a larger fibre size and a significantly lower histochemical staining for SDHase activity (Bone, 1978b). Indeed, quantitative ultrastructural studies by Bone (1978b) have shown ten times as many mitochondria in deep red fibres (Table II) and a more highly developed capillary bed (~ 4.7 capillaries/fibre). Mitochondria in superficial red and white muscle differ from those of the deep red fibres in showing a reduction of cristae development and the presence of lamellar figures interrupting the cristae array (Bone, 1978b). Mg^{2+} Ca^{2+} myofibrillar ATPase activities of Atlantic bluefin tuna (*Thunnus thynnus*) red muscles are twice that of white muscle and somewhat higher than analogous red muscles of less active species (Table I) (Johnston & Tota, 1974). In tuna muscular effort associated with ventilation has been transferred from the buccal to myotomal muscles (ram ventilation) (Dizon, Brill & Yuen, 1978). This feature is associated with continuous high-speed swimming. Significantly it has been suggested that both tuna red and white muscles have a significant capacity of aerobic glucose utilization (Guppy, Hulbert & Hochachka, 1979).

TABLE II

Mitochondrial content (fractional fibre volume %) of fish myotomal muscle fibres

Species	Red	Muscle Type Intermediate	White	Source
ELASMOBRANCHS				
Sharks *Galeus melastomus*	34	16	1	Kryvi (1977)
Etmopterus spinax	30	7	0.5	
CHONDROSTEANS				
Sturgeon *Acipenser stellatus*	30	3.7	0.7	Kryvi, Flood & Guljaev (1980)
TELEOSTS				
Coalfish *Gadus virens*	25	—	1	Patterson & Goldspink (1972)
Atlantic mackerel *Scomber scombrus*	35.5	—	2	Bone (1978a)
Skipjack tuna *Katsuwonus pelamis*	16.4(DR)	3.3 (SR)	2.2	Bone (1978b)
Crucian carp *Carassius carassius*	25	20	4	Johnston & Maitland (1980)
Plaice *Pleuronectes platessa*	25	—	2	Johnston (1981)
Antarctic 'cod' *Notothenia rossii*	30	—	2	Walesby & Johnston (1980a)

Abbreviations: DR, deep red muscle; SR, superficial red muscle. Note that in *Notothenia rossii* the deep pectoral adductor muscle provides the main propulsive thrust for sustained swimming. Red fibres in this muscle have even higher mitochondrial densities ($\sim 38\%$) than the myotomal muscles above.

ULTRASTRUCTURE

There have been a number of recent quantitative studies of the fine structure of fish myotomal (Patterson & Goldspink, 1972, 1973; Kryvi, 1977; Kryvi & Totland, 1978; Bone, 1978b; Walesby & Johnston, 1980a; Johnston & Maitland, 1980) and pectoral muscles (Kryvi & Totland, 1978; Walesby & Johnston, 1980a).

The striking feature of fish red muscle is the high mitochondrial density, which often exceeds that found in the most active mammalian muscles. For example, the fractional volume occupied by mitochondria in red myotomal muscles of various cold-water species is comparable to their volume in the ventricular muscle of the mouse

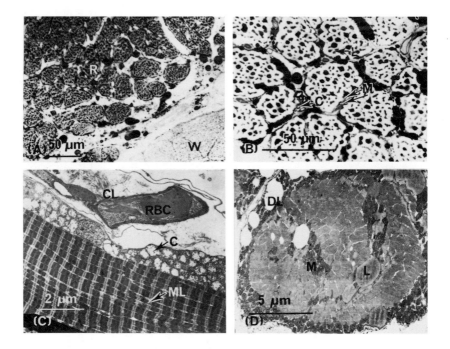

FIG. 3. (A) Brook trout (*Salvelinus fontinalis*) red muscle fibres. Transverse frozen section (10 μm) stained for succinic dehydrogenase as a mitochondrial marker enzyme. Note lipid droplets between red fibres (R) and the higher SDHase staining in red than in white fibres (W) (bottom left). (B) Brook trout red fibres, semi-thin (1 μm) araldite embedded section stained with p-phenylene diamine (PPDA). Note the abundant capillary supply (C) and the high proportion of fibre volume occupied by mitochondria (M) and lipid (L). (C) Longitudinal section of plaice red fibre from a four-month-starved fish. Note loss of cristae from mitochondria (C) also the presence of a distinct M line (ML) and capillary (CL) containing a nucleated red blood corpuscle (RBC). (D) Plaice (*Pleuronectes platessa* L.) red fibre. Electronmicrograph showing high mitochondrial density (M) and numerous lipid droplets (L). This fish has been food deprived for several weeks. Note the depletion of lipid deposits (DL).

(34%) and finch (37%) (Bossen, Sommer & Waugh, 1978) (Table II; Fig. 3). In contrast, approximately 0.5–8% of white fibre volume is occupied by mitochondria, depending on position within the myotome and species (Table II). An interesting exception to this correlation between muscle pigmentation and respiratory capacity occurs in icefishes of the family Channichthyidae. In this group all muscles are pure white in colour owing to the absence of both haemoglobin (Rudd, 1954) and myoglobin (Hamoir, 1978; Walesby, Nicol & Johnston, in press). The superficial myotomal fibres in the pelagic icefish *Champsocephalus gunnarii* have a mitochondrial density of around 45% and are almost entirely surrounded by capillaries (Walesby, Flitney & Johnston, unpublished results).

Another characteristic of fish red or slow fibres concerns the highly developed sarcoplasmic reticulum (SR) and T-tubule system. The fractional volume occupied by SR in red muscle is greater than for tonic and slow twitch muscles in other vertebrates and approaches that of fast twitch fibres (Johnston, 1980b). Quantitative data from studies of fish sarcotubular systems are summarized in Table III. In most myotomal muscles studied T-tubules are located at the junction

TABLE III

Fractional volume (%) occupied by sarcoplasmic reticulum and T-system in some representative vertebrate muscles

| Species | Sarcoplasmic reticulum | | T-system | |
	Muscle Type			
	Red	*White*	*Red*	*White*
ELASMOBRANCHS				
Etmopterus spinax[a]	4.9	6.0	0.33	0.55
Galeus melastomus[a]	4.6	6.8	0.57	0.89
TELEOSTS				
Salmo gairdneri[b]	5.1	13.7	0.10	0.40
	Tonic	*Twitch*	*Tonic*	*Twitch*
AMPHIBIANS				
Rana pipiens[c]	4-5[f]	9.1	0.18[g]	0.32
	Slow Twitch	*Fast Twitch*	*Slow Twitch*	*Fast Twitch*
MAMMALS				
Guinea pig[d,e]	3.2	4.6	0.14	0.27

Data is taken from the following: [a]Kryvi (1977); [b]Nag (1972); [c]Mobley & Eisenberg (1975); [d]Eisenberg & Kuda (1975); [e]Eisenberg, Kuda & Peter (1974); [f]Peachey (1965); [g]Flitney (1971) (*Rana temporaria*).

of the Z disc (Franzini-Armstrong & Porter, 1964; Kilarski, 1966; Nag, 1972; Patterson & Goldspink, 1972; Kryvi, 1977). A similar location of the tubular system is found among the locomotory muscles of lampreys (Terävainen, 1971), urodeles (Totland, 1976) and anurans (Peachey, 1965). In fish extraocular (Kilarski, 1966) swim bladder and drum muscles (Eichelberg, 1976) and in hagfish (Korneliussen & Nicolaysen, 1973) and mammalian muscles the T-tubules are situated at the A-1 boundary. The functional significance of these different positions is unknown.

The fractional volume occupied by myofibrils varies from 80–96% in white muscles to 40–60% in red muscles (see Johnston, 1980b, for a review). Myofibrillar packing is more regular in white muscles. White fibres of teleosts are characterized by elongated peripheral myofibrils (Fig. 4). In small fibres the arrangement of fibrils resembles

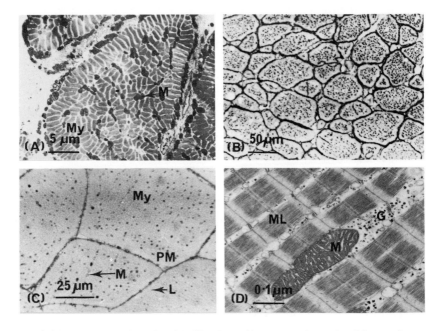

FIG. 4. (A) Transverse section of a pink fibre from the myotomal muscle of the crucian carp (*Carassius carassius* L.). Note intermediate mitochondrial density (M) and irregular myofibrillar packing (My). In this species lipid droplets are not common in any fibre type; the principal stored fuel is glycogen. (B) Brook trout white fibres. Transverse frozen section (10 μm) stained for succinic dehydrogenase. Dark spots within the fibres correspond to lipid and mitochondrial accumulations. Note the wide range of fibre size. (C) Brook trout white fibres. Semi-thin (1 μm) araldite embedded section stained with p-phenylene diamine (PPDA). Note the peripheral ribbon-like myofibrils (PM), the dense myofibrillar packing (My) and the accumulations of mitochondria (M) and lipid (L). (D) Longitudinal section of crucian carp white fibre. Note distinctive M-line (ML), mitochondria (M) with highly developed cristae array and glycogen rosettes (G).

a cart-wheel with spokes radiating to a small central hub of sarco-plasm. This feature is not observed in elasmobranchs, holocephali or higher vertebrates and may be related to the pattern of fibre growth in teleosts.

In contrast to higher vertebrates, both red and white fibres have a distinctive M-line (Figs 3 & 4). Other ultrastructural criteria for dis-tinguishing between fibre types include a somewhat thicker Z-line in white than in red fibres (Patterson & Goldspink, 1972) and differ-ences in the deposition of stored metabolites (Bone, 1978a). Glycogen particles are present either as rosettes or small chains in all fibre types but are particularly abundant in red fibres. With the exception of tuna *Euthynnus pelamis* which has higher glycogen stores in white fibres (Guppy *et al.*, 1979) this is a reversal of the typical vertebrate pattern (Johnston, 1980b). Biochemical determinations of gly-cogen concentrations vary from 350—3600 mg/100 g in red to 9—1440 mg/100 g in white fibres depending on activity level, nutritional status, and species (Love, 1970). Lipid droplets are usually more abundant in red than white fibres constituting 11% of fibre volume in brook trout (*Salvelinus fontinalis* Mitchill) and plaice (*Pleuronectes platessa*) (Johnston & Moon, 1980b; Johnston, 1981) (Fig. 3). In fatty fish such as anchovy (Johnston, unpublished) a layer of fat cells occurs between the skin and red muscle layer, while in other species, for example mackerel (Bone, 1978b) and eel, adipocytes are widely distributed among the white muscle fibres. The extent to which the muscle lipid reserves of such fish represent metabolic stores, or function as part of a buoyancy mechanism, is unknown.

There have been relatively few ultrastructural investigations of the fibre types other than red or white. Intermediate/pink fibres of the carp (Johnston & Maitland, 1980) and sharks (*Galeus melastomus* and *Etmopterus spinax*) (Kryvi, 1977) have an intermediate mito-chondrial density and myofibrillar packing to red and white fibres. The dogfish superficial fibres described by Bone (1978a) are distinct from either red or white fibres in containing an inconspicuous M-line, virtually no lipid, simple mitochondria, and a relatively poorly developed T-system and SR.

INNERVATION AND ELECTROPHYSIOLOGICAL PROPERTIES

In dogfish, red fibres receive a number of terminations of the en-grappe type derived from two separate axons. The motor axons either pass into the fibre from the myoseptal ends or run across the surface of the muscle innervating a number of different fibres (Fig. 5).

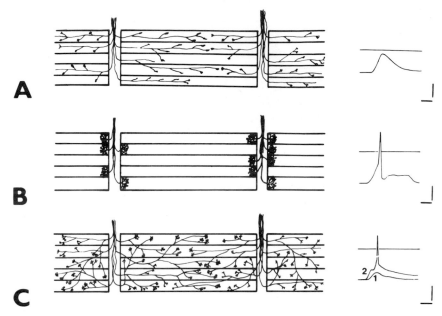

FIG. 5. A diagrammatic representation of the different patterns of motor innervation (left) and modes of activation (right) of fish muscle fibres. Traces of electrical activity show stylised responses to depolarizing pulses or nerve stimulation. Vertical calibration bars represent 20 mV, and horizontal bars 5 ms. Zero potential is indicated with a horizontal line. (A) All fish red fibres examined are multiply innervated and are activated by junction potentials (see Barets, 1961). (B) White fibres in elasmobranchs, dipnoans and certain teleosts (see Fig. 6) are focally innervated at one myoseptal end. Motor terminations are of the en-grappe type. Focally innervated fibres show a propagating action potential overshooting zero potential by around 20 mV (see Hagiwara & Takahashi, 1967). (C) Multiply innervated white fibres are found in all acanthopterygians and certain other fish groups (see Fig. 6). Each muscle fibre is innervated by several different motor axons. Stimulation of the nerve supply results in two distinct kinds of electrical response: (1) junction potentials; and (2) spike potentials (see Hudson, 1969).

From electrophysiological measurements of membrane space constants Stanfield (1972) concluded that motor terminals were around $150-200\,\mu$m apart. Subjunctional folds are present on terminals of both superficial and outer and inner red fibres of *Scyliorhinus* (Bone, 1972). Typically, red fibres do not produce action potentials in response to depolarizing pulses (Stanfield, 1972). Thus in this respect and in relation to their innervation dogfish red fibres resemble the true slow or tonic fibres of amphibians. However, Stanfield (1972) found that eight out of 27 red fibres examined showed a significantly large inward sodium current on depolarization to suggest they were capable of propagating an action potential. Of the remaining fibres six showed no inward Na current and the others only a small inward Na current. Thus although it is likely that red fibres are normally activated by junction potentials the electrophysiology of fish slow fibres warrants further investigation. The pattern of innervation of red muscles in other elasmobranchs and indeed in dipnoans, holosteans, chondrosteans and teleosts would appear to be broadly similar to that described from dogfish (Bone, 1964).

White muscle fibres in dogfish are innervated by large diameter axons which run in the myosepta and give rise to basket-like enplaque end-formations at one end of the fibre (Bone, 1964, 1966). Each muscle fibre is innervated by two separate axons which fuse to form a single end-plate (Bone, 1964, 1972). In dogfish, the two motor terminals contain vesicles of different size ranges, 50 nm and 100 nm (Best & Bone, 1973). The significance of this dual innervation is unknown since only acetyl cholinesterase has been demonstrated in sub-synaptic folds (Pecot-Dechavassine, 1961). Bone (1978a) has calculated that each motor unit in dogfish white muscle contains around 50—100 fibres. In a study of several species of tropical stingray Hagiwara & Takahashi (1967) found that white fibres showed typical over-shooting spike-potentials on depolarization. Thus white fibres of elasmobranchs resemble frog fast twitch fibres. In some elasmobranchs, for example *Torpedo*, the more superficial white fibres receive basket-like endings to the mid-region rather than the ends of the fibre (Bone, 1964).

The innervation of white muscles in chondrosteans, holosteans and dipnoans has been less studied but appears to resemble that of elasmobranchs (Fig. 6) (Bone, 1964). However, among teleosts another type of innervation occurs and focally innervated white muscles are only found among the more taxonomically primitive groups (Fig. 6). Indeed, Bone (1970) has suggested that the type of innervation may even serve as a taxonomic character. No acanthopterygians have focal innervation and of the orders which are considered to be primitive on other grounds, for example the Salmoniformes, only the Salmonidae are multiterminally innervated (Bone, 1970). Interestingly, orders such as the Osteoglossiformes and Ostariophysi also contain families with both types of innervation. For example within the Osteoglossiformes which comprise freshwater butterfly fish (Pantodontidae), feather backs (Notopteridae) and mooneyes (Hiodontidae), only *Hiodon* has terminal innervation (Bone, 1978a).

Thus in the majority of the 20,000 teleosts, including the Perciformes and Cypriniformes which together account for around four-fifths of all the species, the white fibres are multiply innervated. Branches of the spinal nerves run in the myosepta and fan out across the surface of the myotome to form a diffuse network such that each fibre receives numerous nerve terminals (Barets, 1961;

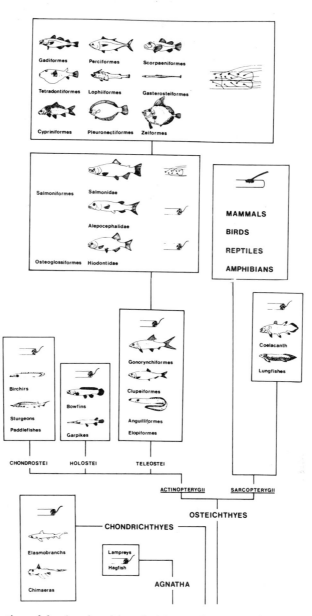

FIG. 6. Distribution of focal and multiterminal innervation among fast muscles of vertebrates. Only representatives of the teleosts have multiterminal innervation. Taxonomically
primitive teleosts (lower box) have focal innervation. The middle box contains orders such
as the Osteoglossiformes and Salmoniformes containing families with both focal and multiply innervated white fibres. This list is not intended to be comprehensive. See Bone (1964,
1970). Teleost orders with taxonomically advanced features have exclusively multiple innervation (upper box). Land vertebrates have single end-plates as do dipnoans, elasmobranchs,
chondrosteans, holosteans, lampreys and hagfish.

Bone, 1964) (Fig. 5). The end-plate formations have been variously described as consisting of small rings or loops as in the tench (Barets, 1961) or as being multiple and made up of numerous separate neuro-fibrillar annuli, e.g. *Myoxocephalus scorpius* (Hudson, 1969). Nerve terminals are usually embedded in the sarcolemma and subjunctional folds are absent (Nishihara, 1967). In *Myoxocephalus*, the short-horned sculpin, Hudson (1969) found that each fibre receives inner-vation from as many as five axons from each of four spinal nerves. The number of terminals on each fibre ranges from eight to 22 with an average of 14 about 0.7 mm apart. Hudson considered that each end-plate is derived from a separate axon while others have con-sidered that fibres are also multiply innervated by the same axon as is the case with red fibres (Barets, 1961). Hudson (1969) calculated that each abdominal myotome in the sculpin consists of 12 motor units all of a considerable size.

There have been relatively few studies of the electrophysiological properties of multiply innervated fast muscles (Barets, 1961; Hidaka & Toida, 1969; Hudson, 1969). Hudson (1969) found that electrical stimulation of spinal nerves elicits two kinds of electrical response, namely spike potentials resulting in a fast twitch, or junction poten-tials leading to a graded local contraction by the muscle. The typical response to supramaximal nerve stimulation is an all-or-none spike potential overshooting zero-potential by as much as 20 mV. Depolar-ization of the membrane occurs in two phases. Duration of the fast phase of the spike potential is of the order of $1.5-2.5 \, m \, s^{-1}$ at 12°C (Hudson, 1969). Variation in stimulus intensity below maximal results in junction potentials the amplitude of which can be quan-titized in discrete steps providing evidence for innervation by several different motor axons. Generally, Hudson found that junction poten-tials did not exceed 35 mV and that in fresh preparations spike po-tentials often arose from a single junction potential. However, in a proportion of cases, particularly in older preparations, only junction potentials could be elicited. At present, it is not possible to conclude whether white fibres can be activated *in vivo* in the absence of spike potentials. Clearly, the electrophysiological properties of polyneuron-ally innervated myotomal muscles are worthy of further study.

The few multiterminally innervated fin muscles that have been investigated appear to be similar to myotomal muscles (Bergman, 1964; Nishihara, 1967). Fin muscles may well provide a convenient nerve—muscle preparation for investigating the combined electrical and mechanical responses of polyneuronally innervated fibres.

MECHANICAL PROPERTIES

In spite of the separation of fish fibre types into different positions within the myotome there have been very few studies of their contractile properties. The complex fibre orientation and myoseptal insertion of fish trunk muscles make such studies technically difficult. In order to obtain preparations in which the fibres run parallel it is necessary to utilize either single or small bundles of fibres. Fin, jaw and sound-producing muscles provide more suitable whole muscle preparations and have been the subject of a limited number of studies (Bergman, 1964; Hidaka & Toida, 1969; Yamamoto, 1972).

Bone & Johnston (in press) have compared the properties of bundles of red and white fibres from the dogfish myotome. Both fibres respond to a single supramaximal stimulus; white fibres giving a fast twitch (half-time peak tension: $t\frac{1}{2} \sim 20$ ms) and red fibres a slow contraction ($t\frac{1}{2} \sim 100$ ms). On multiple stimulation, fused tetani are produced at frequencies above 8 Hz. Twitch-tetanus ratios of around 0.5 occur at 5 Hz and 10 Hz for red and white fibres respectively. Very different stimulation characteristics have been reported for multiply innervated teleost muscle. Flitney & Johnston (1979) studied red and white fibre bundles isolated from adductor operculi muscles of *Tilapia mossambica*. Red fibres from *Tilapia* only respond at stimulation frequencies in excess of 5–10 Hz. Both fibre types produce graded, fused tetani, reaching a maximum at 250–300 Hz. Unloaded speeds of shortening are 2.6 L s^{-1} for white and 1.5 L s^{-1} for red fibres at 18°C (Flitney & Johnston, 1979). However, rate of rise of tension during supramaximal stimulation at 200 Hz was 6.5 times greater for white than for red fibres.

The responses of fibre bundles to different stimulation frequencies for (a) the focally innervated white fibres of the cuckoo ray *Raja naevus* and by (b) the polyneuronally innervated myotomal fibres of the cod *Gadus morhua* L. are shown in Fig. 7. Typical tetanic fusion frequencies on multiple stimulation are 5–10 Hz for skate and 40–50 Hz for cod. Maximal tensions require stimulation in excess of 200 Hz for cod and only 20 Hz for skate. The rate of tension development is also frequency-dependent. Alexander (1969) has calculated that the complex geometry of white fibres results in very little shortening ($\sim 3\% L_0$), even at maximum body flexure. It seems likely that the rate of tension development for multiply innervated fibres *in vivo* may be limited as much by the muscles membrane properties as by its maximum speed of shortening.

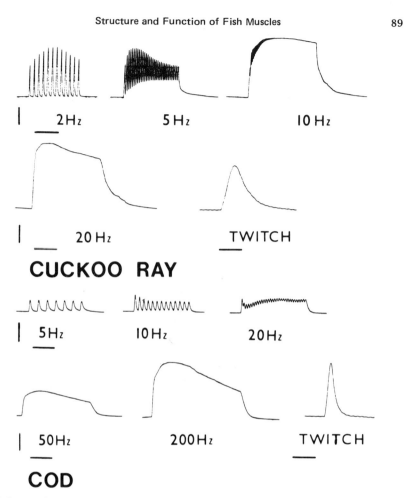

FIG. 7. Isometric contraction of bundles of white muscle fibres at different stimulation frequencies. (Top) Cuckoo ray *Raja naevus* (Muller & Henle) pectoral muscle. (Bottom) Cod *Gadus morhua* dorsal myotomal muscle. Stimulation voltage 6V 1 ms pulse width. Vertical scale bar represents tension 3 kg/m^2. Horizontal scale bars represent 2s (cuckoo ray), 500 ms (cod) (multiple stimulation) and for twitches 100 ms. Temperature 18°C. Note the lower tetanic fusion frequency and lower stimulation frequency required to elicit maximum tension in ray compared to cod muscle. After Johnston (1980a).

A full understanding of these differences in the modes of activation of fast fibres from different groups of fishes must await mechanical studies with innervated single fibre preparations.

REGULATION OF MUSCLE CONTRACTILITY

Some features of the regulation of contractility in fish skeletal muscle are illustrated in Fig. 8. Tension development rises as a function of

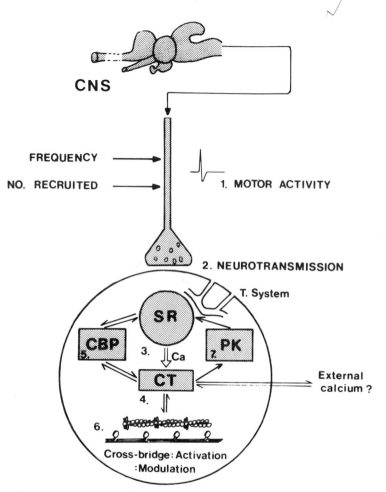

FIG. 8. The control of contractility in fish muscle. Abbreviation: CNS, central nervous system; SR, sarcoplasmic reticulum; CBP, cytoplasmic calcium binding proteins; CT, calcium transient; PK, Ca^{2+}-dependent protein kinase. Factors modulating force production include: (1) The number and frequency of motor units activated. (2) Neurotransmission. (3) The muscle action potential (or distributed depolarization) is transmitted to the SR via the T-tubule system causing release of bound calcium. (4) This results in a transient increase in free calcium from 0.1 to $10\,\mu M$. (5) The duration of the calcium transient is thought to be reduced in fast muscles by the presence of high concentrations of cytoplasmic calcium binding proteins (parvalbumins), see Fig. 9. (6) Calcium ions activate (a) muscle contraction and (b) glycogen breakdown by direct binding effects to troponin C and phosphorylase kinase respectively. There is also evidence that force production can be modulated by myosin phosphorylation through a specific Ca^{2+}-activated myosin light-chain kinase. (7) In slow mammalian muscles calcium uptake into the SR is modulated by a cytoplasmic factor thought to be a Ca^{2+}-dependent protein kinase. Slow muscles of salmonid fishes do not appear to have such a mechanism (McArdle & Johnston, in press). The extent to which this is true of other species is unknown. Incubation of *Tilapia* red and white muscles in ringers containing zero calcium or verapamil ($5 \times 10\,M^{-5}$) leads to a rapid and reversible loss of tension (Flitney & Johnston, 1979). It appears, therefore, that fish muscles have a requirement for extracellular calcium. Whether Ca^{2+} crosses the sarcolemma or remains bound to sites on the outside is unknown.

myoplasmic free calcium concentration within the range 0.1 to 10 μM. In multiply innervated fibres the release of calcium from the SR is dependent on the degree of depolarization of the muscle membrane which is in turn determined by the number of motor neurones activated and their firing frequency. In fish white but not red fibres cytoplasmic calcium binding proteins (parvalbumins) play an important role in regulating free calcium concentrations (Gerday & Gillis, 1976; Pechère, Derancourt & Harech, 1977).

Parvalbumins are found in all vertebrate fast muscles but occur in particularly high concentrations in fish white fibres ($\sim$15% soluble proteins) (Le Peuch, Demaille & Pechère, 1978). Characteristically they are acidic proteins of low molecular weight (11,000–12,000 daltons) which can bind 2-g atoms Ca^{2+}/mole (Pechère, Capony & Demaille, 1973). Sequence studies have shown strict conservation of the amino acid residues at the Ca^{2+}-binding sites and a high degree of homology with other muscle calcium binding proteins, troponin C, myosin P light chain and calmodulin (Collins, 1976; Perry, 1979). In the metal-free form parvalbumins are able to inhibit Mg^{2+} Ca^{2+} myofibrillar ATPase activity by chelating Ca^{2+} ions bound to troponin C (Pechère, Derancourt et al., 1977). Calcium bound to parvalbumins can be exchanged and accumulated by SR vesicles (Gerday & Gillis, 1976). The relative binding constants of Ca^{2+} for troponin C and parvalbumins are Kd's 10^{-6} M and 10^{-7} M respectively (Benzonana, Capony & Pechère, 1972; Potter & Gergely, 1975). The current view is that parvalbumins function in aiding rapid relaxation (Gerday & Gillis, 1976; Pechère, Derancourt et al., 1977). Since parvalbumins occur in molar excess over TNC in fish muscles Ca^{2+}-released by the SR causes only a transient activation of crossbridges as Ca^{2+} rapidly becomes bound within the cytoplasm. Thus relaxation occurs before all the Ca^{2+} is resequestered within the SR (see Fig. 9).

In some fish the sarcoplasmic reticulum occupies a somewhat similar proportion of fibre volume in both fast and slow fibres (e.g. Patterson & Goldspink, 1972; see also Table III).

Evidence has been obtained that Ca^{2+}-uptake in mammalian slow but not fast muscles is regulated by a Ca^{2+}-dependent protein kinase which phosphorylates components of the pump necessary for transport. Although sarcoplasmic reticulum isolated from red and white muscles of rainbow trout has a somewhat different protein composition and it does not appear to require protein kinases or cAMP for full activation (McArdle & Johnston, in press). The extent to which this is true for other species and represents a phylogenetic difference between fishes and mammals is unknown.

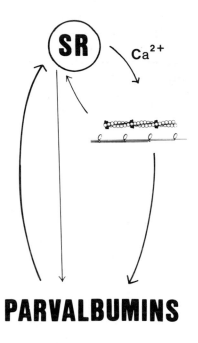

PARVALBUMINS

FIG. 9. Diagrammatic representation of the role of parvalbumins in Ca^{2+} movements in fish fast muscle fibres. The width of the lines represents the relative magnitude of the Ca^{2+} fluxes. After Pechère, Derancourt et al. (1977).

Activation of myosin crossbridges in fish muscle is achieved through direct effects of Ca^{2+}-binding to troponin C as in other vertebrates (Lehman & Szent-Györgyi, 1975). In addition, phosphorylation of myosin through a specific Ca^{2+}-activated myosin light-chain kinase may lead to a modulation of force production (see Perry, 1979).

FIBRE RECRUITMENT DURING SWIMMING

The recruitment of different fibre types with increasing speed has been studied in dogfish (Bone, 1966), rainbow trout (Hudson, 1973), carp (Johnston, Davison et al., 1977), herring (Bone et al., 1978), skipjack tuna (Rayner & Keenan, 1967; Brill & Dizon, 1979), bass (Freadman, 1979, saithe (Johnston & Moon, 1980a) and brook trout (Johnston & Moon, 1980b). This represents only a small proportion of the 20,500 species of living fishes. However, at present the results are consistent with a division in the pattern of fibre recruitment between different groups of fishes according to the type of innervation of the fast motor system. Focally innervated fast fibres appear only to be recruited for burst activity. For example, electromyographical studies in the Pacific herring show that only red fibres are recruited at speeds that can be sustained for several hours. Higher speeds (5 body lengths/s) result in the recruitment of white fibres and fatigue of the fish within a further 1—2 min swimming (Bone *et al.*, 1978). Qualitatively similar results have been obtained for the dogfish (*Scyliorhinus canicula*) which also has focally innervated white fibres (Bone, 1966). In contrast a number of studies have shown that both red and white fibres are recruited at sustainable swimming speeds in species with multiply innervated fast muscles (Fig. 10) (Hudson, 1973); Johnston, Davison *et al.*, 1977; Bone *et al.*, 1978).

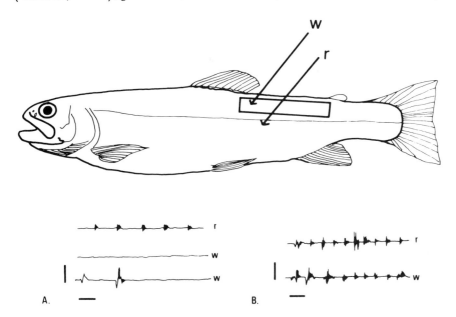

FIG. 10. Recruitment of fibre types of the brook trout (*Salvelinus fontinalis* Mitchill) during steady swimming. (Top) Shows insertion of EMG electrodes in red (r) and white (w) muscles. (Bottom A) Top two traces, EMGs recorded from red and white muscle regions during steady swimming at 1 length/s. Note potentials are only recorded from the red muscle layer. Lower trace, a burst of electrical activity in white muscle associated with a change of position in the swimming chamber during the same experiment. (Bottom B) Recordings of EMGs from red and white muscles at a steady swimming speed of 3 lengths/s. Brook trout are able to maintain this speed indefinitely. Horizontal scale represents 500 ms. Vertical scale represents 1 mv. (From Johnston & Moon, 1980b.)

For example the threshold speed for recruitment of white fibres is
0.8—1.9 lengths/s in saithe (Johnston & Moon, 1980a), and 2.0
lengths/s in carp (Johnston, Davison *et al.*, 1977). Somewhat higher
speeds for recruitment of white fibres have been reported in striped
bass (*Morone saxatilis*), 3.2 lengths/s, and bluefish (*Pomatomus
sultatrix*), 4.5 lengths/s (Freadman, 1979). Thus in some species with
multiply innervated fast muscles there may be a secondary reversion
to the more primitive pattern of fibre recruitment such as is found in
the dogfish. Recent studies have shown a wide range of aerobic capacities for teleost white muscles between species, presumably reflecting different degrees of involvement of this fibre type in sustained
swimming (Johnston & Moon, 1980b).

Only in carp have electromyographs (EMGs) been recorded from
the pink muscle layers. In this study it was shown that the order of
recruitment of fibre types with increasing speed is red muscle > pink
muscle > white muscle (Johnston, Davison *et al.*, 1977).

METABOLISM

Enzyme Activities of Red and White Muscle Fibres

Determinations of maximal *in vitro* activities of enzymes can be used
to assess the metabolic capacities of different muscles. In the case of
non-equilibrium enzymes such information gives a useful semi-
quantitative estimate of maximal metabolic flux (see Newsholme &
Start, 1973; Newsholme, Zammit & Crabtree, 1978). Enzymes
catalysing non-equilibrium reactions can be identified by measurements of mass action ratios and they usually have among the lowest
activities in the pathway, being regulatory steps in metabolism.
Enzymes catalysing equilibrium reactions only provide qualitative
information about the relative importance of particular metabolite
pathways and the principal fuels supporting activity. Enzyme studies
of fish muscle have recently been reviewed (Zammit & Newsholme,
1979; Johnston, 1980b). Table IV gives the maximal *in vitro* activities
of some enzymes of energy metabolism from the fast and slow
muscles of the brook trout measured under optimal conditions of
pH, substrates and co-ions (Johnston & Moon, 1980b). Typically
white muscles have a higher glycolytic potential (phosphorylase,
PFK activities) and lower oxidative capacity (citrate synthetase) than
red muscles.

Bárány (1967) has demonstrated for a wide range of vertebrate
muscles that myofibrillar ATPase activities parallel the unloaded

TABLE IV

*Activities of some key enzymes of energy metabolism and metabolite concentrations in the red and white muscles of brook trout (*Salvelinus fontinalis Mitchill*). Data from Johnston & Moon (1980b) and Walesby & Johnston (1980b)*

Enzyme Activities (μmol substrate/g dry wt./min)	Red Muscle	White Muscle
$Mg^{2+} Ca^{2+}$ myofibrillar ATPase[+]	0.25	0.73
Creatine kinase	166	538
Adenylate kinase	145	380
5'AMP aminohydrolase	38	176
Glycogen phosphorylase	3.3	12
Hexokinase	1.1	2.8
Phosphofructokinase	39	67
Pyruvate kinase	271	701
Lactate dehydrogenase	687	1651
Cytochrome oxidase	7.9	1.8
Citrate synthetase	10.5	3.2
3-OH Acyl coA dehydrogenase	0.3	0.05
Metabolite concentrations (μmol/g dry wt.)		
ATP	2.9	6.1
ADP	0.4	0.6
AMP	0.2	0.2
Pi	3.8	7.1
Phosphoryl Creatine	9.3	25.7

Activity expressed as μmol Pi released/mg myofibrillar protein/min.

speeds of shortening. On this basis, white muscles of the trout and other species studied (see Table I) are around two to five times as fast as red muscle. Activities of enzymes responsible for maintaining ATP supply during contraction such as creatine kinase, adenylate kinase and 5'AMP amino-hydrolase parallel those of myofibrillar ATPase. 5'AMP amino-hydrolase is particularly important in fast muscles since it catalyses the deamination of AMP to IMP thus shifting the equilibrium of adenylate kinase in favour of ATP production. It has been calculated that in fatigued muscles up to half of the ATP for contraction is supplied by this pathway.

In a study of a large number of invertebrate and vertebrate species Beis & Newsholme (1975) found that muscles with the lowest ATP/AMP ratios usually have the lowest rate of ATP utilization and the least variation in energy requirement. The ATP/AMP ratios in the red and white muscles of brook trout (R, 18.6; W, 38.3) and tuna (R, 26.3; W, 61.1) reflect the relative activity levels of these muscles and lend support to this concept (Table IV).

Both metabolite concentrations (Freed, 1971; Walesby & Johnston, 1980b) and enzyme activities can be modified by environmental factors such as temperature, activity and nutritional status (see Johnston, 1980b).

Burst Swimming

The type of locomotory activity best understood is burst swimming. Maximum swimming speeds can only be maintained for relatively few tail-beat cycles and are approximately 26 lengths/s for 10 cm fish and 4 lengths/s for 1 m fish (Wardle, 1975). Burst swimming is accompanied by an extremely rapid activation of glycolysis in white muscle. In trout this results in the utilization of around 50% of muscle glycogen stores in 15 s which is equivalent to a flux of 40 μmol glycogen derived glucose/kg/s (Stevens & Black, 1966). There is near-quantitative conversion of muscle glycogen to lactate which may be retained in the muscle for several hours following the cessation of exercise (Wardle, 1978). Lactate levels of 59 (trout) and 69 mmol/kg (skipjack tuna) have been recorded in white muscles following 1—2 min burst swimming (Black, Robertson & Parker, 1961; Guppy et al., 1979). Recovery of white muscle lactate to pre-activity states may require up to 18 h in some species (Black, Robertson et al., 1961) but is usually complete within 30—60 min in red muscle (Johnston & Goldspink, 1973a).

Changes in metabolite concentrations between rest and maximal activity in tuna white muscle are illustrated in Fig. 11. Burst swimming is associated with a large decrease in phosphoryl creatine (90%) and ATP (53%) and a small increase in AMP concentrations (22%) (Guppy et al., 1979: table 2).

In the carp Driedzic & Hochachka (1976) observed a 48% decrease in the total adenylate pool and an essentially 1:1 increase in 1MP and NH_4 due to the deamination of AMP by 5′ amino-hydrolase (see above). it is possible that this is one source of the anaerobic NH_4^+ production which has been frequently observed following enforced exercise in fish (Kutty, 1972).

The regulation of glycogenolysis in fish white muscle is illustrated in Fig. 12. Non-equilibrium or regulatory steps in anaerobic glycolysis include phosphorylase, phosphofructokinase and pyruvate kinase (Newsholme & Start, 1973). There is an interesting phylogenetic difference in the control of glycogen breakdown between fish and

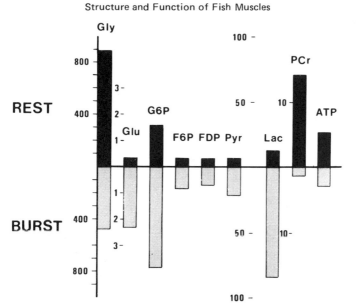

FIG. 11. Metabolite concentrations in freeze-clamped white muscles of skipjack tuna *Euthynnus pelamis* during rest and burst swimming activity (data from Guppy *et al.*, 1979: table 2). Abbreviations: Gly, glycogen; Glu, glucose; G6P, glucose-6-phosphate; F6P, fructose-6-phosphate; FDP, fructose 1, 6-diphosphate; Pyr, pyruvate; Lac, lactate; PC, phosphoryl creatine; ATP, adenosine 5'-triphosphate. Units of concentration μmol g wet weight muscle^{-1}. Dark shading represents data from resting fish and light shading burst swimming.

mammalian muscles (Fischer, Blum *et al.*, 1975). While rabbit muscle phosphorylase kinase is rapidly activated or inhibited by cAMP dependent protein kinases or specific phosphatases, the dogfish enzyme depends only on Ca^{2+} for its activity (Cohen, Duewer & Fischer, 1971; Fischer, Blum *et al.*, 1975; Fischer, Alaba *et al.*, 1978). Indeed, there is no production of cAMP in response to noradrenaline in dogfish white muscle (Fischer, Blum *et al.*, 1975). The lack of a hormonal mechanism for activating glycogen breakdown in dogfish white muscle may be related to the relatively poor circulation through this tissue during maximal activity. *In vitro* phosphofructokinase is known to be activated by substrates, AMP, ADP, Pi and NH_4 and inhibited by phosphoryl creatine, ATP and citrate (Mansour, 1972). Substrate cycling between F-6-P and FDP by the coupling of the PFK and FDPase reactions is thought to provide a mechanism for increasing the sensitivity of F-6-P phosphorylation to changes in AMP (Newsholme, 1976) (Fig. 12).

Increases in AMP concentration activate PFK and inhibit FDPase resulting in a proportionally larger increase in glycolysis than is possible in the absence of substrate cycling. According to this scheme

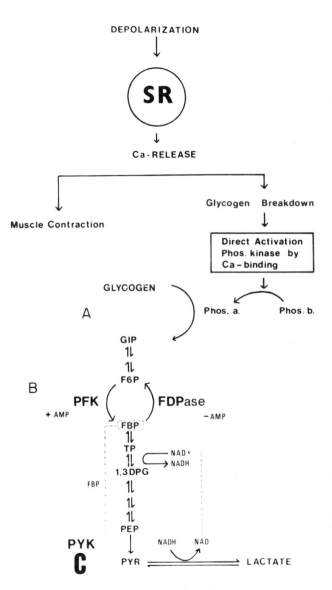

FIG. 12. Control of anaerobic glycogenolysis in fish white muscle. Regulatory steps include (A) glycogen phosphorylase, (B) phosphofructokinase and (C) pyruvate kinase (see text for details). Abbreviations: Phos, phosphorylase; G1P, glucose-1-phosphate; PFK, phospho-fructokinase; FBP, fructose 1,6-biphosphate; AMP, adenosine 5′ monophosphate; TP, triose phosphates; NAD, β-nicotinamide adenine dinucleotide; NADH, β-nicotinamide adenine dinucleotide reduced form; 1, 3 DPG, 1, 3 diphosphoglyceric acid; PEP, phospho(enol) pyruvate; PYK, pyruvate kinase. Other abbreviations are given in the legend to Fig. 11.

small changes in ATP lead to proportionately larger changes in AMP through the adenylate kinase reaction. Thus, in general, the higher the initial ATP:AMP ratio the greater the increase in AMP for a given breakdown of ATP and the greater the stimulation of glycolysis (Newsholme *et al.*, 1978). An interesting feature of the control of glycolysis in fish muscles is the presence of a regulatory pyruvate kinase. *In vitro* pyruvate kinase is inhibited by ATP and alanine and activated by substrates and FDP (Somero & Hochachka, 1968; Johnston, 1975). Regulatory pyruvate kinases are associated with muscles in which there is a rapid and dramatic increase in glycolysis between standard and burst activity. For example modulator-sensitive pyruvate kinase occurs in the locomotory muscles of diving mammals such as whales but not the limb muscles of terrestrial mammals (Hochachka & Storey, 1975). The observed changes in metabolites and Ca^{2+} release from SR during burst swimming are sufficient to cause a rapid activation of glycolysis through the known regulatory mechanisms described above (Figs 11, 12).

Steady Swimming

The metabolism of muscles during steady swimming is much less well understood. One recurring problem with studies of metabolism of sustained swimming has been the failure of many investigators to control the activity state of fish either before or during the exercise period. It is essential to know the exact swimming speed in lengths/s and to control as far as possible the condition of the fish prior to its introduction to the swimming chamber (e.g. temperature, nutritional status, level of stress etc.). In addition the adaptive variation observed within fish populations and particularly between species means that it is often impossible to reach general conclusions about metabolism.

In general, fish can increase their oxygen uptake by a factor of 10–15 times between rest and maximal activity (Bennett, 1978). Values for maximum oxygen consumption are comparable to the basal levels for birds and mammals and are in the range 190–644 ml $kg^{-1} h^{-1}$ (see Jones & Randall, 1978). Unfortunately, only fragmentary data are known for tuna which might be expected to have aerobic activity levels more comparable to homeotherms. It seems likely that red muscle receives a significant proportion of the cardiac output during steady swimming. This is particularly true for species in which red muscles alone support sustained activity (e.g. elasmobranchs, dipnoans, primitive teleosts). Since red fibres constitute a small fraction of the trunk musculature (5–10%) they undoubtedly possess a highly active aerobic metabolism. Indeed, the mitochondrial densities of fish red fibres are greater than those of similar fibres in

TABLE V

Maximal activities of enzymes of fatty acid and ketone body oxidation in fish red muscle. Activities expressed as μmol substrate utilized, g wet weight muscle^{-1} min^{-1}

Species	Hydroxybutyrate dehydrogenase	Carnitine palmitoyl transferase	Triacylglycerol lipase
Plaice	< 0.01	0.07	3.7
Bass	< 0.01	0.75	1.8
Mackerel	< 0.01	0.60	4.5
Dogfish	1.3	< 0.01	0.2
Ray	0.37	< 0.01	0.54
Spurdog	0.64	< 0.01	0.1

Data from Zammit & Newsholme (1979).

mammalian limb muscles (see Table II). Little is known about the regulation of the tricarboxylic acid cycle in fish (for a review see Driedzic & Hochachka, 1978). Lipid is known to be an important fuel during steady swimming particularly in those species which undertake a spawning migration (see Bilinski, 1974). There are indications that both lipolysis and the transport of fat differ significantly between fish and mammals (see Bilinski, 1974). For example, it has been reported that the activation of lipolysis in fish does not proceed via cAMP-dependent protein kinases (Farkas, 1969). There are also differences in the sites of storage and metabolism of lipid between different fish groups. In elasmobranchs triacyl glycerol is stored in the liver, whereas in teleosts a significant proportion is deposited either in discrete stores in the viscera or dispersed throughout the muscle fibres in adipocytes. Red muscles of teleosts contain high activities of carnitine palmitoyl transferase and triacyl glycerol lipase but not 3-hydroxybutyrate dehydrogenase. The opposite situation is found in elasmobranchs which have high activities of enzymes of ketone body but not fatty acid oxidation (Zammit & Newsholme, 1979) (see Table V). Triacyl glycerol and non-esterified fatty acids also occur in much higher concentrations in the plasma of teleosts than elasmobranchs (Zammit & Newsholme, 1979). These differences are augmented during starvation. Thus it seems likely that the most important fat fuels are ketone bodies in elasmobranchs and fatty acids in teleosts. Red muscles in salmonids can oxidize long chain saturated fatty acids to CO_2 at around ten times the rate of that of white muscles (Jonas & Bilinski, 1964; Bilinski, 1974). Similarly, carnitine palmitoyl transferase activities of teleost white muscles are generally only 5% of that of red muscles (Crabtree & Newsholme,

1972). Thus it seems likely that oxidation of fats as fuels is much more important in red than in white muscles even when the latter are recruited for sustained activity.

Numerous studies have shown glycogen depletion during sustained swimming in both the red and white muscles of teleosts (Pritchard, Hunter & Lasker, 1971; Johnston & Goldspink, 1973a,b). However, the pathways of carbohydrate metabolism utilized by multiply innervated white fibres at sustained speeds are not known for any species. Two basic possibilities exist: ATP generation could be aerobic using either local glycogen stores or blood glucose, or largely anaerobic resulting in the production of lactic acid. For the latter case, a number of schemes have been proposed involving lactate-exchange with other tissues to keep the fish in overall aerobic balance.

In a quantitative study of glycogen utilization in crucian carp Johnston & Goldspink (1973c) found that glycogen utilization at three lengths/s was two to three times higher in red than in white muscles. However, since red fibres only comprise 7% of the trunk musculature they only account for 15–20% of the total glycogen utilized at this swimming speed.

Crabtree & Newsholme (1972) have demonstrated a good correlation between hexokinase activities in various muscles and calculated maximum capacities for glucose oxidation. Table VI gives values for red and white muscle hexokinase activities in various species. It can be seen that the importance of aerobic glucose utilization is likely to vary considerably between species for both myotomal muscle types. White fibres in two species of trout and the skipjack tuna have high hexokinase activities comparable to those of red muscle (Table VI). Other evidence for a significant aerobic capacity in the white muscles of these species is as follows. In rainbow trout the ratio of mitochondria between red and white muscles is 8:3 by number (Nag, 1972). Activities of citrate synthetase and cytochrome oxidase in brook trout white muscle are 25–35% of that of red muscle (Johnston & Moon, 1980b). Thus in at least trout, and possibly tuna, the aerobic capacity of the white muscle is probably sufficient to support sustained activity and blood glucose may be an important fuel.

A number of early studies reported lactate accumulation two to three times resting levels in various salmonids forced to swim at sustained swimming speeds (Black, Bosomworth & Docherty, 1966). These results may well reflect the presence of a generalized stressed reaction in acutely exercised fish (see Wardle, 1978). Studies of trout trained to swim in an exercise chamber for several weeks prior to experimentation show no net accumulation of lactate over rested

TABLE VI

Activities of hexokinase in the red and white muscles of fish. Red fibres were all isolated from myotomal muscles except for Notothenia rossii (labriform locomotion) where values represent activities in the deep pectoral adductor muscle

Species	Temperature	Red muscle	White muscle	Source
Dogfish (Scyliorhinus canicula)	10°C	0.5	0.1	Crabtree & Newsholme (1972)
Herring (Clupea harengus)	10°C	0.1	–	Newsholme et al. (1978)
Rainbow trout (Salmo gairdneri)	20°C	1.3	0.6	Johnston (1977)
Brook trout (Salvelinus fontinalis)	15°C	0.3	0.6	Johnston & Moon (1980b)
Bass (Dicentrarchus labrax)	10°C	0.2	0.2	Newsholme et al. (1978)
Coal fish (Pollachius virens)	15°C	0.7	0.1	Johnston & Moon (1980a)
Plaice (Pleuronectes platessa)	15°C	0.4	0.06	Moon & Johnston (1980)
Common carp (Cyprinus carpio)	20°C	1.2	0.1	Johnston (1977)
Skipjack tuna (Katsuwonus pelamis)	25°C	1.2	0.8	Guppy et al. (1979)
Red mullet (Mullius surmeletus)	10°C	2.1	0.1	Newsholme et al. (1978)
Antarctic cod (Notothenia rossii)	4°C	0.8*	0.1	Walesby & Johnston (1980a)
Sculpin (Myoxocephalus scorpius)	15°C	1.8	0.4	Fitch & Johnston (unpublished results)

Values are expressed as μmol substrate utilized, g wet weight muscle^{-1} min^{-1}. Details of numbers of animals used and standard errors of the mean values are given in the original publications. Unpublished results for Myoxocephalus scorpius represent determinations for six fish, red muscle 1.84 ± 0.27 (SE) white muscle 0.39 ± 0.02 (SE). Assay temperatures are given above.

fish for speeds up to at least four body lengths/s (Johnston & Moon, 1980b).

However, in many species it would appear that anaerobic pathways are important in providing ATP for sustained activity. For example, Smit *et al.* (1971) have calculated that goldfish (*Carassius auratus*) obtain 80% of their energy requirements anaerobically during high speed sustained swimming. Further evidence for the importance of anaerobic pathways at sub-maximal swimming speeds comes from measurements of key enzyme activities following endurance exercise training (see Johnston & Moon, 1980a,b). In contrast to mammals, endurance exercise training in fish leads to an increase in glycolytic enzyme activities with little change in overall aerobic capacity of the muscles (Johnston & Moon, 1980a,b).

Metabolic Fate of Lactate Produced by the Swimming Muscles

The metabolic fate of lactate produced during sustained and following burst swimming, has been the subject of a large number of papers. It is likely that a major part of the lactate produced during exercise is oxidized to pyruvate (Bilinski, 1974). Bilinski & Jonas (1972) have demonstrated that gills, liver, red muscle and kidney are all able to oxidize ^{14}C-lactate at high rates *in vitro*. Bone (1975) has suggested that fish may be able to maintain overall aerobic balance by transferring lactate produced by anaerobic glycogenolysis in the white muscle to the gills and other peripheral sites for subsequent oxidation. The extent to which fish liver can utilize lactate as a gluconeogenic substrate is unclear. Black, Bosomworth *et al.* (1966) concluded that the Cori cycle is of little importance in trout since liver glycogen levels are not restored to normal even 24 h following burst swimming. However, recently it has been demonstrated that eel hepatocytes convert lactate to glucose at high rates *in vitro* (Renaud & Moon, 1980).

Wittenberger and co-workers have suggested that red muscle operates a direct glucose—lactate exchange with white muscles (Wittenberger & Diacuic, 1965; Wittenberger, 1973; Wittenberger, Coprean & Morar, 1975). This idea developed from a suggestion by Braekkan (1956) that red muscle in fish is homologous to the liver in higher vertebrates. The demonstration of a contractile function for red muscle and the large diffusion distances between red and white muscle masses rules out this hypothesis at least in its more extreme forms. However, on the basis of a variety of *indirect evidence* it has been proposed that red muscle does have a significant role in synthesizing glucose from circulating lactate (see Driedzic & Hochachka,

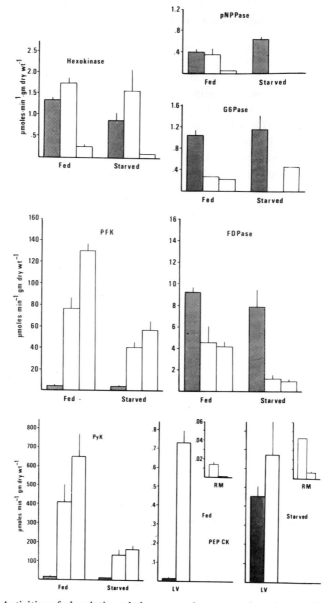

FIG. 13. Activities of glycolytic and gluconeogenic enzymes in red and white muscles and liver of fed and four-month-starved plaice (*Pleuronectes platessa*) (from Moon & Johnston, 1980). (Top) Activities (μmol min^{-1} g dry weight^{-1}) of hexokinase, glucose-6-phosphatase (G-6-Pase) and p-nitrophenyl phosphatase (pNPPase). (Centre) Activities (μmol min^{-1} g dry weight^{-1}) of phosphofructokinase (PFK) and fructose-1,6 bisphosphatase (FDPase). (Bottom) Activities (μmol min^{-1} g dry weight^{-1}) of pyruvate kinase (PyK). Phosphoe-enol pyruvate carboxykinase (PEP Ck) activities are given for the soluble (left) and bound (right) fractions of liver (LV) and red muscle (RM). Note the very low activities in red muscle. No activities of PEP Ck were detected in white muscle. In each case bars represent enzyme activities in liver (left), red muscle (centre) and white muscle (right). Data represent mean ± SE of six fish.

1975; Wittenberger *et al.*, 1975; Hulbert & Moon, 1978; Batty & Wardle, 1979). For example, Batty & Wardle (1979) have calculated, on the basis of *in vivo* ^{3}H-glucose turnover experiments in plaice, that the transport of glucose into the muscle following recovery from exercise is insufficient to account for the observed rates of glycogen formation. However, a significant gluconeogenic role for red muscle is not supported by measurements of key enzyme activities (Johnston & Moon, 1979; Moon & Johnston, 1980). Activities of pyruvate carboxylase, PEP carboxykinase and glucose-6-phosphatase in plaice are either absent or present at very low levels in red muscle compared to liver, even under conditions of starvation which would be expected to augment gluconeogenesis (Fig. 13). Whether gluconeogenesis occurs in the red muscle of other species or whether the very low enzyme levels in plaice red muscle are sufficient to account for some glycogen synthesis during a long recovery period (12 h) remains to be seen.

Utilization of Muscle Proteins

Many species of fish undergo periods of seasonal starvation linked to fluctuations in food supply (see Love, 1970). In fish that do not store large quantities of lipid, white muscle myofibrillar proteins serve as an important source of metabolic precursors during starvation. For example, in plaice, after four months starvation white muscle water contents increase from around 75 to 95% with a corresponding loss of contractile proteins (Johnston & Goldspink, 1973b; Johnston, 1981). It seems likely that amino acids released from the muscles during starvation are converted to glucose via gluconeogenic pathways in the liver and hence constitute an important fuel for muscular activity under these conditions (Moon & Johnston, 1980).

Finally a number of authors have suggested that fish muscles are a major site of amino acid oxidation. For example, glycine and histidine together constitute around 50% of the free amino acid pool in the muscles of salmonid fishes (Fontaine & Marchelidon, 1971). The initial enzyme in histidine degradation is histidase, which is found in high titres in the muscles of some fish but not mammals and has a Km for histidine which is five times lower than the liver analogue (see Driedzic & Hochachka, 1978). Wood, Duncan & Jackson (1960) have suggested that histidine is oxidized *in situ* as an energy source by salmon muscle during spawning migration. There is, however, as yet little direct evidence for amino acid oxidation in fish muscles (for a review see Driedzic & Hochachka, 1978).

ACKNOWLEDGEMENTS

The author's work is supported by grants from the Natural Environment Research Council, Science Research Council and Wellcome Trust.

REFERENCES

Alexander, R. McN. (1969). The orientation of muscle fibres in the myomeres of fishes. *J. mar. biol. Ass. U.K.* 49: 263–290.

Bárány, M. (1967). ATPase activity of myosin correlated with speed of muscle shortening. *J. gen. Physiol.* 51: 197–216.

Barets, A. (1961). Contribution à l'étude des systemes moteur lent et rapide du muscle lateral des teleostéens. *Archs. Anat. Morph. exp.* 50: (Suppl.): 91–187.

Batty, R. S. & Wardle, C. S. (1979). Restoration of glycogen from lactic acid in the anaerobic swimming muscle of plaice, *Pleuronectes platessa* L. *J. Fish Biol.* 15: 509–519.

Beis, I. & Newsholme, E. A. (1975). The contents of adenine nucleotides, phosphogens and some glycolytic intermediates in resting muscles from vertebrates and invertebrates. *Biochem. J.* 152: 23–32.

Bennett, A. F. (1978). Activity metabolism of the lower vertebrates. *An. Rev. Physiol.* 40: 447–469.

Benzonana, G., Capony, J.-P. & Pechère, J. F. (1972). The binding of calcium to muscular parvalbumins. *Biochim. Biophys. Acta* 278: 110–116.

Bergman, R. A. (1964). Mechanical properties of the dorsal fin musculature of the marine teleost *Hippocampus hudsonius*. *Bull. John Hopkins Hosp.* 114: 344–353.

Best, A. C. G. & Bone, Q. (1973). The terminal neuro-muscular junctions of lower chordates. *Z. Zellforsch. mikrosk. Anat.* 143: 495–504.

Bilinski, E. (1974). Biochemical aspects of fish swimming. In *Biochemical and biophysical perspectives in marine biology* 1: 239–288. Malins, D. C. & Sargent, J. R. (Eds). London & New York: Academic Press.

Bilinski, E. & Jonas, R. E. E. (1972). Oxidation of lactate to carbon dioxide by rainbow trout (*Salmo gairdneri*) tissues. *J. Fish. Res. Bd Can.* 29: 1467–1471.

Black, E. C., Bosomworth, N. J. & Docherty, G. E. (1966). Combined effects of starvation and exercise on glycogen metabolism of rainbow trout, *Salmo gairdneri*. *J. Fish. Res. Bd Can.* 23: 1461–1463.

Black, E. C., Robertson, A. C. & Parker, R. R. (1961). Some aspects of carbohydrate metabolism in fish. In *Comparative physiology of carbohydrate metabolism in heterothermic animals*. 89–124. Martin, A. W. (Ed.) Seattle: Univ. Washington.

Boddeke, R., Slijper, E. J. & van der Stelt, A. (1959). Histological characteristics of the body musculature of fishes in connection with their mode of life. *Proc. K. ned. Akad. Wet.* (C.) 62: 576–588.

Bokdawala, F. D. & George, J. C. (1967). A histochemical study of the red and white muscles of the carp, *Cirrhina mrigala*. *J. anim. Morph. Physiol.* 14: 60–68.

Bone, Q. (1964). Patterns of muscular innervation in the lower chordates. *Int. Rev. Neurobiol.* 6: 99—147.

Bone, Q. (1966). On the function of the two types of myotomal muscle fibres in elasmobranch fish. *J. mar. biol. Ass. UK* 46: 321—349.

Bone, Q. (1970). Muscular innervation and fish classification. *Simp. Int. Zoofil. Ist. Univ. Salamanca* 1970: 369—377.

Bone, Q. (1972). The dogfish neuromuscular junction: Dual innervation of vertebrate striated muscle fibres ? *J. Cell Sci.* 10: 657—665.

Bone, Q. (1975). Muscular and energetic aspects of fish swimming. In *Swimming and flying in nature.* 2: 493—528. Wu, T. Y.-T., Brokaw, C. J. & Brennen, C. (Eds). New York: Plenum.

Bone, Q. (1978a). Locomotor muscle. In *Fish physiology* 7: 361—424. Hoar, W. S. & Randall, D. J. (Eds). New York & London: Academic Press.

Bone, Q. (1978b). Myotomal muscle fibre types in *Scomber* and *Katsuwonus.* In *The physiological ecology of tunas*: 183—205. Sharp, G. D. & Dizon, A. E. (Eds). New York & London: Academic Press.

Bone, Q. & Chubb, A. D. (1978). The histochemical demonstration of myofibrillar ATPase in elasmobranch muscle. *Histochem. J.* 10: 489—494.

Bone, Q. & Johnston, I. A. (In press). Contractile properties of dogfish muscle fibres. *J. mar. biol. Ass. UK.*

Bone, Q., Kicznuik, J. & Jones, D. R. (1978). On the role of the different fibre types in fish myotomes at intermediate swimming speeds. *Fish. Bull., U.S.* 76: 691—699.

Bossen, E. J., Sommer, J. R. & Waugh, R. A. (1978). Comparative steriology of the mouse and finch left ventricle. *Tissue & Cell* 10: 773—784.

Braekkan, D. R. (1956). Function of the red muscle in fish. *Nature, Lond.* 178: 747—748.

Brill, R. W. & Dizon, A. E. (1979). Red and white muscle fibre activity in swimming skipjack tuna, *Katsuwonus pelamis* (L.). *J. Fish Biol.* 15: 679—685.

Cohen, P., Duewer, T. & Fischer, E. H. (1971). Phosphorylase from dogfish skeletal muscle. Purification and a comparison of its physical properties to those of rabbit muscle phosphorylase. *Biochemistry* 10: 2683—2694.

Collins, J. H. (1976). Structure and evolution of troponin C and related proteins. *Symp. Soc. exp. Biol.* 9: 303—334.

Crabtree, B. & Newsholme, E. A. (1972). The activities of phosphorylase hexokinase, phosphofructokinase, lactate dehydrogenase and glycerol 3-phosphate dehydrogenases in muscles from vertebrates and invertebrates. *Biochem. J.* 126: 49—58.

Davison, W., Johnston, I. A. & Goldspink, G. (1976). The division of labour between fish myotomal muscles during swimming. *J. Physiol., Lond.* 263: 185—186.

Dizon, A., Brill, R. & Yuen, H. (1978). Respiratory physiology of skipjack tuna. In *Physiological ecology of tuna*: 233—259. Sharp, G. & Dizon, A. (Eds). London & New York: Academic Press.

Driedzic, W. R. & Hochachka, P. W. (1975). The unanswered question of high anaerobic capabilities of carp white muscle. *Can. J. Zool.* 53: 706—712.

Driedzic, W. R. & Hochachka, P. W. (1976). Control of energy metabolism in fish white muscle. *Am. J. Physiol.* 230: 579—582.

Driedzic, W. R. & Hochachka, P. W. (1978). Metabolism of fish during exercise. In *Fish physiology* 7: 503—543. Hoar, W. S. & Randall, D. J. (Eds). New York & London: Academic Press.

Eichelberg, H. (1976). The fine structure of the drum muscles of the trigger fish, *Therapon jarbua*, as compared with the trunk musculature. *Cell Tissue Res.* **174**: 453—463.

Eisenberg, B. R. & Kuda, A. M. (1975). Steriological analysis of mammalian skeletal muscle II. White vastus muscle of the adult guinea pig. *J. Ultrastruct. Res.* **51**: 176—187.

Eisenberg, B. R., Kuda, A. M. & Peter, J. B. (1974). Steriological analysis of mammalian skeletal muscle. I. Soleus muscle of the adult guinea pig. *J. Cell Biol.* **60**: 732—754.

Farkas, T. (1969). Studies on the mobilization of fats in lower vertebrates. *Acta Biochim. Biophys. Acad. Sci. Hung.* **4**: 237—249.

Fischer, E. H., Alaba, J. O., Brautigan, D. L., Kerrick, W. G. L., Malencik, D.A., Moeschler, H. J., Picton, C. & Pocinwong, W. (1978). Evolutionary aspects of the structure and regulation of phosphorylase kinase. In *Versatility of proteins*: 133—149. Choh, Hao Li (Ed.). London & New York: Academic Press.

Fischer, E. H., Blum, H. E., Byers, B., Heizmann, C., Kerrick, G. W., Lehky, P., Malencik, D. A. & Pocinwong, S. (1975). Concerted regulation of glycogen metabolism and muscle contraction. In *Metabolic interconversion of enzymes*: 1—8. Shaltiel, S. (Ed.). Heidelberg: Springer-Verlag.

Flitney, F. W. (1971). The volume of the T-system and its association with the sarcoplasmic reticulum in slow muscle fibres of the frog. *J. Physiol., Lond.* **217**: 243—257.

Flitney, F. W. & Johnston, I. A. (1979). Mechanical properties of isolated fish red and white muscle fibres. *J. Physiol., Lond.* **295**: 49—50P.

Focant, B., Huriaux, F. & Johnston, I. A. (1976). Subunit composition of fish myofibrils: the light chains of myosin. *Int. J. Biochem.* **7**: 129—133.

Fontaine, M. & Marchelidon, J. (1971). Amino acid contents of the brain and the muscle of young salmon (*Salmo salar* L.) at parr and smolt stages. *Comp. Biochem. Physiol.* (A.) **40**: 127—134.

Franzini-Armstrong, C. & Porter, K. R. (1964). Sarcolemmal invaginations constituting the T-system in fish muscle fibres. *J. Cell Biol.* **22**: 675—696.

Freadman, M. A. (1979). Role of partitioning of swimming musculature of striped Bass, *Morone saxatilis* Walbaum and Bluefish, *Pomatomus saltatrix* L. *J. Fish Biol.* **15**: 417—423.

Freed, J. M. (1971). Properties of muscle phosphofructokinase of cold- and warm-acclimated *Carassius auratus. Comp. Biochem. Physiol.* **39**B: 747—764.

Gerday, C. & Gillis, J. M. (1976). The possible role of parvalbumins in the control of contraction. *J. Physiol., Lond.* **258**: 96—97P.

Greer-Walker, M. (1970). Growth and development of the skeletal muscle fibres of the cod (*Gadus morhua* L.). *J. Cons. perm. int. Explor. Mer* **33**: 228—244.

Greer-Walker, M. & Pull, G. A. (1975). A survey of red and white muscle in marine fish. *J. Fish Biol.* **7**: 295—300.

Guppy, M., Hulbert, W. C. & Hochachka, P. W. (1979). Metabolic sources of heat and power in tuna muscles II. Enzyme and metabolite profiles. *J. exp. Biol.* **82**: 303—320.

Hagiwara, S. & Takahashi, K. (1967). Resting and spike potentials of skeletal muscle fibres in salt-water elasmobranch and teleost fish. *J. Physiol., Lond.* **190**: 499—518.

Hamoir, G. (1978). Différentiation protéinique des muscles striés blancs, jaunes et cardiaques d'un Poisson antarctique exempt d'hemoglobine, *Champsocephalus gunnari. C. r. hebd. Seánc. Acad. Sci., Paris* (D) 286: 145–148.

Hidaka, T. & Toida, N. (1969). Biophysical and mechanical properties of red and white muscle fibres in fish. *J. Physiol., Lond.* 201: 49–59.

Hochachka, P. W. & Storey, K. B. (1975). Metabolic consequences of diving in animals and man. *Science, Wash.* 187: 613–621.

Hudson, R. C. L. (1969). Polyneuronal innervation of the fast muscles of the marine teleost *Cottus scorpius* L. *J. exp. Biol.* 50: 47–67.

Hudson, R. C. L. (1973). On the function of the white muscles in teleosts at intermediate swimming speeds. *J. exp. Biol.* 58: 509–522.

Hulbert, W. C. & Moon, T. W. (1978). The potential for lactate utilization by red and white muscle of the eel *Anguilla rostrata* L. *Can J. Zool.* 56: 128–135.

Johnston, I. A. (1975). Pyruvate kinase from the red skeletal musculature of the carp (*Carassius carassius* L.). *Biochem. Biophys. Res. Commn* 63: 115–122.

Johnston, I. A. (1977). A comparative study of glycolysis in red and white muscles of the trout (*Salmo gairdnerii*) and mirror carp (*Cyprinus carpio*). *J. Fish Biol.* 11: 575–588.

Johnston, I. A. (1980a). Contractile properties of fish fast muscle fibres. *Marine Biol. Letters.* 1: 323–328.

Johnston, I. A. (1980b). Specializations of fish muscle. In *Development and specializations of muscle*: 123–148. (Soc. exp. Biol. Seminar Series 7). Goldspink, D. F. (Ed.). Cambridge: University Press.

Johnston, I. A. (1981). Quantitative analyses of muscle breakdown during starvation in the marine flat-fish (*Pleuronectes platessa*). *Cell Tissues Res.* 214: 369–386.

Johnston, I. A., Davison, W. & Goldspink, G. (1977). Energy metabolism of carp swimming muscles. *J. comp. Physiol.* 114: 203–216.

Johnston, I. A., Frearson, N. & Goldspink, G. (1972). Myofibrillar ATPase activities of red and white muscles of marine fish. *Experientia* 28: 713–714.

Johnston, I. A. & Goldspink, G. (1973a). A study of the swimming performance of the crucian carp *Carassius carassius* (L.) in relation to the effects of exercise and recovery on biochemical changes in the myotomal muscles and liver. *J. Fish Biol.* 5: 249–260.

Johnston, I. A. & Goldspink, G. (1973b). Quantitative studies of muscle glycogen utilisation during sustained swimming in crucian carp (*Carassius carassius* L.). *J. exp. Biol.* 59: 607–615.

Johnston, I. A. & Goldspink, G. (1973c). A study of glycogen and lactate in the myotomal muscles and liver of the coalfish (*Gadus virens* L.) during sustained swimming . *J. mar. biol. Ass. UK* 53: 17–26.

Johnston, I. A. & Maitland, B. (1980). Temperature acclimation in crucian carp: A morphometric analysis of muscle fibre ultrastructure. *J. Fish Biol.* 17: 113–125.

Johnston, I. A. & Moon, T. W. (1979). Glycolytic and gluconeogenic enzyme activities in the skeletal muscles and liver of a teleost fish (*Pleuronectes platessa*). *Trans. Biochem. Soc.* 7: 661–663.

Johnston, I. A. & Moon, T. W. (1980a). Endurance exercise training in the fast and slow muscles of a teleost fish (*Pollachius virens*). *J. comp. Physiol.* (B) 135: 147–156.

Johnston, I. A. & Moon, T. W. (1980b). Exercise training in skeletal muscle of brook trout (*Salvelinus fontinalis*). *J. exp. Biol.* 87: 177—194.

Johnston, I. A. & Moon, T. W. (In press). Fine structure and metabolism of multiply innervated fast muscle fibres in teleost fish. *Cell Tissue Res.*

Johnston, I. A., Patterson, S., Ward, P. S. & Goldspink, G. (1974). The histochemical demonstration of myofibrillar adenosine triphosphatase activity in fish muscle. *Can. J. Zool.* 52: 871—877.

Johnston, I. A. & Tota, B. (1974). Myofibrillar ATPase in the various red and white trunk muscles of the tunny (*Thunnus thynnus* L.) and the Tub Gurnard (*Trigla lucerna* L.). *J. comp. Biochem. Physiol.* 49: 367—373.

Johnston, I. A., Ward, P. S. & Goldspink, G. (1975). Studies on the swimming musculature of the rainbow trout. I. Fibre types. *J. Fish Biol.* 7: 451—458.

Jonas, R. E. E. & Bilinski, E. (1964). Utilization of lipids by fish III. Fatty acid oxidation by various tissues from sockeye salmon (*Oncorhynchus nerka*). *J. Fish. Res. Bd Can.* 21: 653—656.

Jones, D. R. & Randall, D. J. (1978). The respiratory and circulatory systems during exercise. In *Fish physiology* 7: 425—492. Hoar, W. S. & Randall, D. J. (Eds). New York & London: Academic Press.

Kilarski, W. (1966). The organisation of the sarcoplasmic reticulum in fish skeletal muscles. III Pike (*Esox lucius* L.). *Bull. Acad. Pol. Sci.* (Biol.) 8: 575—579.

Korneliussen, H., Dahl, H. A. & Paulsen, J. E. (1978). Histochemical definition of muscle fibre types in the trunk musculature of a teleost fish (cod, *Gadus morhua* L.). *Histochemistry* 55: 1—16.

Korneliussen, H. & Nicholaysen, K. (1973). Ultrastructure of four types of striated muscle fibres in the Atlantic hagfish (*Myxine glutinosa*, L.). *Z. Zellforsch. mikrosk. Anat.* 143: 273—290.

Kryvi, H. (1977). Ultrastructure of the different fibre types in axial muscles of the sharks *Etmopterus spinax* and *Galeus melastomus*. *Cell Tissue Res.* 184: 287—300.

Kryvi, H., Flood, P. & Guljaev, D. (1980). The ultrastructure and vascular supply of the different fibre types in the axial muscles of the sturgeon, *Acipenser stellatus*. *Cell Tissue Res.* 212: 117—126.

Kryvi, H. & Totland, G. K. (1978). Fibre types in locomotory muscles of the cartilaginous fish *Chimaera monstrosa*. *J. Fish Biol.* 12: 257—265.

Kutty, M. N. (1972). Respiratory quotient and ammonia excretion in *Tilapia mossambica*. *Mar. Biol.* 16: 126—133.

Lehman, W. & Szent-Györgyi, A. G. (1975). Regulation of muscular contraction. Distribution of actin control and myosin control in the animal kingdom. *J. gen. Physiol.* 66: 1—30.

Le Peuch, C. J., Demaille, J. & Pechère, J. F. (1978). Radioelectrophoresis: a specific microassay for parvalbumins, application to muscle biopsies from man and other vertebrates. *Biochim. Biophys. Acta* 537: 153—159.

Lindsey, C. C. (1978). Form function and locomotory habits in fish. In *Fish physiology* 7: 1—100. Hoar, W. S. & Randall, D. J. (Eds). New York & London: Academic Press.

Lorenzini, S. (1678). *Observazioni intorno alle Torpedini*. Onofri.

Love, R. M. (1970). *The chemical biology of fishes*. London & New York: Academic Press.

Lowey, S. & Risby, D. (1971). Light chains from fast and slow muscle myosins. *Nature, Lond.* 234: 81—85.

Mansour, T. E. (1972). Phosphofructokinase. *Curr. Top. Cell Regul.* 5: 1—46.

McArdle, H. J. & Johnston, I. A. (In press). Ca^{2+}-uptake by tissue sections and biochemical characteristics of sarcoplasmic reticulum isolated from fish fast and slow muscles. *Eur. J. Cell Biol.*

Mobley, B. A. & Eisenberg, B. R. (1975). Sizes of components in frog skeletal muscle measured by methods of stereology. *J. gen. Physiol.* 66: 31—45.

Moon, T. W. & Johnston, I. A. (1980). Starvation and the activities of glycolytic and gluconeogenic enzymes in skeletal muscles and liver of the plaice, *Pleuronectes platessa. J. comp. Physiol.* 136: 31—38.

Mosse, P. R. L. (1978). The distribution of capillaries in the somatic musculature of two vertebrate types with particular reference to teleost fish. *Cell Tissue Res.* 187: 281—303.

Mosse, P. R. L. (1979). Capillary distribution and metabolic histochemistry of the lateral propulsive musculature of pelagic teleost fish. *Cell Tissue Res.* 203: 141—660.

Mosse, P. R. L. & Hudson, R. C. L. (1977). The functional roles of different muscle fibre types identified in the myotomes of marine teleosts: a behavioural, anatomical and histochemical study. *J. Fish. Biol.* 11: 417—430.

Nag, A. C. (1972). Ultrastructure and adenosine triphosphatase activity of red and white muscle fibers of the caudal region of a fish, *Salmo gairdneri. J. Cell Biol.* 55: 42—57.

Newsholme, E. A. (1976). The role of the fructose 6-phosphate/fructose 1,6 diphosphate cycle in metabolic regulation and heat generation. *Trans. Biochem. Soc.* 4: 978—984.

Newsholme, E. A. & Start, C. (1973). *Regulation in metabolism.* New York: Wiley.

Newsholme, E. A., Zammit, V. A. & Crabtree, B. (1978). The role of glucose and glycogen as fuels for muscle. *Trans. Biochem. Soc.* 6: 512—520.

Nishihara, H. (1967). Studies on the fine structure of red and white fin muscles of the fish (*Carassius auratus*). *Archvm histol. jap.* 28: 425—447.

Patterson, S. & Goldspink, G. (1972). The fine structure of red and white myotomal muscle fibres of the coalfish (*Gadus virens*). *Z. Zellforsch. mikrosk. Anat.* 133: 463—474.

Patterson, S. & Goldspink, G. (1973). The effect of starvation on the ultrastructure of the red and white myotomal muscles of the Crucian carp (*Carassius carassius*). *Z. Zellforsch. mikrosk. Anat* 146: 375—384.

Peachey, L. D. (1965). The sarcoplasmic reticulum and transverse tubules of the frog sartorius. *J. Cell Biol.* 25: 209—231.

Pecot-Dechavassine, M. (1961). Etude biochemique, pharmacologique et histochemique de cholinesterase des muscles striés chez les poissons, les batraciens, et les mammifères. *Archs Anat. microsc. Morph. exp.* 50(Suppl.): 341—438.

Pechère, J. F., Capony, J. P. & Demaille, J. (1973). Evolutionary aspects of the structure of muscular parvalbumins. *Syst. Zool.* 22: 533—548.

Pechère, J. F., Derancourt, J. & Harech, J. (1977). The participation of parvalbumins in the activation—relaxation cycle of vertebrate fast skeletal muscle. *FEBS Lett.* 75: 111—114.

Perry, S. V. (1979). The regulation of contractile activity in muscle. *Trans. Biochem. Soc.* 7: 593—617.

Potter, J. D. & Gergely, J. (1975). The calcium and magnesium binding sites on troponin and their role in the regulation of myofibrillar adenosine triphosphatase. *J. biol. Chem.* 250: 4628—4633.

Pritchard, A. W., Hunter, J. R. & Lasker, R. (1971). The relation between exercise and biochemical changes in red and white muscles and liver in the jack mackerel *Trachurus symmetricus*. *Fish. Bull. US Fish Wildl. Serv.* **69**: 379–386.

Rayner, M. D. & Keenan, M. J. (1967). Role of red and white muscles in the swimming of the skipjack tuna. *Nature, Lond.* **214**: 392–393.

Renaud, J. M. & Moon, T. W. (1980). Characterization of the isolated American eel (*Anguilla rostrata* Le Sueur) hepatocyte. *J. comp. Physiol.* **135**: 115–127.

Rudd, J. T. (1954). Vertebrates without erythrocytes and blood pigment. *Nature, Lond.* **173**: 848–850.

Sharp, G. D. & Pirages, S. (1978). The distribution of red and white swimming muscles, their biochemistry and the biochemical phylogeny of selected scombrid fishes. In *The physiological ecology of tunas*: 41–78. Sharp, G. D. & Dizon, A. E. (Eds). New York & London: Academic Press.

Smit, H., Amelink-Koutstall, J. M., Vigverberg, J. & von Vaupel-Klein, J. C. (1971). Oxygen consumption and efficiency of swimming goldfish. *Comp. Biochem. Physiol.* A**30**: 1–28.

Somero, G. N. & Hochachka, P. W. (1968). The effect of temperature on catalytic and regulatory functions of pyruvate kinases of rainbow trout and the Antarctic fish, *Trematomus bernacchii*. *Biochem. J.* **110**: 395–400.

Stanfield, P. R. (1972). Electrical properties of white and red muscle fibres of the elasmobranch fish *Scyliorhinus canicula*. *J. Physiol., Lond.* **222**: 161–186.

Stevens, E. D. (1968). The effects of exercise on the distribution of blood to various organs in rainbow trout. *Comp. Biochem. Physiol.* **25**: 615–625.

Stevens, E. D. & Black, E. C. (1966). The effects of intermittent exercise on carbohydrate metabolism in rainbow trout, *Salmo gairdneri*. *J. Fish. Res. Bd Can.* **23**: 471–495.

Terävainen, H. (1971). Anatomical and physiological studies on muscles of lamprey. *J. Neurophysiol.* **34**: 954–973.

Totland, G. K. (1976). Three muscle fibre types in the axial muscles of Axolotl (*Ambystoma americanum*, Shaw). A quantitative light and electron microscopic study. *Cell Tiss. Res.* **168**: 65–78.

Walesby, N. J. & Johnston, I. A. (1980a). Fibre types in the locomotory muscles of an Antarctic teleost, *Notothenia rossii*: A histochemical, ultrastructural and biochemical study. *Cell Tissue Res.* **208**: 143–164.

Walesby, N. J. & Johnston, I. A. (1980b). Temperature acclimation in Brook trout muscle: Adenine nucleotide concentrations, phosphorylation state and adenylate energy charge. *J. Comp. Physiol.* **139**: 127–133.

Walesby, N. J., Nicol, C. J. M. & Johnston, I. A. (In press). Metabolic differentiation of muscle fibres from a haemoglobinless (*Champsocephalus gunnari*) and a red-blooded (*Notothenia rossii*) Antarctic fish. *Bull. Br. Antarct. Surv.* No. 53.

Wardle, C. S. (1975). Limit of fish swimming speed. *Nature, Lond.* **255**: 725–727.

Wardle, C. S. (1978). Non-release of lactic acid from anaerobic swimming muscle of plaice, *Pleuronectes platessa* L., a stress reaction. *J. exp. Biol.* **77**: 141–155.

Webb, P. W. (1973). Kinematics of pectoral fin propulsion in *Cymatogaster aggregata*. *J. exp. Biol.* **59**: 697–710.

Webb, P. W. (1975). Hydrodynamics and energetics of fish propulsion. *Bull. Fish. Res. Bd Can.* No. 190: 1—158.

Wittenberger, C. (1973). Metabolic interaction between isolated white and red muscles. *Revue roum. Biol.* (Zool.) 18: 71—76.

Wittenberger, C., Coprean, D. & Morar, L. (1975). Studies of the carbohydrate metabolism of the lateral muscles in carp (influence of phlorizin, insulin and adrenaline). *J. comp. Physiol.* 101: 161—172.

Wittenberger, C. & Diacuic, I. V. (1965). Effort metabolism of lateral muscles in carp. *J. Fish. Res. Bd Can.* 22: 1397—1406.

Wood, J. D., Duncan, D. W. & Jackson, M. (1960). Biochemical studies on sockeye salmon during spawning migration XI. The free histidine content of the tissues. *J. Fish. Res. Bd Can.* 17: 347—351.

Yamamoto, T. (1972). Electrical and mechanical properties of the red and white muscles in the silver carp. *J. exp. Biol.* 57: 551—567.

Zammit, V. A. & Newsholme, E. A. (1979). Activities of enzymes of fat and ketone-body metabolism and effects of starvation on blood concentrations of glucose and fat fuels in teleost and elasmobranch fish. *Biochem. J.* 184: 313—322.

Symp. zool. Soc. Lond. (1981) No. 48, 115–136

The Organization of the Nervous System of Fishes in Relation to Locomotion

B. L. ROBERTS

The Laboratory of the Marine Biological Association, Plymouth, England

SYNOPSIS

This chapter summarizes present knowledge of how the nervous system of fishes generates motor activity that is appropriate for locomotion. The fundamental requirements for undulatory movement are met almost entirely by the spinal cord, acting under the general supervision of the brain. The timing and content of the motor output of the spinal cord result from the intrinsic properties of networks of nerve cells. It is unclear how these networks operate but to function correctly they require signals from brain centres as well as from sense organs that are excited by body movements.

Some of the brain centres match the locomotory sequences to the requirements of posture and equilibrium, while others trigger specific body movements. A good example of the latter is provided by the teleost Mauthner cell which has specialized connections that bring about a rapid unilateral movement that is the basis of the "startle response" of these fishes. These hindbrain centres are under the general control of the cerebellum.

INTRODUCTION

The main type of locomotion utilized by fishes is bodily undulation and this mode of progression was evidently inherited from some invertebrate ancestor. Studies made over the last few years on the neuronal basis of locomotion in several invertebrates have shown that their movements rely on the activity of groups of nerve cells within the nervous system that produce regular motor patterns (the motor program) and we shall see that there is some evidence for a similar arrangement in fishes.

Comparisons of cinefilms of fish moving in water and of newts walking on land, as provided by Gray (1968), give us good reason to suppose that the movements of the limbs of terrestrial vertebrates have been superimposed on undulatory movements of the body. There have, of course, been changes in neuronal organization associated with the new patterns of locomotion, particularly in the elaboration of the cerebral cortex and, with this, the acquisition of skilled movements. But many of the actions of the nervous system of higher vertebrates can be attributed to the control requirements of numerous separate muscles needed to move the limbs and to the problems of posture and balance posed by terrestrial life.

These changing neuronal demands have been superimposed on a basic plan established by the fishes so we may hope in an examination of the stereotyped locomotory movements of these animals to find model systems that reveal to us the principles of organization that are basic to vertebrate locomotion.

THE MOTOR PROGRAM

The Muscle Systems and Their Neural Requirements

The various types of muscle fibre that comprise the body myotomes of fishes (Johnston, this volume), with their differing arrangements of innervation, have distinctive mechanical and electrical properties and require different patterns of neural activation which must be incorporated within the motor program. The focally innervated ("white") muscle fibres produce overshooting action potentials, and a single stimulus to the nerve fibres that supply them will bring about contraction; in contrast, the multiterminally innervated ("red") muscle fibres begin to contract only when they have received several triggering impulses (see Bone, 1978).

Because of these properties gradation of the forces exerted by blocks of white muscle fibres is brought about in the usual vertebrate manner by a variation in the number of supplying nerve fibres that are active and in their frequency of discharge (Roberts, 1969b), whereas gradation of contraction of the superficial fibres can be achieved within individual fibres because spatial and temporal facilitation will enhance the amplitude of the end-plate potentials and hence the resulting contraction.

Patterns of Muscle Activity During Locomotion

Our knowledge of the spatial and temporal patterns of activation for these different muscle systems has come from electromyographic recordings from the various muscle types in several species of fish swimming in water tunnels or from swimming spinal elasmobranch preparations. An important result from this work was the recognition (Bone, 1966) that the steady swimming movements ("cruising") of

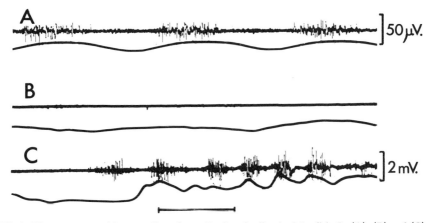

FIG. 1. Electromyographic recordings from "swimming" spinal dogfish. In (A), (B) and (C) the top trace is the electromyogram and the lower trace is the transducer record of the movement. Recordings from the superficial red fibres (A) and white fibres (B) during "cruising" locomotion; in (C), white fibres become active in response to a tail pinch. (From: Roberts, 1969b.)

the dogfish, *Scyliorhinus*, were produced by activity of only the superficial muscle fibres and that the bulk of the musculature was quiescent, becoming active only when more rapid movements were required (Fig. 1).

This conclusion has now been reported by several authors using various species of fish and is also supported by biochemical findings, and appears to apply to species that have superficial fibres and terminally-innervated white muscle fibres. The situation is more complex in the higher teleosts where evidence from biochemistry (reviewed by Bone, 1978) and from electromyographical studies indicates that, even at cruising speeds, activity is not limited to the superficial fibres. In the trout, for example, superficial fibres are active at all swimming speeds but the deeper fast fibres become active at 75% of the sustainable speed, producing electrical records similar to those of the superficial fibres, while at "burst" speeds these muscle fibres generate much larger potentials which are most probably action potentials (Hudson, 1973). Similar findings for the carp have been obtained by Bone, Kiceniuk & Jones (1978).

The duration of the burst of impulses recorded in the electromyogram, depends on the recording conditions, myotome position, species of fish and swimming speed but remains fairly constant as a proportion of the tail-beat frequency and gives a linear relationship when plotted against cycle period (*Scyliorhinus*: Roberts, 1969b; Williamson & Roberts, 1980; *Squalus*: Grillner, 1974; trout and eel: Grillner & Kashin, 1976).

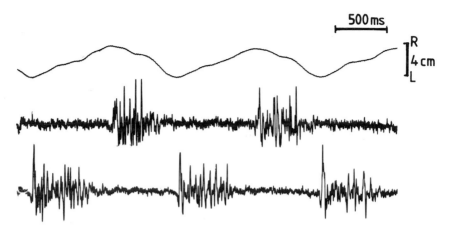

FIG. 2. Electromyographic recordings from "swimming" spinal dogfish. The top trace is a transducer record of the movement; the other traces are records from the superficial muscles of the left (middle) and right (bottom) side of the same body segment. (From: Williamson & Roberts, 1980.)

In undulatory swimming the period during which no potentials are seen in the electromyogram coincides with a burst of activity of the muscle fibres on the opposite side of the body (Figs 2 and 3). In Fig. 2 the muscle fibres on both sides of the same segmental level are seen not to be simultaneously active nor is there any overlap, an arrangement that is quite different from that found in terrestrial vertebrates where overlap between the contraction times of the protagonist and antagonist muscles is important for limb deceleration; the dense medium surrounding a fish presumably makes this an unnecessary requirement. The discharges recorded from the most rostral segments are longer than those obtained more caudally, and rostrally there may be some overlap in contralateral contraction. Blight (1976), who reported this in the tench, has pointed out that this will tend to limit the excursions of the head to the axis of overall progression.

Another finding provided by electromyographical studies is that myotomes along the same side of the body contract at different times, the posterior segments lagging behind the more anterior segments (Fig. 3). The temporal relationship between the ipsilateral segments has been studied in the dogfish *Squalus* by Grillner (1974) and in the eel and trout by Grillner & Kashin (1976). They found for all three fish that the time lag progressively increases in the rostro-caudal direction and that because the value of this lag is inversely proportional to the swimming speed it represents a constant proportion of the swimming period (i.e. the phase lag is constant). Of

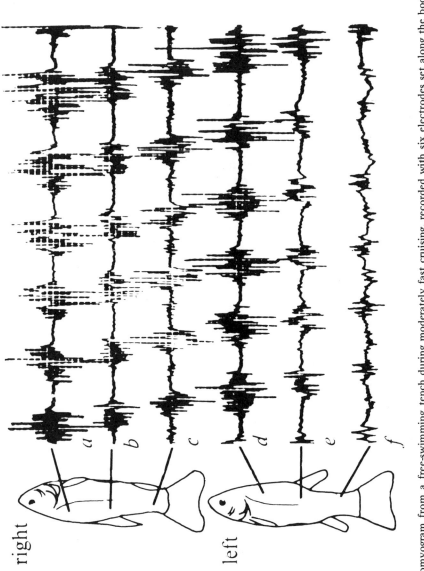

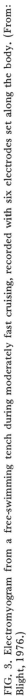

FIG. 3. Electromyogram from a free-swimming tench during moderately fast cruising, recorded with six electrodes set along the body. (From: Blight, 1976.)

course the size of the phase lag for different fish will determine just how many body segments on one side of the body will be contracting in unison and so will set the wavelength of the body undulation.

In a fish like the eel most of the body is bent into large-amplitude backward-moving waves whereas in shorter, thicker fish waves of propulsion are more limited to the tail region. This spectrum of body form, from multi-waved undulation to exclusive caudal fin oscillation (i.e. the familiar categories of anguilliform, carangiform and so on) reflects the differences in timing between the contraction of body segments and whether there is any overlap between activity on both sides of the body.

We know from observations made by Bainbridge (1958) and others using water tunnels that swimming fish increase their forward velocity by making tail movements of enhanced frequency and amplitude. Tail beat frequency depends on the timing of muscle activation while the amplitude of movement is a result of the proportion of the musculature that is contracting at any one time. Changes in wavelength, apparently a less frequently modified parameter, would be brought about by changing the relative activation of different myotomes along the body.

All these requirements, as well as the specific demands of the individual types of muscle fibre, must be met by the appropriate compilation of a motor program by the central nervous system.

The Motor Program is Determined by the Spinal Cord

Although the overall equilibrium and pathway of a swimming fish are determined by neural circuits within the brain, we are confident that the basic form and rhythm of the locomotory movements are factors that are intrinsic to the organization of the spinal cord and emerge from its internal connections.

The importance of the spinal cord in determining the fundamental locomotory pattern is now recognized throughout the vertebrates (Graham Brown, 1911; Grillner, 1975) but the fishes provide some particularly interesting and illustrative experimental conditions. It has been recognized for a long time in the case of the dogfishes that spinal preparations — where the spinal cord has been totally severed close to the brain — can perform continuous locomotory movements that are essentially of normal form (Steiner, 1888). Only a few other species of fish show regular spinal locomotion (e.g. chronically spinal *Conger* (Gray & Sand, 1936) and *Anguilla* (von Holst, 1936)) although a few swimming cycles can usually be elicited from the other spinal species by cutaneous stimulation or by unpatterned electrical stimulation of the spinal cord.

In view of the significance of fishes for the understanding of spinal locomotion it is particularly unfortunate that current knowledge of the structural organization of the spinal cord in these animals is still restricted predominantly to those findings obtained from comparative anatomical studies that were fashionable more than 50 years ago (reviewed by Nieuwenhuys, 1964).

Light microscope sections of both teleost and elasmobranch spinal cord show an organization which is immediately familar to anyone acquainted with the mammalian cord because of the arrangement of the grey matter as dorsal and ventral horns (Fig. 4). The motor neurones within the ventral horn are the final common pathway for effecting movement and interact continually with several other neural systems — local intermediate neurones and sensory fibres and descending and ascending tracts. The motor neurones have large cell bodies with extensive dendrites that spread dorsally and laterally to provide a surface for interaction with other axons. There is some evidence that motor neurones that supply red muscle fibres are smaller than those that supply the white fibres (lamprey: Teravainen & Rovainen, 1971).

The sensory fibres from the body enter via the dorsal roots to terminate on intermediate neurones within the dorsal funiculus and elsewhere. However, some are thought to end monosynaptically on motor neurones although there is at present supporting physiological evidence for this only for rays (Leonard, Rudomin & Willis, 1978) and not for teleosts (Bando, 1975).

The main axons that descend from the brain to interact on spinal neurones come from the reticular and vestibular centres within the hindbrain; the more rostral brain centres apparently obtain access to the spinal cord via these two descending systems.

The Generation of the Motor Program

It has become increasingly clear from studies on a variety of invertebrates that the basic neural pattern required for locomotion (i.e. the motor program) is derived from a pattern generator that resides within the nervous system (see Stein, 1978). In some species this generator is probably capable of developing most if not all of the motor program needed for locomotion, but in others information from other parts of the nervous system and from sensory activity is normally essential.

Amongst fishes, the spinal elasmobranch with its spontaneous locomotion provides excellent experimental material for the study of the generator's properties. When this preparation is immobilized with

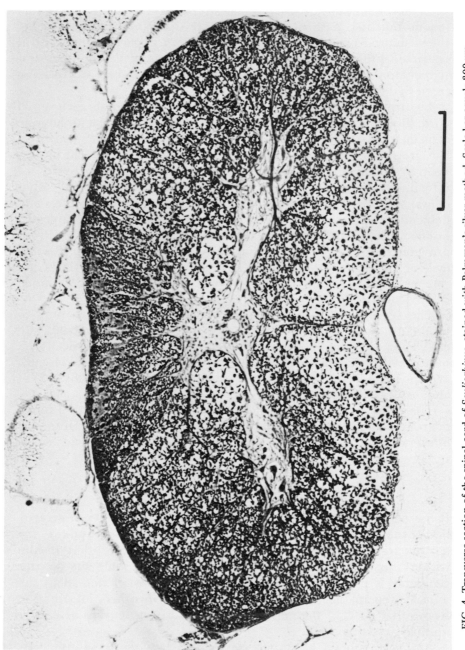

FIG. 4. Transverse section of the spinal cord of *Scyliorhinus*, stained with Palmgren's silver method. Scale bar equals 300 μm.

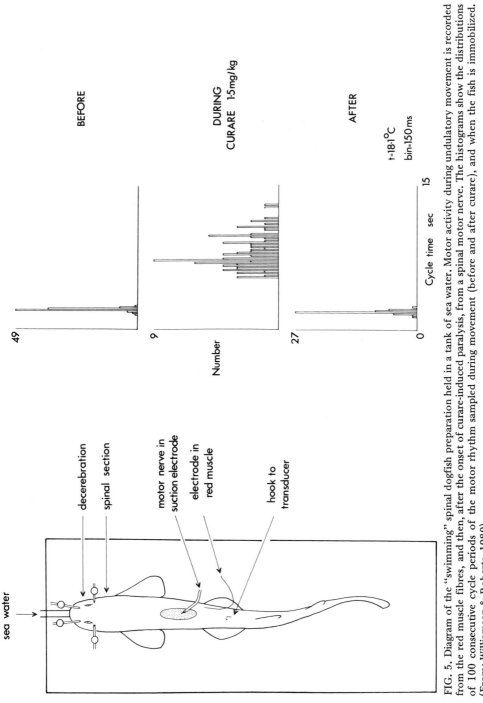

FIG. 5. Diagram of the "swimming" spinal dogfish preparation held in a tank of sea water. Motor activity during undulatory movement is recorded from the red muscle fibres, and then, after the onset of curare-induced paralysis, from a spinal motor nerve. The histograms show the distributions of 100 consecutive cycle periods of the motor rhythm sampled during movement (before and after curare), and when the fish is immobilized. (From: Williamson & Roberts, 1980).

suitable paralytic drugs it is possible to isolate the spinal cord circuits from timed proprioceptive activity evoked by undulatory body movements and so provide some idea of the operation of the intrinsic generator (Roberts, 1969a; Grillner, Perret & Zangger, 1976; Williamson & Roberts, 1980).

In the experiment illustrated in Fig. 5 a spinal dogfish is "swimming", with its head held firmly in a holder, in a tank of sea water. Electromyographic recordings from the red muscle monitor the persistent undulatory movements which can be subsequently prevented by an injection of curare. In the paralysed fish recordings are then made from the motor bundles of the spinal nerves; the paralytic effects of the curare eventually wane and the fish once again makes rhythmical, swimming-type movements.

The histograms of Fig. 5 show the distributions of the cycle period of the motor rhythm obtained from one fish before, during, and on recovery from curare-induced paralysis. During swimming (before curare) the rhythm of the movements is very steady with cycle period values around 1.8 s, but these lengthen and become more variable when all movements have ceased. In this study (Williamson & Roberts, 1980) the motor rhythms obtained during paralysis were always found to be longer (about twice as long) than the swimming rhythms. Motor activity generated at this rate would be quite inappropriate for locomotion, for a free-swimming intact *Scyliorhinus* undulates its tail at about 50 beats/min (i.e. cycle period of 1.2 s) and would "stall" and sink if it produced tail beats of lower frequency.

Although the motor output of the isolated spinal cord occurs at incorrect frequencies it shows many of the features of organization essential for locomotion. Thus the motor discharges on opposite sides of the same segment are completely out of phase (Roberts, 1969a; Grillner & Wallen, 1977) and the activity of rostral segments precedes that of more caudal nerves (Grillner *et al.*, 1976). The burst duration also changes appropriately with changing cycle period (Williamson & Roberts, 1980).

The fact that the generator in *Scyliorhinus* operates below normal discharge rates (in *Squalus* the discharges are apparently at normal rate, Grillner *et al.*, 1976) is presumably an expression of the overall excitability of the spinal cord which is normally regulated by supraspinal influences as well as by input from body proprioceptors. The supraspinal effects can be discounted in the spinal dogfish and so the changes in timing seen in curarized *Scyliorhinus* can be attributed to the absence of appropriate sensory feedback from the body movements.

Preliminary microelectrode studies in *Scyliorhinus* (Roberts & Williamson, in press) indicate that many of the neurones of the spinal cord in the paralysed spinal preparation discharge in rhythmical bursts that are in time with, but bear various phase relations to, the motor output. Until these various neurones are morphologically characterized and their interconnections determined it is not possible to draw up feasible networks for the nerve cells that contribute to the generator.

Although amongst invertebrates there are known examples of single "pacemaker" neurones it is more probable that for the fishes the motor rhythm results from the emergent property of a network of neurones. The term "spinal generator" implies that a discrete set of neurones, distinguishable from other spinal nerve cells, is solely responsible for the locomotory pattern; however it may well be that we shall find that such a distinction cannot be made.

PROPRIOCEPTORS AND LOCOMOTION

As well as providing the muscle fibres with a motor innervation the segmental nerves also carry sensory nerve fibres to the spinal cord. In contrast to terrestrial vertebrates where there is a considerable range of morphologically complex proprioceptors situated amongst muscle fibres, in tendons and joints, to register muscle length and tension and angle of the limbs, the fishes have only a very limited number of locomotory sensors. Two morphologically differentiated endings have been described in elasmobranch fishes: the Wunderer corpuscle, a coiled ending lying predominantly in the dermis of the skin; and the ending of Poloumordwinoff, an elongate ending found amongst the muscle fibres of the fins of skates and rays. For teleosts, up to the present time, no definite morphologically distinct type of ending has been reported, although free-nerve endings have been seen. The properties of these various types of ending have been reviewed recently (Bone, 1978; Roberts, 1978); all the evidence suggests that they could provide the fish with some feedback during locomotion.

Some idea of the impact of the information generated by these sense organs during locomotion has come from studies in which locomotory performance has been examined after the sensory (dorsal) roots to the spinal cord have been cut ("deafferentation") or from observations on motor activity after the body waveform has been modified so as to alter the sensory inflow.

Deafferentation experiments have been carried out on elasmobranchs by several authors (Gray & Sand, 1936; Lissmann, 1946; Grillner *et al.*, 1976) and on teleosts by Gray (1936) and von Holst (1936); in all cases the results have led to conflicting interpretations. The studies by Gray & Sand (1936), based on cinephotography, and those by Grillner and his co-workers, using electromyography, show that deafferentated portions of the spinal cord of the dogfish can still produce organized movements that differ only slightly from the normal pattern. Lissmann (1946) also found this to be the case but only if spinal segments in other body regions were receiving proprioceptive input, whereas Grillner *et al.* (1976) observed that even in totally deafferentated preparations the motor discharges, although markedly reduced in amplitude, were of essentially normal rhythm. The differences between these results may possibly be explained by the very marked reduction in the amplitude of the movements that follows deafferentation.

Although motor rhythms persist in the isolated cord it is clear that the sensory input does influence the timing of the body movements, for when the body of a swimming spinal dogfish is subjected to a forced oscillation that changes the waveform motor-neuronal recordings show (Fig. 6) that the motor output immediately becomes entrained to the applied new rhythm (Roberts, 1969b; Grillner & Wallen, 1977). Recent studies (Roberts & Williamson, in press), in which body movements of *Scyliorhinus* were weakened, and so reduced in amplitude, by the injection of low dosages of curare, have also shown that over a certain range of body movements the frequency of tail beats is related to the size of the movements, and presumably, therefore, to the amount of sensory inflow.

The neurones of the spinal cord of fishes, when suitably excited, are evidently able to generate bursts of motor discharges that form the patterns basic to undulatory motion. The role of the proprioceptors appears to be to complement this central effect and to raise the excitability of the cord neurones to appropriate levels, to stop the generator being destabilized by local perturbations and to adapt the standardized output produced by the generator to changes in body form that result, for example, from body growth or from the use of different muscle systems.

THE BRAIN AND MOVEMENT

The Mesencephalic Locomotor Region

Stimulation of the spinal cord has been reported to establish and change the form of locomotory movements in spinal fish (*Anguilla*:

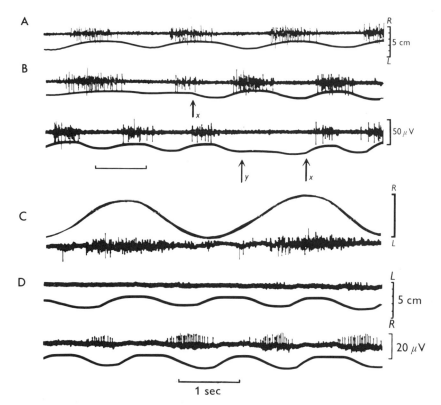

FIG. 6. The effect of activity of body sense organs on the timing of motor discharges. (A) Recordings taken during steady "swimming"; (B) the same fish when at *x* an oscillation was imparted to the body. At *y* the body was released; (C) the application of a slow oscillation to the body of a swimming fish; (D) application of body oscillation to a non-swimming spinal fish. Note the appearance of rhythmical motor discharge. (From: Roberts, 1969b.)

Gray, 1936; *Squalus*: Lissmann, 1946; Grillner *et al.*, 1976). We do not know if this results from a direct action on local spinal intermediate neurones, whose action then reverberates along the cord, or whether the stimulus is exciting specific long descending pathways, originating from nerve cell bodies in the brain. But there are several well worked examples from invertebrates of such "command fibre" systems (arthropod swimmeret movements; wing beat of insects; swimming in *Tritonia* (reviewed in Stein, 1978)), and in mammals it has been shown that locomotion can be initiated and regulated by stimulation of a midbrain centre — the mesencephalic locomotor region (MLR) (Shik & Orlovsky, 1976).

In the fish *Cyprinus* and *Hexogrammus* Kashin, Feldman and Orlovsky (1974) found that continual electrical stimulation of a limited region in the caudal part of the midbrain tegmentum would evoke co-ordinated swimming movements of the caudal and other fins. The type of locomotion could be altered by changing the strength of stimulation — weak stimulation triggered movement of the pectoral fins while stronger stimulation brought about movements of the caudal fin as well. The frequency of the fin movements was also dependent on the strength of stimulation (Fig. 7).

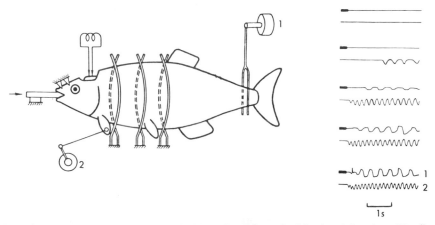

FIG. 7. Locomotory movement evoked by stimulation of midbrain of *Cyprinus*. The five recordings from the caudal fin (top trace) and pectoral fin (bottom trace) show the effect of increasing strength of stimulation (modified from Kashin, Feldman & Orlovsky, 1974).

Electrical stimulation of the midbrain tegmentum in *Opsanus* (Demski & Gerald, 1974) produced complex locomotory movements and Peter (1977) found that damage to a similar region in *Carassius* produced marked changes in swimming behaviour. These results provide support for the concept of an MLR in these fish; in the elasmobranchs the picture is less clear, for although locomotion can be established in decerebrate preparations of *Dasyatis* by means of continuous midbrain stimulation (Droge & Leonard, 1979), in sharks Demski (1977) obtained locomotion by stimulation in other centres as well.

Certainly not all the descending pathways are essential for locomotion for goldfish can still swim satisfactorily when only 50% of the descending fibres are present (Bernstein, 1964; Bernstein & Gelderd, 1970).

The Mauthner Cell in Locomotion

A strikingly large pair of axons seen in transverse sections cut through the spinal cord of some fishes and amphibians, arises from specialized medullary neurones, the Mauthner cells. These conspicuous neurones have for a long while been implicated in the generation of body movement and were initially held responsible for the rhythm, muscle tone and for equilibrium. Bartelemez (1915) was the first person to associate these cells with the "startle response", a characteristic rapid movement produced by these animals in response to sudden visual and vibratory stimuli, and this view of their function is now widely held (see Faber & Korn, 1978). The best evidence for excluding the Mauthner cell from a part in steady locomotion comes from the observation that swimming behaviour eventually returns after the spinal cord of goldfish has been transected even though the Mauthner cell axons do not regenerate past the transection point (Bernstein, 1964). Eaton and his colleagues, working with zebrafish (Eaton, Bombardieri & Meyer, 1977), and Zottoli (1977), using goldfish, have shown with electrophysiological procedures that these cells become active just before certain types of startle movements are made (Fig. 8a). The startle response in a subcarangiform fish such as the goldfish starts with either a C or an S-shape configuration of the body and all the data now suggest that the C-type start — usually called the "tail flip" — is associated with Mauthner cell activity.

The response has a latency as short as 5—8 ms and commences with a C posture of the body, followed by a strong propulsive stroke of the tail in a direction opposite to the initial contraction, which carries the animal away from its starting position, accelerating the animal to peak velocity within 20 ms (Fig. 8c).

Figure 8b provides an analysis of a startle response of a goldfish based on high-speed cine-photography and for six trials charts the pathway of the head during the movement. For the initial period, when the body bends into a C-shape, the pathways followed by the head in different trials are identical, but later the pathways differ. It is the initial stereotyped phase that is believed to be the result of Mauthner cell activation.

Several features of this cell can be seen as specializations for its role in this type of movement. Thus the cell does not discharge repetitively and has a high behavioural threshold. It receives afferent excitation and inhibition from ipsi- and contralateral VIIIth nerves and posterior lateral-line nerves and some of these fibres make synapses that are rapidly transmitting electrotonic connections. The overall excitability of the Mauthner cell is regulated by inhibitory intermediate neurones that are driven by the Mauthner cell collaterals and by primary afferent fibres and this inhibition is uniquely of two forms: a classical chemical inhibition and an electrical inhibition that depends on extracellular currents hyperpolarizing the cell (see Faber

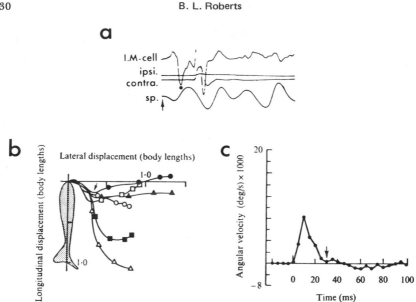

FIG. 8. The Mauthner cell and the "startle response" of the goldfish. (a) Activity obtained
from a Mauthner cell (top trace, dot) and from the body musculature (ipsi- and contra-
traces) during a startle response evoked by sound (sp. trace); (b) plot of the pathway of the
head of a goldfish during a *C*-type startle response; (c) plot of angular velocity of the head
during the *C*-type start. The arrow indicates the start of the return phase. (Modified from:
(a) Zottoli, 1977; (b and c) Eaton *et al.*, 1977.)

& Korn, 1978). The Mauthner cell axons cross in the brain-stem to
the contralateral side and then pass along the spinal cord to activate
the motor neurones of that side, an arrangement that ensures that
the head is moved away from the side of stimulation. The axons are
large, lack nodes, conduct very rapidly (70–100 m/s) and make
chemical synapses onto the ventral dendrites of ipsilateral motor
neurones and onto intermediate neurones that prevent concurrent
motor-neuronal activity. The properties of these synapses are unknown but they may be similar to those described in the hatchet fish
which "flies" by means of pectoral fin movements. In this fish the
Mauthner cell connects with an intermediary giant cell which, in
turn, excites the fin motor neurones via electrotonic synapses. The
junctions between the Mauthner cell and the giant fibre have a high
safety factor but fatigue easily (Highstein & Bennett, 1975).

The Cerebellum

The nerve cells that send their axons into the spinal cord are themselves in receipt of signals from the spinal cord and from other brain
centres, one of the most important of which, the cerebellum, is a

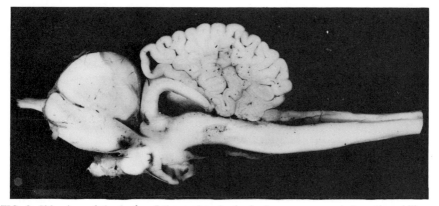

FIG. 9. Side view of the brain of the porbeagle shark, *Lamna nasus*, cut along the midline. Note the highly convoluted cerebellum.

structure that is strikingly similar in form throughout the vertebrates. The photograph (Fig. 9) of a longitudinal section through the brain of the porbeagle shark, *Lamna*, emphasizes the resemblance to the mammalian cerebellum. The cerebellum is rarely so large in bony fishes although the cellular content of Purkinje cells, stellate cells, parallel fibres and so on, is fully established.

The earliest attempts to relate cerebellar function in fish to locomotion were based on ablation and gave very confusing results (reviewed by Healey, 1957), not only because of differences in techniques but also because some workers have expected the profound locomotory disturbances of the decerebellated mammal to be seen also in fishes. In fact closer examination of the mammalian results suggest that the underlying cause of the locomotory disturbances is the disruption of postural control and of course this problem is considerably simplified for aquatic animals because of the support supplied by the dense medium. Luciani (1891) had observed that a decerebellate dog, though unable to stand, when placed in water could swim normally and other workers have also found that mammalian swimming movements are little disturbed by cerebellectomy. Therefore, in this context, the reports of limited effects of cerebellectomy in fishes are not too surprising.

A recent study on the consequences of cerebellar ablation for a reflex movement in *Scyliorhinus* gives some insight into the significance of the cerebellum (Paul & Roberts, 1979). In these experiments reflex elevating movements of the pectoral fins are evoked by electrical stimulation through fine implanted wires (Fig. 10A). In this reflex the fin is lifted, held briefly and then relaxed again and produces two phases in the electromyogram (Fig. 10B) recorded

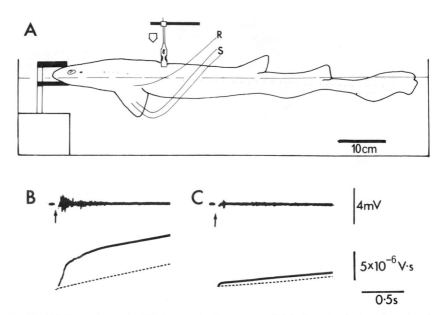

FIG. 10. (A) Decerebrate dogfish in a tank of sea water. S, bipolar stimulating wires placed subdermally in pectoral fin; R, electromyographic electrode in levator muscle. (B and C) The effect of cerebellectomy on the pectoral fin reflex. In each record the top trace is the myogram, the middle trace is the integrated myogram and the dotted line indicates the integrator output when there was no stimulation. Records in (B) from decerebrate, in (C) from decerebellated fish. (Modified from: Paul & Roberts, 1979.)

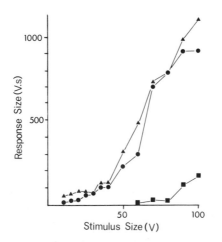

FIG. 11. Size of the pectoral fin reflex, measured from the integrated myogram (in volts seconds) plotted against stimulus strength in volts. Circles — decerebrate; squares — decerebellate; triangles — spinal preparation. (Modified from: Paul & Roberts, 1979.)

from the levator muscle. Removal of the cerebellum produces marked changes in the form of this reflex (Fig. 10C). In the decerebellate preparation the pattern of the reflex is quite different from that of either the decerebrate or spinal animal for the magnitude is considerably reduced and the threshold elevated (Fig. 11). In contrast, in the spinal fish the reflex is essentially normal, a result that suggests that, as for body undulation, the pattern of the movement is determined entirely within the spinal cord. After cerebellar ablation there is an increase in inhibitory drive, presumably generated by the hindbrain neurones, that operates on the spinal circuits to control their inherent excitability.

On the basis of these results Paul & Roberts (1979) suggest that the cerebellum modulates the inhibitory drive to the spinal cord so that during a movement the total cerebellar output to the hindbrain is sculptured to modify selectively its influence on the spinal cord. The cerebellum is to be seen, therefore, not as deriving the pattern of a particular movement but as determining how much of the pattern, derived elsewhere in the nervous system, should be expressed.

Clearly, on this view of cerebellar function, different regions of the body will be regulated by selected areas of the cerebellum which will effectively hold a map of the body's motor activities; animals with complex movements accordingly require elaborated cerebellar cortices. This interpretation may provide an explanation of why the elasmobranch cerebellum is relatively so much larger than that found in teleosts, because those teleosts that possess neutral buoyancy do not have the problem of matching thrust against lift which confronts the swimming elasmobranch whose equilibrium is essentially a dynamic one. A useful analogy may perhaps be made between the complex manoeuvres that are required to fly an aeroplane as compared to the simpler control systems needed to operate a barrage balloon!

Our knowledge of the interconnections between the cerebellum, other brain centres and the spinal cord is far too incomplete at the moment for us to develop any kind of detailed "circuit diagram" of the locomotory system in fishes. The block diagram of Fig. 12 summarizes the main relationships we have been considering and focuses attention on the spinal cord as the main source of locomotory design. The motor programs generated there are appropriately transformed to meet behavioural and environmental demands; first, by commands generated within the brain centres and second, by signals received from body sense organs and adjacent segments of the cord.

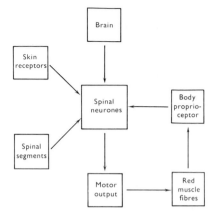

FIG. 12. Diagram of main components involved in locomotory control. (From: Roberts, 1969b.)

ACKNOWLEDGEMENT

I would like to thank Dr R. M. Williamson for many helpful discussions and for his comments on the manuscript. The work with Dr Williamson is funded by the Wellcome Trust and that with Dr D. H. Paul by the SRC.

REFERENCES

Bainbridge, R. (1958). The speed of swimming of fish as related to size and to the frequency and amplitude of the tail beat. *J. exp. Biol.* 35: 109–133.

Bando, T. (1975). Synaptic organization in teleost spinal motoneurons. *Jap. J. Physiol.* 25: 317–331.

Bartelemez, G. W. (1915). Mauthner's cell and the nucleus motorius tegmenti. *J. comp. Neurol.* 25: 87–128.

Bernstein, J. J. (1964). Relation of spinal cord and regeneration to age in adult goldfish. *Expl Neurol.* 9: 161–174.

Bernstein, J. J. & Gelderd, J. B. (1970). Regenerative capacity of long spinal tracts in the goldfish. *Brain Res.* 19: 21–26.

Blight, A. (1976). Undulatory swimming with and without waves of contraction. *Nature, Lond.* 264: 352–354.

Bone, Q. (1966). On the function of the two types of myotomal muscle fibre in elasmobranch fish. *J. mar. biol. Ass. U.K.* 46: 321–349.

Bone, Q. (1978). Locomotor muscle. In *Fish physiology* 7: 361–424. Hoar, W. S. & Randall, D. K. (Eds). New York & London: Academic Press.

Bone, Q., Kiceniuk, J. & Jones, D. R. (1978). On the role of different fiber types in fish myotomes at intermediate swimming speeds. *Fish. Bull. U.S.* 76: 691–699.

Demski, L. S. (1977). Electrical stimulation of the shark brain. *Am. Zool.* 17: 487–500.

Demski, L. S. & Gerald, J. W. (1974). Sound production and other behavioral effects of midbrain stimulation in free-swimming toadfish, *Opsanus beta. Brain, Behav. & Evol.* 9: 41—59.

Droge, M. H. & Leonard, R. B. (1979). Fictive locomotion in the stingray *Dasyatis sabina. Neurosci. Abst.* 5: 368.

Eaton, R. C., Bombardieri, R. A. & Meyer, D. L. (1977). The Mauthner-initiated startle response in teleost fish. *J. exp. Biol.* 66: 65—81.

Faber, D. S. & Korn, H. (1978). *Neurobiology of the Mauthner cell.* New York: Raven Press.

Graham Brown, T. (1911). The intrinsic factors in the act of progression in the mammal. *Proc. R. Soc.* (B) 84: 308—319.

Gray, J. (1936). Studies in animal locomotion: IV. The neuromuscular mechanism of swimming in the eel. *J. exp. Biol.* 13: 170—180.

Gray, J. (1968). *Animal locomotion.* London: Weidenfeld & Nicholson.

Gray, J. & Sand, A. (1936). The locomotory rhythm of the dogfish (*Scyllium canicula*). *J. exp. Biol.* 13: 200—209.

Grillner, S. (1974). On the generation of locomotion in the spinal dogfish. *Expl Brain Res.* 20: 459—470.

Grillner, S. (1975). Locomotion in vertebrates: central mechanisms and reflex interaction. *Physiol. Rev.* 55: 247—306.

Grillner, S. & Kashin, S. (1976). On the generation and performance of swimming in fish. In *Neural control of locomotion*: 181—201. Herman, R. M., Grillner, S., Stein, P. S. G. & Stuart, D. G. (Eds). New York: Plenum.

Grillner, S., Perret, C. & Zangger, P. (1976). Central generation of locomotion in the spinal dogfish. *Brain Res.* 109: 255—269.

Grillner, S. & Wallen, P. (1977). Is there a peripheral control of the central pattern generators for swimming in dogfish? *Brain Res.* 127: 291—295.

Healey, E. G. (1957). The nervous system. In *The physiology of fishes.* 9: 1—119. Brown, M. E. (Ed.). London & New York: Academic Press.

Highstein, S. M. & Bennett, M. V. L. (1975). Fatigue and recovery of transmission at the Mauthner fiber-giant fiber synapse of the hatchetfish. *Brain Res.* 98: 229—242.

Hudson, R. C. L. (1973). On the function of the white muscles in teleosts at intermediate swimming speeds. *J. exp. Biol.* 58: 509—522.

Kashin, S. M., Feldman, A. G. & Orlovsky, G. N. (1974). Locomotion of fish evoked by electrical stimulation of the brain. *Brain Res.* 82: 41—47.

Leonard, R. B., Rudomin, P. & Willis, W. D. (1978). Central effects of volleys in sensory and motor components of peripheral nerve in the stingray, *Dasyatis sabina. J. Neurophysiol.* 41: 108—125.

Lissman, H. W. (1946). The neurological basis of the locomotory rhythm in the spinal dogfish. II. The effect of deafferentation. *J. exp. Biol.* 23: 162—176.

Luciani, L. (1891). *Il cerveletto: Nuovi studi di fisiologia normale e patologica.* Firenze: Le Monnier.

Nieuwenhuys, R. (1964). Comparative anatomy of the spinal cord. *Progr. Brain Res.* 11: 1—57.

Paul, D. H. & Roberts, B. L. (1979). The significance of cerebellar function for a reflex movement of the dogfish. *J. comp. Physiol.* 134: 69—74.

Peter, R. E. (1977). Effects of midbrain tegmentum and diencephalon lesions on swimming and body orientation in goldfish. *Expl Neurol.* 57: 922—927.

Roberts, B. L. (1969a). Spontaneous rhythms in the motoneurons of spinal dogfish. *J. mar. biol. Ass. U.K.* 49: 33—49.

Roberts, B. L. (1969b). The co-ordination of the rhythmical fin movements of dogfish. *J. mar. biol. Ass. U.K.* **49**: 357–425.

Roberts, B. L. (1978). Mechanoreception and the behavior of elasmobranch fishes with special reference to the acoustico-lateralis system. In *Sensory biology of sharks, skates and rays*: 331–390. Hodgson, E. S. & Mathewson, R. F. (Eds). Arlington: Office of Naval Research.

Roberts, B. L. & Williamson, R. M. (In press). The activity of cord neurones in the dogfish during fictive locomotion. *J. Physiol. Lond.*

Shik, M. L., & Orlovsky, G. N. (1976). Neurophysiology of locomotor automatism. *Physiol. Rev.* **56**: 465–501.

Stein, P. G. (1978). Motor systems, with specific reference to the control of locomotion. *A. Rev. Neurosci.* **1**: 61–81.

Steiner, J. (1888). *Die Functionen des Centralnervonsystem und ihre Phylogenese.* No. 2. *Die Fische.* Braunschweig: Vieweg.

Teravainen, H. & Rovainen, C. M. (1971). Fast and slow motoneurons to body muscle of the sea lamprey. *J. Neurophysiol.* **34**: 990–998.

von Holst, E. (1936). Erregungsbildung und Erregungsleitung im Fischruckenmark. *Pflüg. Arch. ges. Physiol.* **235**: 345–359.

Williamson, R. M. & Roberts, B. L. (1980). The timing of motoneuronal activity in the swimming spinal dogfish. *Proc. R. Soc. Lond.* (B) **211**: 119–133.

Zottoli, S. J. (1977). Correlation of the startle reflex and Mauthner cell auditory responses in unrestrained goldfish. *J. exp. Biol.* **66**: 243–254.

Symp. zool. Soc. Lond. (1981) No. 48, 137–172

Flight Adaptations in Vertebrates

J. M. V. RAYNER

Department of Zoology, University of Bristol,
Woodland Road, Bristol, England

SYNOPSIS

The presence of flying animals in each of the five vertebrate classes testifies to the success of flight as a locomotor adaptation in a wide variety of ecological conditions. This chapter explores the mechanisms by which flight can prove advantageous. Most flight adaptations are concerned with energetics, either through reducing the energy cost of locomotion or by enabling economic foraging. Many flying animals are able to glide but cannot propel or sustain themselves in the air. Nonetheless their locomotion is interesting, and illustrates well many of the principles governing flying locomotion. Gliding vertebrates have had relatively few mechanical problems to solve; animals which have developed flapping flight have tackled successfully daunting physiological, anatomical and aerodynamic problems to reach their present sophisticated degree of adaptation. The most obvious change as flapping flight evolved was the lengthening of the forelimbs to form wings sufficiently long to produce sufficient lift and thrust. Additionally, muscles strong enough to drive the wing and metabolism able to provide sufficient power for flight are essential, and must have developed before sustained flapping flight as we know it today could have evolved. Various explanations of how flight evolved have been advanced, but none seems fully satisfactory. Flapping flight has allowed birds and bats to radiate to a wide variety of flight adaptations. Each of these, considered alongside the animal's ecological status, permits the animal effectively to balance its energy budget. Present research is directed towards quantifying flight energy, so that we may better understand the relations between ecology and morphology in flying animals, and towards investigating the aerodynamic mechanisms by which flapping flight takes place. A comparison of the flight morphology and behaviour of birds and bats illustrates well many of the constraints within which a flight adaptation must be designed.

INTRODUCTION

Flight as a form of locomotion requires a high degree of morphological adaptation, but brings substantial rewards to those animals that adopt it. That it is a highly successful and versatile adaptation is indicated by the presence of flying forms in each of the five vertebrate

classes; indeed, true flapping flight requiring an even greater degree
of specialization has evolved in four of them. This chapter is intended
to explore the range of flight adaptations among vertebrates, and to
indicate how important an animal's locomotion habits are as a factor
in its ecology and energetics. By studying the more primitive flight
of the various gliding animals we can understand the factors which
determine the advantages to be gained from flight, and can explore
the mechanisms favouring the evolution of flapping powered flight in
birds and bats. Any flight adaptation is a compromise between a
variety of ecological, physiological, behavioural and mechanical
demands on an animal's design. That these constraints can be viably
balanced is indicated by the success of flying animals: the variety of
flight adaptations indicates that there are a number of ways of
achieving this.

The most important adaptation required by any flying animal is a
large surface area to support its weight. By spreading wings or membranes the animal can take advantage of aerofoil action (Fig. 1) to
influence the airflow around its body, thereby generating lift to
increase the horizontal distance travelled during its flight. There are
only a limited number of types of anatomical adaptation which
permit enlargement of the flight surfaces. The most obvious strategies
are, briefly, to enlarge the feet (flying frogs), to lengthen the ribs
(flying lizards), to develop membranes around the body and limbs
(flying geckos, flying squirrels, bats), and to relinquish entirely other
use of the forelimbs, as is required for flapping flight. The animal
must also have a certain degree of control over its flight surfaces:
they must be held rigid during steady flight, but must be moved
rapidly to counter sudden crises and in take-off and landing. It must
also have a sensory system which can obtain and interpret information while flying; this can be a significant problem if the animal
flies fast (perhaps because its flight adaptations are not sufficiently
well developed for it to be capable of controlled slow flight) or if, as
in the case of flying mammals, it is nocturnal. If the animal is to be a
flapping flier it will also need to develop a skeletal and muscular
system capable of tolerating the stresses involved and of driving the
wings. More importantly, it must also acquire the necessary physiological and biochemical adaptations to provide a sufficient oxygen
supply to the main flight muscles and to enable the optimum conversion of that oxygen to produce mechanical work. Flapping flight
is a strenuous mode of locomotion and makes high demands on
metabolism, especially at extreme high and low flight speeds. Lift
generation becomes harder as the animal flies slower, since all air
movements necessary for propulsion and lift must be generated by

the wing movements alone. On the other hand in fast flight drag of the wings and body is substantial, and considerable thrust forces are required for propulsion. Although costly in its demands on energy, flapping flight is most efficient at relatively high speeds, and the cost of transport (energy consumed transporting unit weight of the animal through unit distance) is relatively low compared with terrestrial locomotion.

Power supply is a major limiting factor for flying animals, and the majority of flight adaptations are related to the minimization of energy consumption, at least through permitting economic foraging. Wing morphology and wing beat kinematics are correlated with life style and flight habits, so that both the amount of energy needed for a particular flight strategy is minimized and the amount of energy or food likely to be obtained from the strategy is maximized. If an animal needed simply to minimize its energy expenditure ignoring other constraints it would certainly not choose to fly; flight (or indeed any locomotor strategy) must give it access to a better supply of food to provide sufficient energy to offset the higher cost of flying; it must also ensure a reasonable degree of security from predation. Each design of flight morphology represents a viable ecological adaptation giving access to a successful ecological niche. It is unfortunate that ecological studies of animal life histories rarely consider locomotion as an important feature: the energy cost of flying is substantial, and is a very significant feature of an energy budget. The problems involved in estimating flight energy are substantial, for although we are able to estimate theoretically the aerodynamic power consumption of a flying animal in the simplest situations, it is far less straightforward to relate this to the actual energy used, metabolic or fuel, for the time spent in flying, the effort used in manoeuvring or in display, and the efficiency of conversion of metabolic energy into mechanical energy while in flight must all be determined. It is the quantification of these factors, as they apply to the life history of an individual animal, which is the greatest challenge in the study of flight.

AERODYNAMICS AND MORPHOLOGY

The aerodynamic principles governing flying animals are well known (Pennycuick, 1972a; Lighthill, 1974), but have important consequences for the morphology of flight adaptations. The principle of aerofoil action by which all flight surfaces operate is that the shape of the surface cross-section modifies the air flow in order to

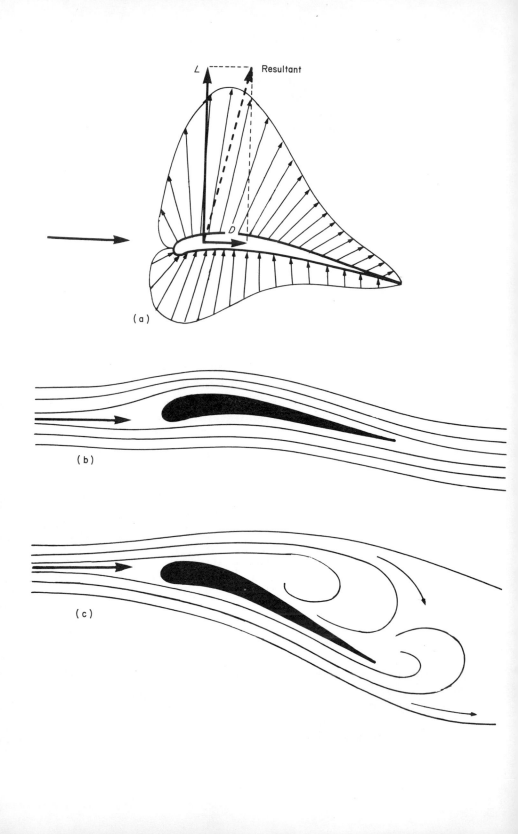

(a)

(b)

(c)

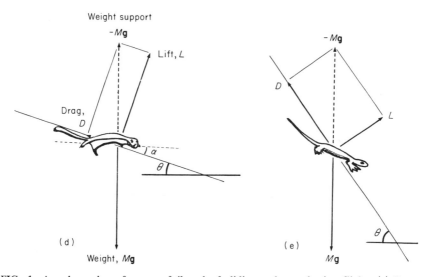

FIG. 1. Aerodynamics of an aerofoil and of gliding and parachuting flight. (a) Pressure distribution and resultant forces for a typical aerofoil. (b) Airflow around an aerofoil in normal flight. (c) Stalled airflow around an aerofoil at high incidence. Balance of forces on an animal in gliding ((d) $\theta < 45°$) and parachuting ((e) $\theta > 45°$). Incidence angle of aerofoil α: in gliding mammals and *Draco* $\alpha \approx 10°$.

generate a pressure difference between the two sides of the wing or flight membrane (Fig. 1); this pressure difference provides the force which supports the animal's weight and propels it through the air. This aerodynamic force is conventionally resolved into two components: the lift, which acts perpendicular to the relative air flow, and the drag which acts parallel to the air flow and must retard the animal. It must be remembered that these are not two separate forces, but are simply components of the resultant of the pressure difference generated by the wings. It is also possible to describe the effect of the wing in terms of the vortex distribution it produces: the "pressure" and "vortex" descriptions of aerofoil action are equivalent (Rayner, 1980). For any aerofoil the lift developed is proportional to the angle of incidence (Fig. 1d); if this angle is too large the air flow over the upper surface breaks away (separates; Fig. 1c), lift falls off rapidly, and the aerofoil is said to stall. Natural wings are generally capable of far better aerodynamic performance in this respect than man-made wings, but the lift available can be an important constraint.

The advantage gained by an animal which can generate even a small amount of lift while moving through the air is great, and represents a strong selective pressure in favour of the development of flight membranes and the ability to use them, provided that the animal's other faculties are not substantially hindered. Lift generation by aerofoil action must always be accompanied by drag retarding the animal; drag is composed partly of the frictional drag of the animal's body and wings (conventionally termed parasite and profile drag) and partly of the necessary drag associated with lift generation (induced drag). Unless the animal can flap its wings and propel itself by generating air movements in the wake whose reaction balances the drag, the kinetic energy necessary to overcome drag can only be obtained at the expense of potential energy, and the glide path must descend relative to the air. The shape and size of the flight surfaces are usually described by two parameters. The wing loading is defined as the weight supported by unit area of the membrane, and is a measure of the pressure difference in flight. The aspect ratio describes the shape of the wing, and is the ratio of wing span to mean wing chord.

If the surface area of the wing is large the rate of descent will be smaller, since the available lift will be greater. Glide speed is proportional to the square root of wing loading, and is not greatly influenced by other flight parameters. Glide angle depends on the geometric properties of the wing rather than on scale. Large aspect ratio enables the animal to have good gliding performance with a shallow glide angle, but large wing span increases the energy cost of flapping flight and can make take-off difficult or even impossible. Increasing wing span with a decrease in wing breath to keep the area constant does not alter glide speed, but reduces the glide angle so that the rate of descent is less and the horizontal distance covered is greater. Glide angle is determined by the relative magnitudes of the lift and drag forces generated by the flight surface (Fig. 1 d and e). The resultant of these two forces must exactly balance the weight in a steady glide. There has been a certain amount of confusion over the terms "parachuting" and "gliding" as applied to passive flying vertebrates, the two words often being used indiscriminately. The two modes of flight should be distinguished by the force responsible for the majority of weight support. Most flying animals rely on lift for this, and in passive flight should be described as "gliding". If, as in a man-made parachute, the weight is supported largely by drag, then the animal can be said to be "parachuting". This distinction can also be phrased as flying at a shallower or steeper angle than 45° to the horizontal, for lift and drag are merely the components of the single vertical force. Oliver (1951) proposed the same distinction on arbitrary grounds, but did not consider the mechanism of weight support. Most flying animals are capable of gliding, but it should be remembered that a gliding animal may choose to descend more steeply and will appear to parachute; to say that an animal is able to glide refers to its optimum performance. Soaring, according to English usage, refers to gliding in non-stationary air, and is a highly specialized flight adaptation found in large birds.

Although long, narrow wings give good glide parameters, the animal must tolerate some lack of non-flying manoeuvrability: the adaptation occurs mainly among seabirds where there is no hindrance to broad wing spreads. If shallow glide angle is not of prime importance there can be distinct advantages in small aspect ratio wings with large area, as are found in the gliding mammals, for stall can be delayed and lift is generated even at high incidence. By flexing its wings a flying animal can increase its flight speed, but its performance is limited by the minimum speed it can achieve, and hence by the area of the flight surfaces; minimum speed is important because it reflects take-off ability, and because a slow speed is important in gliding or manoeuvring in dangerous situations. Sample values of wing loading are shown for flying animals of different body mass in Fig. 2. Birds and bats show the expected increase in loading with body mass approximately as the one-third power (Lighthill, 1977),

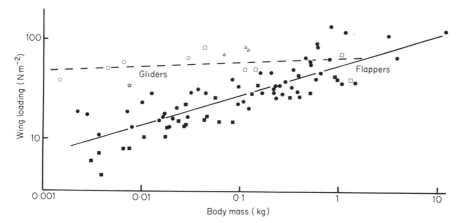

FIG. 2. Wing loadings for flying vertebrates against body mass on logarithmic coordinates. ●, birds; ■, bats; ○, *Draco* (Reptilia; Agamidae); ☆, *Rhacophorus* (Anura; Rhacophoridae); △, flying fish; □, gliding mammals. The regression line for flapping fliers (solid symbols) indicates that wing loadings rise with size, whereas for gliding fliers (open symbols) there is comparatively little increase in wing loading.

bats having rather lower loadings than birds owing to the light build of their membranes and their need for agility. Larger animals have to fly faster in order to generate sufficient lift, and wing loading must increase with size. By contrast, the wing loadings of gliding animals do not increase with size: the smaller animals can afford to fly fast, and their morphology prevents any further enlargement of the flight surfaces. Larger gliding animals cannot tolerate high speeds, and the gliding option is only a valid adaptation if the membrane is large and well developed.

To explore the principles which have governed the development of flight adaptations in vertebrates, I wish to look first at the various passive fliers among the vertebrates: those species of amphibians, reptiles, fish and mammals which have acquired the ability to glide, but have no powers of flapping flight. These interesting animals are valuable in that they can help us to understand a great deal about the ecological benefits of flight and the forces which have encouraged the evolution of flapping flight, for the ancestors of flapping animals must have possessed primitive adaptations possibly similar to those we can trace in gliding vertebrates today.

GLIDING VERTEBRATES

Amphibians

Gliding is represented in the Amphibia by the celebrated "flying frogs" of south-east Asia, Australasia and South and Central America. Tree frogs are widespread, and are distinguished from other Anura by their slight build, small size and by the adhesive pads on the ends of their digits; in the tropics they often live high in the forest canopy, and it is not surprising that several species have a limited form of flight. Flying frogs have enlarged their flight surfaces by lengthening their digits and growing membranes between them. Many are able to flatten their bodies so that the entire animal contributes to lift generation. They are readily distinguished by the webbing of their front feet; although the webbing also camouflages the animal against bark or foliage (and is usually coloured accordingly), its main function is to act as a lifting surface during flight. The most famous record of a gliding frog is Wallace's (1869) observation of *Rhacophorus rheinwardtii* (Rhacophoridae); a more detailed discussion of the same animal has been given by Siedlecki (1908). The animals' flight is very agile, and they glide freely; its purpose is usually claimed to be escape from predation, but Siedlecki argues that they also use it to feed, for their stomachs contained insects which were only active at night and which were never found on the trees. Flight is used by several other species of rhacophorid and hylid frogs. Cott (1926) describes experiments on the flying ability of *Phrynohyas* (=*Hyla*) *venulosa*. He found that it was able to reduce its rate of descent, but did not glide well and should be considered as a "parachutist". This animal possesses little adaptation for flight, and by comparison the membranes of the similarly-sized *Rhacophorus rheinwardtii* are considerably larger. Other neotropical and Australasian hylid frogs have more obvious webbing, and can be classed as gliding. Duellman (1970) reports a glide by *Hyla miliaria* at an angle of about 18°; this is the largest of a group of hylids which also have membraneous flanges along the hind edge of each limb as an additional flight surface. Scott & Starrett (1974) discuss the flight performance of a number of species of flying frogs, and in most of the observations cited glide angles as shallow as 30° were found.

In flight flying frogs hold their limbs close to the body, with feet splayed (Fig. 3). At the beginning of the flight the animal leaps with its hind legs stretched out; these are brought into the gliding position when a steady glide has been achieved. The majority of weight support is provided by the hindlimbs, which possess the largest webs. The lift force from these acts behind the centre of gravity, and stability is ensured by the webs on the front feet. The glide performance will be similar to that of a "Canard" aircraft, which is exceptionally stable to pitching; lateral stability is controlled by movements of the forelimbs. It seems reasonable to suggest that these adaptations together with the low wing loading (Siedlecki, 1908; Fig. 2) permit the animal a slow and shallow glide.

It is not easy to speculate on the importance of gliding in the biology of arboreal frogs. There are several species with weakly developed webbing on their forefeet, suggesting that it is fairly common. There is a definite trend for gliding frogs to be larger than their non-gliding relatives, but this may reflect the lack of any need for distinct gliding abilities in smaller species which are less likely to be harmed by a long fall. Escape from predation undoubtedly plays a significant role, but the ability to move around the forest canopy to find food and water will be of value to the animals; to what extent these and other features influence the appearance of gliding in frogs will not be clear until their habits are much better understood.

Reptiles

The most celebrated flying reptiles are the pterosaurs, of the Jurassic and Cretaceous periods. These have been the subject of a great deal of attention (see, for example, Bramwell & Whitfield, 1974; Desmond, 1975). Some were by present-day standards enormous: the largest known at present had a wingspan of 15.5 m (Lawson, 1975).

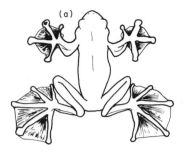
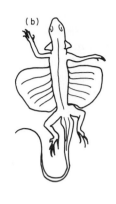

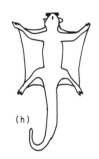
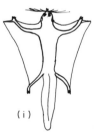

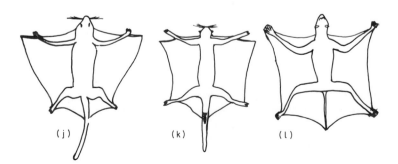

Undoubtedly many had the capacity for powered flight much as we know it in birds and bats. The wing design in which the entire flight surface is supported by a bony forelimb along the leading edge cannot have given the animal a great deal of agility, but does illustrate the principle that flapping flight can only be achieved if other uses of a limb are sacrificed. Whether the largest pterosaurs were capable of flapping flight as opposed to gliding and soaring remains open to argument. By analogy with present-day birds the larger creatures would have been unable to provide the power output required for sustained flapping, and the wings are unlikely to have been able to support the stresses involved. It seems most likely that they relied on soaring to gain height, perhaps leaping from cliffs to gain initial momentum.

There are no living reptiles using powered flight, and it seems unlikely that poikilothermic reptile metabolism could support it even if the necessary adaptations were to appear. However, the gliding reptiles found in the tropics are of some interest, for they have evolved a simple but effective means of locomotion. Flying reptiles and amphibians have been described by Oliver (1951) and Mertens (1960).

The flying "dragon" genus *Draco*

The best known gliding reptile is the flying dragon of south India and south-east Asia; there are about 20 species in the genus *Draco* (Agamidae) found throughout this region. These animals have developed a flight surface by lengthening between five and seven of their ribs, and using these to support a thin membrane (patagium) entirely independent of the limbs. Their flight has been described by Klingel (1965), Colbert (1967) and Scholey (in preparation c), who report a characteristic flight pattern. On leaping from a point high up on the trunk of a tree, the animal descends in a rapid dive for about two metres at an angle as steep as 70 or 80°. When sufficient kinetic energy has been built up the animal twists its tail and raises the fore edge of the patagium to generate more lift and slow the descent. It then travels in a shallow glide, at an angle of about 15°, until just

FIG. 3. Gliding adaptations in vertebrates. Drawings not to scale; *approximate* mass and head-body length of mature individuals shown after each animal. (a) *Rhacophorus rheinwardtii* (Anura; Rhacophoridae) 8 g, 7 cm. (b) *Draco maculatus* (Reptilia; Agamidae) 6 g, 8 cm. (c) *Ptychozoon kuhli* (Reptilia; Gekkonidae) 15 g, 10 cm. (d) *Exocoetus volitans* (Pisces; Exocoetidae) 50 g, 30 cm. (e) *Hirundichthys rondeletii* (Pisces; Exocoetidae), plan view. (f) *Gasteropelecus levis* (Pisces; Gasteropelecidae) 10 g, 10 cm. (g) *Gasteropelecus levis* (Pisces; Gasteropelecidae), front view. (h) *Petaurus breviceps* (Marsupialia; Phalangeridae) 110 g, 16.5 cm. (i) *Pteromys volans* (Rodentia; Sciuridae) 140 g, 16.5 cm. (j) *Petaurista petaurista* (Rodentia; Sciuridae) 1.3 kg, 40 cm. (k) *Anomalurops beecroftii* (Rodentia, Anomaluridae) 650 g, 31 cm. (l) *Cynocephalus volans* (Dermoptera) 1.1 kg, 34 cm.

before landing on another tree when the angle of incidence of the membrane is increased and the animal lands gently. In flight it has a great deal of control over its movements, largely by the long tail; its vision must be acute, for it can fly through small gaps in foliage and past obstacles with little noticeable change in speed.

Flight in *Draco* is a feeding adaptation; to a lesser extent it permits relatively cheap movement around the forest. The lizards need to obtain a continual supply of ants, and occupy an ecological niche at mid-height in the rain forest where there are few horizontal branches; they spend most of their feeding time on vertical tree surfaces. Hunting downwards appears to be difficult, and the animal usually feeds while climbing (Scholey, in preparation c); by gliding down to the foot of another tree (or indeed of the same tree), and then feeding while climbing, the lizard ensures a continuous food supply with the minimum of energy expenditure. The construction of the joints between the ribs and vertebrae prevents the patagium from folding upwards under the effect of air pressure, and therefore little energy is consumed in flight. Without the ability to fly the lizard could not occupy this niche, and although in flight it can be conspicuous to predators the advantage in ease of feeding outweighs this.

Draco is the only living vertebrate with a flight surface formed from lengthened ribs, but a number of prehistoric reptiles were morphologically very similar. *Icarosaurus* (Colbert, 1970) was about twice the size of *Draco* but had very much longer wings; its flight was probably much the same as that of *Draco* (although the long wings would have given it a shallower glide angle), and Colbert infers that its habits were similar. The adaptation has also been noted in *Daedalosaurus* from Madagascar (Carroll, 1978) and a similar but as yet unnamed animal from north-east England (Pettigrew, 1979); both are broadly similar to *Icarosaurus* and *Draco*, but have about 20 elongated ribs. It seems reasonable to conclude that all these animals were convergent on the same highly-specialized insectivorous niche now occupied by *Draco*. Although the tropical forests in other parts of the globe would appear to offer the same advantages to a gliding animal the adaption is not found elsewhere today.

Other flying lizards

Gliding has been reported in other species of lizards, in which the flight adaptation is much less distinct. There has been a considerable amount of controversy attached to the question of flight in these animals; while it is certain that they do possess the faculty, at least to some limited degree, the question of its importance will be finally resolved only when the animals concerned have been closely observed in their natural habitats. The most widely discussed are the "flying" geckos, *Ptychozoon*. These lizards are larger and more solidly constructed than *Draco*; the most remarkable features are the broad flaps of skin on either side of the body which are folded under the animal when at rest, and spread sideways in flight by stretching the limbs. In addition there are similar small flaps on either side of the neck, the body and tail are flattened, and the feet are webbed. The adaptations of these animals, in particular of *P. lionatum*, have been considered in detail by Russell (1979), who also reviews the various observations and opinions advanced about the gliding ability. Russell suggests that the flaps are primarily concerned with crypsis, and only later acquired their function as a flight surface. *Ptychozoon* is very inconspicuous when pressed against a tree trunk, and it is hard to see how the presence or absence of the flaps could alter its chances of survival. *Ptychozoon* certainly does fly, although less frequently and with less proficiency than *Draco*: it is too heavy and the membranes are too small for it to have a shallow glide angle: all observations (see Russell, 1979) estimate glide angle as about 45° or steeper, and the animal should therefore be classed as "parachuting"; the outspread skin folds certainly give the appearance of a true parachute. Flight is not as important in the ecology of *Ptychozoon* as in *Draco*, and is probably used primarily for escape. Anti-predator behaviour alone is unlikely to have justified the development of the flight membranes without their added camouflage advantage. If *Ptychozoon* used flight as a normal form of locomotion it would be lighter and would have better glide performance; the morphology of the skin flaps prevents this because the span of the membranes cannot be enlarged beyond the length of the limbs.

A number of other species of gekkonid lizards possess similar adaptations to *Ptychozoon*, although the membranes are rarely so well developed (Mertens, 1960; Russell, 1979). Mertens has suggested, on the basis of their similarity with *Ptychozoon*, that these animals should be able to fly, but no observations support this. Various other lizards flatten their bodies while leaping (Oliver, 1951; Mertens, 1960), and probably represent a primitive but little developed flight adaptation.

Snakes

The snakes are remarkably successful animals which have adapted to a wide variety of habitats with the minimum of changes in morphology. Nonetheless, it is still surprising that there is a flying snake. Traveller's tales had told of such a creature for many years, but these were not confirmed until 1906 when Shelford reported that a number

of snakes in the jungles of Borneo fall from tall trees, and that experiments with a specimen of *Chrysopelea* (Colubridae) had confirmed
that this snake was able to "glide" remarkably well. He found that
it drew in the skin on the ventral surface of its body to form a deep
concave hollow, and by this means retarded its fall. Shelford also
remarked that a number of related species from all over south-east
Asia have similar morphological features, and by analogy are also
able to fly. There do not appear to be any photographs available of
these snakes in flight, but a glide of a *Chrysopelea* species was filmed
for the BBC television series "Eastward with Attenborough" (*c.*
1968). The angle of descent appears to be about 30°, suggesting most
remarkable performance for an animal which at first sight is so ill-
suited for this mode of locomotion. The most striking feature of the
glide is the serpentine coil into which the body is configured; once
the steady glide is attained the body is held rigidly coiled in a tri-
angular-shaped plane, with the tail used to control the glide. The
transverse coils of the body are close to one another, and it is tempt-
ing to suggest that in this configuration the snake operates as some
kind of "slotted delta wing". I have no confirmation for this idea,
but the glide performance is sufficiently impressive for there to be
some more effective lifting mechanism than can be explained solely
by the concave ventral surface. Though this may have some aerodynamic effect, it is more likely to be used for landing, with the
impact being taken by the ribs rather than by the soft ventral tissues.
Flight in snakes probably owes its origin to the pursuit of prey
(mainly small lizards which leap or even fly), but once acquired the
ability is useful in that it enables the snake to move around freely.
The animal's ecological niche cannot be greatly different from that
of other arboreal colubrid snakes, for the morphological adaptations
required for flight are minimal, and little selective pressure would be
needed to encourage their appearance. The more significant changes
needed would have been in behaviour, in developing the confidence
and balance needed for a glide.

Flying Fish

Marine flying fish

A number of species of marine fishes have developed passive gliding
flight. These belong to the Exocoetidae (including the two-winged
species *Exocoetus* and the four-winged *Cypselurus*) and the related
Hemirhamphidae. For descriptions of the species concerned and a
review of the extensive literature on the fishes and their flight see
Klausewitz (1960) and Aleyev (1977: 169–175). Flying fishes occur

throughout tropical oceans, and are all similar in design. They inhabit the upper layer of the ocean, feeding on plankton, and are well adapted to breaking through the water surface and to flight. The pectoral fins are much enlarged, and are situated near to the centre of gravity. In flight they are stretched out beside the body, and appear in plan (Fig. 3) very similar to the swept-back wings of hirundine birds. Because of the position of the fins there is nothing to hinder their enlargement in order to improve flight; nonetheless, flying fishes have relatively high wing loadings. Stability is not a serious problem, for a fish losing control during a glide will merely fall into the water; some exocoetids can use their ventral fins for balance and control when necessary. The other flight adaptations include a thin cylindrical body, to reduce energy loss when passing through the water surface, and an enlarged lower lobe of the caudal fin to generate lift before take-off. The total length of the fish ranges from about 20 cm in *Exocoetus* to 50 cm in *Cypselurus*; generally, the four-winged species are larger. Members of the family Dactylopteridae, the flying gurnards, have often been described as flying fish owing to their resemblance to the Exocoetidae. Like many other fish they leap out of the water when pursued, but there is no particular reason to believe that these animals do fly, and if they in fact do they will certainly be less adept than the Exocoetidae.

Once it has taken off a flying fish glides over the water surface, but owing to the drag of body and wings gliding must be associated with height loss or with deceleration, and gliding is only effective if the take-off is sufficiently fast for the fish to travel some distance. The average length of flight is about 50 m at a speed of 25 m/s, with maximum height above the water of about 6 m (Aleyev, 1977). To "fly" still further the fish allows its tail to touch the water surface to give a few strokes to produce further thrust and lift in order to prolong the glide; total lengths of 400 m can be achieved in this way. It is usually considered that flight in flying fishes arose as a response to predation. However, predation is unlikely to provide sufficient selective pressure for the evolution of significant locomotor adaptations: animals unable to escape before the adaptation is fully developed are likely to be caught, and a half-developed adaptation probably makes the animal more vulnerable. There is another explanation for flight in fish which I suggest deserves some attention. Owing to the greater density of water, the drag experienced by a body underwater is of the order of 800 times greater than it is in air. If the fish could achieve sufficient thrust while under water, and then break the surface with little energy loss, it would be able to travel further than if it remained underwater, because deceleration owing to air drag would be comparatively negligible. This is exactly what flying fish have learned to do. Immediately on breaking the surface the fins are spread — they must be folded under water owing to the high drag — and the glide begins. Provided that the tail beats can provide sufficient thrust for the surface to be broken at a speed which will enable lift generation in flight and will take the fish far enough, the cost of transport will be lower than that for swimming continuously, so that flying carries significant energetic advantages. The principle is identical to that advanced by Weihs (1974) to explain burst swimming, and which also applies to intermittent flight of birds (Rayner, 1977) and to gliding flight in mammals (see p. 155): it can be better to build up a store of energy (in these cases kinetic and potential) to be used to overcome drag, than to expend more energy working continuously against drag as in steady locomotion. For flying fish the energy savings are probably large, since a fish swimming near to the surface, as plankton-feeding fish must, experiences considerably more drag than if it were in deep water. To be most effective this mode of locomotion will impose limits on the range of sizes within which it is useful, and this no doubt explains the narrow range of sizes of marine flying fish. Flying is valuable, but probably not vital, in escaping predators, for many fish leap from the water to confuse predators which do not have air—water vision. I consider the primary reason for the adaptation to gliding in these fishes is the energetic saving, and the accompanying reduction of cost of transport.

It is possible that the same mechanism explains the locomotor patterns of a number of other marine animals, which have not evolved adaptations for flight, but still benefit energetically from breaking the surface. Penguins, dolphins, squid, several species of fish and some small whales all regularly jump through the surface for short distances. A major factor is the need to breathe, but it could be advantageous energetically to "fly", even for short periods.

Freshwater flying fish
There are two types of freshwater flying fish. The butterfly fish *Pantodon buchholzii* from West African rivers is similar in appearance to the flying gurnards, but certainly does fly. It glides on outstretched pectoral fins for short distances to catch flying insects. Under water the fishes' movements are slow and sluggish, but the flights are rapid; the metabolism of these animals would be unsuited, owing to their small size, to anything more than sudden bursts of activity. The only fish using flapping flight are the South American hatchet fishes of the family Gasteropelecidae. These small (10 cm)

animals also use flight to catch insects above the surface, but can propel themselves for distances of about 5 m. They owe their unusual hatchet-like shape to the extensive breast musculature comprising about 25% of the body mass and used to flap the pectoral fins. The adaptation is highly specialized, but the pectoral fins are also used in water for control and propulsion. It is unlikely that the flights could be sustained, for the fish cannot obtain sufficient oxygen underwater to power the muscles for any length of time, and they would require rest after a brief insect-catching flight.

Mammals

Gliding flight has appeared in mammals of three different orders by the relatively simple expedient of developing a fur-covered membrane spanning the fore- and hindlimbs. This adaptation is found in three genera of marsupials (in each of which it evolved independently) in the Dermoptera (an order containing only one animal, the colugo or "flying lemur"), and in two families of rodents, the Sciuridae (with about 12 genera) and the Anomaluridae (with three genera). The habits and ecology of these animals have been discussed elsewhere (Rayner & Scholey, in preparation), and their flight and its significance are only briefly described here. These creatures are distributed in arboreal habitats throughout the globe, and are only absent from South America, south and west Europe and certain zoologically-isolated islands; as with other gliding adaptations, they are most diverse and common in south-east Asia. The size range (10 g—1.5 kg) is broadly similar to that of bats. The adaptation must be extremely successful, since it has arisen independently at least seven times within the recent past and has radiated so widely. Unrelated species often appear very similar to each other, for the viable range of design compromises in which gliding is successful is narrow, and has enforced a remarkable amount of convergence. The extant gliding mammals have a number of important features in common: they are all predominantly arboreal, they are all nocturnal, and are all largely vegetarian; each of these characteristics together with their wide use of gliding flight is vital to the ecology of these interesting animals. They are essentially tropical forest species, and the flying squirrels distributed through North America and the Holarctic have radiated from former tropical regions, to become adapted for temperate life.

The flight membrane stretches from the rear edge of the forelimb to the fore edge of the hindlimb, and is controlled by a complicated system of muscles (Johnson-Murray, 1977); in some species (see Fig. 3) there are further membranes connecting the forelimbs to the neck and the hindlimbs to the tail. Although the membrane is not broad, it is large owing to the very low aspect ratio (about $1-1.5$), and wing loadings are comparable with those of birds (Fig. 2). The membrane is most well developed in the colugo, one of the largest of these animals, which is helpless on the ground and hangs sloth-like from branches; this is the only species in which no part of the body extends beyond the membrane, and as in bats even the fingers are included in it. In all gliding mammals the membrane can be controlled accurately by adjustments to the position of the four limbs, and the animals have great control over their glides; the smaller species are extraordinarily agile. Gliding rodents have a cartilaginous spine attached to the forelimbs to support the membrane, springing from the wrist in Sciuridae and from the elbow in Anomaluridae; this stretches the membrane, and allows some increase in span without excessive lengthening of the limbs; on rotation of the forelimbs the spine alters the angle of incidence of the membrane, thereby varying the pressure distribution and the glide performance. Spanwise enlargement of the membrane by lengthening of the digits and forelimbs, as has occurred in bats, would reduce the glide angle, but would raise problems of stability unless flapping flight were developed: this would involve present-day gliding mammals in (?unsuccessful) competition with bats. Longer forelimbs would prove a hindrance in climbing, and would undoubtedly reduce agility. The design of gliding mammals is a compromise, but a very successful one, between a number of independent constraints.

Remarkably little has been reported on the flight performance of these animals; although common they are nocturnal and exceptionally hard to observe in all but the most fortunate conditions. Nachtigall, Grosch & Schultze-Westrum (1974) and Nachtigall (1979a,b) describe the glide performance of the gliding marsupial *Petaurus breviceps* and describe various experiments on the aerodynamic properties of a mounted specimen in a wind tunnel. These measurements suggest excellent control performance, with a glide angle between 11 and 27°. These are very steep glides compared with birds and man-made gliders, but are remarkable in view of the low aspect ratio of the membrane. The flight path is very similar to that of *Draco*, with a steep descent followed by a shallow glide when the velocity is sufficiently high, and a slight rise while braking before landing by increasing the membrane incidence; by comparison with birds, the gliding mammals can sustain very high incidence angles, since the low aspect ratio of the membrane and the hairs on the upper surface delay stall by preventing separation of the air flow.

Borodulina & Blagosklonov (1951) and Polyakova & Sokolov (1965) report similar findings with the European flying squirrel *Pteromys volans*. The largest of the gliding sciurid rodents, *Petaurista petaurista*, has been observed in the wild by Scholey (in preparation b), who has recorded glides of up to 130 m from tall trees, with an optimum glide angle of about 11°. Glide velocities are approximately constant in all these species in the range 8–14 m/s; there is little variation with size owing to the near-constant wing loadings (Fig. 2).

It has often been argued that gliding arose in mammals in response to predation. The advantages of being able to leap safely from a high branch are obvious, but there are much greater benefits from being able to move around a forest safely and quickly. It has been shown (Rayner & Scholey, in preparation) that gliding mammals enjoy great energetic advantages compared with other non-gliding creatures. They can move fast compared with running owing to the high glide speed, and provided that they are not heavier than about 1.5 kg, they enjoy significant savings of energy in travelling long distances if they climb a tree and then glide, compared with the alternative of level running. The resulting saving in cost of transport allows an enhanced foraging radius, which is vital to animals which may have to cover large distances each night in order to find the correct food. If the additional cost of running to circumvent obstacles on the ground or in the canopy and of climbing trees to find fruit is included, the comparative advantage to the gliding animal becomes greater still, and explains the great success and wide distribution of gliding mammals.

It has been suggested (see, for example, Smith, 1976; Scholey, in preparation a) that gliding was an intermediate stage in the evolution of bats. The energetic advantages of gliding confirm that the fruit bat ancestor would have been ecologically successful, and the control performance would favour the assumed insectivorous habits of the precursor of the microchiropteran bats. If the proto-bat lengthened its forelimbs to enlarge the membrane to a true wing, learnt to hang with its hind legs while roosting, and at the same time acquired the necessary muscles and metabolism, it is easy to see how the bats developed. The same ecological constraints would have forced the proto-bats to be nocturnal as affect the present-day gliding mammals, but there were then no competitors preventing gliding from developing into flapping (Scholey, in preparation a).

FLAPPING FLIGHT — BIRDS AND BATS

Evolution of Bird Flight

Since the first fossil of *Archaeopteryx lithographica* was discovered, many opinions have been advanced as to the origin of birds and the evolution of flight. Much of the contention has concerned the phylogenetic relationships between birds, *Archaeopteryx* in particular, and the Archosauria and Reptilia, and need only concern us here in that the descent from the coelorusaurian dinosaurs which is now widely accepted is more likely to have allowed development of the metabolic adaptations needed for powered flight (principally warm-bloodedness) than is the alternative of descent from the reptiles. Discussion has also revolved around the so-called "arboreal" and "cursorial" theories, according to which the animals which were the precursors of *Archaeopteryx* were either tree- or ground-dwelling. It is now accepted that *Archaeopteryx* was capable of powered flight at least to some degree (Yalden, 1971). The long tail and short blunt wings suggest a bird making short flights among trees or bushes and using the tail for control and stability. (This cannot be conclusive support for the arboreal theory, however, for these features may be associated with the close relationship with bipedal dinosaurs.) Convincing evidence that *Archaeopteryx* was a powered flier is provided by the well developed furcula, which must have been the origin of a powerful pectoral muscle (Olson & Feduccia, 1979), and by the highly asymmetric primary feathers, a characteristic of birds using strong powered flight (Feduccia & Tordoff, 1979), which can only be explained by the aerofoil action of separated primary feathers in flapping flight.

The details of the arboreal — cursorial argument are well known, and are reviewed by De Beer (1954) and Ostrom (1974), but despite many well argued suggestions no convincing description of the habits of *Archaeopteryx* consistent with the available evidence has yet been advanced. According to the arboreal theory, proto-birds lived among trees, were probably bipedal, and learnt to fly by first leaping between branches, then gliding with outstretched forelimbs, finally developing the wings and muscles necessary for powered flight; at this stage some birds were able to radiate to non-arboreal niches. Bock (1965, 1979) has explained in detail how this process could have taken place, with the animal fully adapted at each stage for its mode of life. The process would have been very similar to that by which bats have evolved from insectivorous arboreal mammals (Smith, 1976; Scholey, in preparation a). The "arboreal" argument is convincing in its simplicity, but recent study shows that it may not be supported by the palaeontological evidence. The cursorial theory as originally put forward envisaged the proto-bird running and flapping its forelimbs in order to obtain short leaps from the ground, which gradually lengthened as the wings grew stronger. This argument has been discounted on the grounds that the main source of thrust, and at first also of lift, is removed as soon as the legs leave the ground, and therefore the selective pressure favouring the development of wings is weak. Recent research by Ostrom (1974, 1979) has led to a modified version of this idea, which attempts to explain cursorial selection for flight. According to this proposal, the proto-bird was cursorial, but used its feathered forelimbs to catch insects as it ran: the waving movements of these limbs enabled it to develop lift while jumping after flying insects. (It is now believed that feathers evolved independent of flight, probably for thermoregulation, and only later acquired their function as flight surfaces (Regal, 1975).) The palaeontological evidence, as interpreted by Ostrom, favours a cursorial habit for *Archaeopteryx*; his theory takes this into account, and also avoids the need for a bipedal animal to colonize an arboreal niche successfully, which the more logical arboreal theory apparently demands. Although persuasive, Ostrom's arguments seem to suffer from the same problem as the original cursorial theory: bipedal running while chasing insects means that the main source of thrust vanishes on take-off. Flapping to catch insects demands horizontal movements of the wings rather than the vertical movements needed to fly, and is unlikely to have favoured development of the musculature needed for flapping flight, which was apparently possessed by *Archaeopteryx*. This design, with a pectoralis muscle much larger than the supracoracoideus, could be explained if the animal first learnt to glide, and developed a pectoral muscle to hold the wings in place. Further, the cursorial argument gives no indication of how either the morphology of the supracoracoideus muscle or the distinctive configuration of flight feathers might have arisen: both are associated with flight and appear to confer no other advantage.

By analogy with the energetic advantages of gliding flight in mammals as a fully adapted stage in the evolution of bats, and since if *Archaeopteryx* was arboreal and a poor flier it did not need to be as "successful" as modern birds for it had no aerial competitors, it seems feasible that *Archaeopteryx* was to some degree arboreal. It is not easy to explain what energetic or ecological advantages were gained from leaping up from a level substrate. There is a great need for a new theory which can reconcile the apparent ecological claims of the arboreal theory with the palaeontological evidence embodied in Ostrom's. It may be that there is some value in a compromise, such as a predominantly ground-dwelling creature gliding from the tops of cliffs. Without doubt future research and discussion will help us to understand the origin of birds and of powered flight in greater detail.

Flapping Flight Kinematics

Bird flight has been the subject of a great deal of attention in recent years, and it is impossible to review all the relevant literature. The following notes indicate some of the more important papers, but should by no means be considered exhaustive! Descriptions of flight kinematics and of the fundamental aerodynamic properties of birds and wings have been given by Brown (1963), Lighthill (1974, 1977), Nachtigall (1975), Bilo & Nachtigall (1977) and Rayner (1979a, 1980). Flight energetics and its importance in migration have been discussed by Pennycuick (1969, 1975). Lorenz (1933), Herzog (1968) and Pennycuick (1969, 1972a, 1975) are valuable general discussions of flight biology, and Berger & Hart (1974) is a detailed review of physiological aspects of flight. A detailed survey of kinematics and morphology has been undertaken by Oehme (see Oehme & Kitzler, 1974, 1975a,b; Oehme, Dathe & Kitzler, 1977; Dathe & Oehme, 1978), and a substantial collection of morphological data has been assembled and analysed by Greenewalt (1962, 1975). Various more specific aspects of flight biology have also been described. The function of flocking and formation flying has long been of interest, and some recent discussions are the subject of a review by May (1979). The importance of flight for feeding in small passerine birds has been emphasized by Norberg (1979), and the use of soaring in foraging has been described by Pennycuick (1972b).

Interest in flapping flight has naturally centred on the kinematics and aerodynamics of the wing beat and on the rate of energy consumption in flight: the basic mechanical principles by which flight is possible. These have been explored in many of the papers referred to above. Although the detailed kinematics of a bird's wing stroke vary between species, the basic pattern of the beat in steady forward flight is dictated by the aerodynamic constraint of maximum lift and thrust generation with minimum drag and energy, and cannot vary to any degree. Figure 4 shows the wing beat cycle for a pigeon flying at about 6 m/s. Approximately 40% of the time occupied by the stroke is taken by the downstroke, which is responsible for most, if

FIG. 4. Wing beat of a pigeon *Columba livia* flying at about 6 m/s. Composite sequence of photographs at approximately equal intervals through the stroke cycle. Photographs by G. R. Spedding, Bristol University.

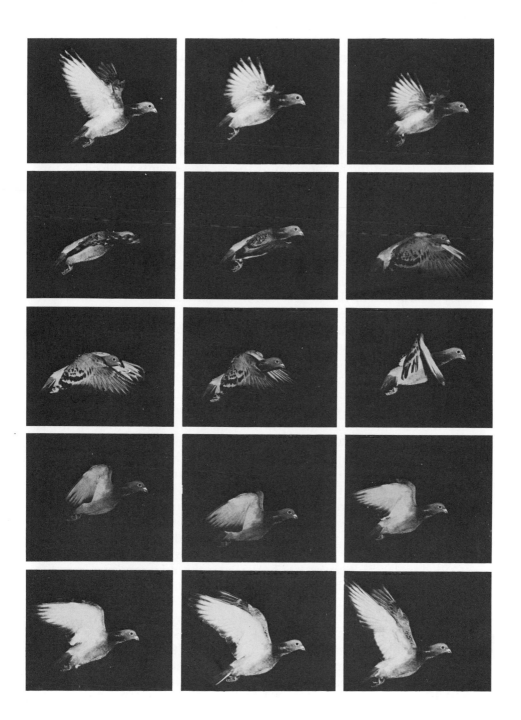

not all, of the lift and thrust generation. During the downstroke the wings are widely spread (the breadth to which they are spread is related to the flight velocity), and move downwards in a plane inclined to the horizontal; this is to ensure that the force generated by each section of the wing has some forward component to overcome drag. It has been described elsewhere (Rayner, 1979b, 1980) how the downstroke gives rise to the air motions necessary to propel and support a flying bird. Each beat sheds a vortex distribution into its wake, which deforms to a closed loop approximately elliptic in shape. The reaction of the momentum of the air movements induced by this vortex loop provides the force necessary for steady level flight, and is identical to the sum of the forces (usually, but misleadingly, resolved into components of lift and drag) generated by aerofoil action at each section of the wing, as in Fig. 1. Experiments by Kokshaysky (1979) confirm this concept of the wake structure, and make it clear that in small birds the upstroke has no aerodynamic effect. It serves as the "return" portion of the stroke, and is remarkably complicated; the details differ considerably between different species, but the pigeon is representative. The wing tip is moved very close to the body, and only a small portion of the wing is presented to the airflow. The primary feathers are often well spread and twisted, and in the latter half of the upstroke the fore edge of the wing twists back over the hind portion and the wing then flicks up to its highest position. The exact purpose of these movements is not clear, but detailed study of high speed films shows that the bird decelerates throughout the upstroke; no appreciable thrust or lift is generated, and the wing motions are associated with drag reduction, and with the generation of air flows around the wing which improve the action of the downstroke. In birds adapted for gliding the upstroke in fast flight takes place with the wings spread and lift is generated; these birds generally have long thin wings with low drag, and can rely on the upstroke for lift without incurring too great a drag penalty. In many smaller species wing drag is high (this is the primary reason for the use of bounding flight in small passerines (Csicsáky, 1977; Rayner, 1977, 1979a)); the drag resulting from lift generation during the upstroke would be large, and the optimum strategy is for the upstroke to be passive. Owing to the design of a feathered wing, birds, unlike bats, do not have the option of generating thrust and negative lift during the upstroke.

It is known (Norberg, 1975; Rayner, 1979a,b) that "unsteady" effects beyond normal aerofoil action are used in force generation by the wings, at least in smaller birds and in the more strenuous activities such as hovering, take-off and landing. We have little indication as yet of what influence they may have on wake generation, or indeed of what unsteady aerodynamic mechanisms may be used. Wing motions similar to the Weis-Fogh clap and fling (Weis-Fogh, 1976) occur frequently in pigeons, and my colleagues and I have recently filmed them in a number of other species, including kestrels and barnacle geese (during landing). They presumably serve to enhance lift and may well be employed as required by many birds. Even if only the secondary feathers are touched the animal will benefit. These feathers are near to the humeral joint; they move less fast than the wing tips, and cannot develop lift as rapidly as the primaries. Many bats clap their wings dorsally, and presumably use the same mechanism. To explain the flight performance of many animals we must assume that other "unfamiliar" aerodynamic mechanisms than have yet been described or classified are employed. Future research will clarify the motions of the wings and help us to understand the mechanisms of wake generation better.

Flapping Flight Energetics

The most important factor in relating the use of flight to a bird's ecological status is the amount of energy consumed in flight, and there have been various attempts to measure or calculate this. Pennycuick (1968) used aerodynamic blade-element and helicopter theory to estimate the mechanical work done by a pigeon in flight. The power components he uses correspond to the rate of working against each of the three components of drag experienced by the animals: parasite drag (body), profile drag (wings) and induced drag (lift-induced aerofoil drag). Some of the methods Pennycuick used to derive power are based on aerodynamic practice and their use for describing flapping flight cannot be justified, but the U-shaped curve he derives to relate aerodynamic power to flight velocity has proved most instructive, and has led to many of our current ideas of flight performance. In later work (1969, 1975, 1978) Pennycuick relates this model to migratory performance and flight behaviour, and shows (1979) how the concepts of cost of transport and foraging radius can help our understanding of flight and foraging behaviour. Pennycuick's model is based on observed measurements of flight kinematics in the pigeon, which are not necessarily representative of other species, particularly when the upstroke is active, and hence may give misleading estimates for other birds. To derive a more realistic aerodynamic model of flapping flight I have used the vortex ring structure of the wake to estimate flight power within the same basic framework as the Pennycuick model (Rayner, 1979a,b). The resulting

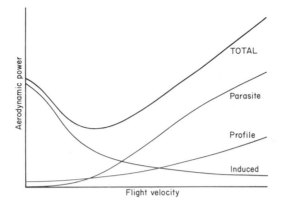

FIG. 5. Aerodynamic power as a function of flight velocity for a small passerine bird. Calculated for gait used in hovering: the high power in fast flight is offset by gait changes and the use of bounding flight.

power against velocity curve (Fig. 5) is similar to Pennycuick's, but by comparison has a rather more pronounced minimum at a slightly lower velocity. This model is more sensitive to morphology and kinematics, but has the disadvantages of requiring detailed kinematic data and of needing large amounts of computer time. It can be used to indicate how the morphology of different species is determined by a balance between ecological/behavioural factors and the constraint, in essence also ecological, of reducing flight power consumption as far as possible; morphological and metabolic adaptations are combined to permit the most efficient use of fuel sources while still retaining the advantages of flight.

Any animal has an obvious need to reduce energy consumption, since those individuals which can perform the same task for a lower energy cost will carry a strong selective advantage. In addition to determining the broad correlation between morphology, habits and habitat, this constraint also dictates the wing beat kinematics, or gait, used in flying. The term "gait" is borrowed from the study of terrestrial locomotion to describe the pattern of wing movements. It is perhaps not used in its strictest sense: the discontinuities which distinguish terrestrial gaits are rarely found in flapping flight, for the biomechanical features which determine them are absent from flying animals. The influence of stroke frequency and downstroke period and amplitude on flight power are illustrated in Fig. 6. At low speeds there is a definite energetic advantage in the use of a fast downstroke of large amplitude, while at high speeds a slow, short downstroke consumes less energy. The range of stroke kinematics available is limited by physiological and biomechanical features of the flight

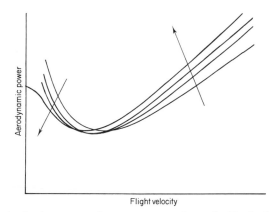

FIG. 6. Effect of gait changes on the power curve shown in Fig. 5. Arrows indicate the effect of increasing the downstroke amplitude and velocity, advantageous at low speeds but detrimental at high. The bird has access to the envelope of the available power curves; these are determined by a range of physiological, mechanical and aerodynamic constraints.

apparatus which we cannot yet accurately quantify, and also by the aerodynamic constraint of avoiding wing stall. The optimum gait pattern will vary between species, but the pattern of wing movements suggested as most advantageous is indeed observed in many birds. The ultimate strategy available to large birds at high speeds is complete reduction of the downstroke, that is gliding. This is not fully comparable for no thrust is generated by a beating wing and the bird can only progress by losing height, but the strategy becomes economic if the bird can climb easily. Many species use rising air masses in order to gain height, and the only work needed to remain airborne keeps the pectoral muscles under tension holding the wing steady. Albatrosses have become so well adapted to soaring flight over the sea that they have a "lock" on the humeral joint which prevents the wing from rising above the horizontal when it is fully stretched (C. J. Pennycuick, personal communication), and have thereby almost eliminated energy expenditure in flight.

To understand fully the ecology of a flying animal we must be able to quantify the energy needed for flight and the fuel which it must be able to acquire. The majority of flight adaptations in both gliding and flapping animals are associated with foraging, in that they permit the animal to obtain a source of food which is rich enough to offset the metabolic cost of flying. If a suitable food source is not guaranteed the flight adaptations would not enjoy the obvious selective advantage which encourages them. The major disadvantage of the aerodynamic models of flight power is that they estimate the mechanical power in flying, that is the mean rate of increase of kinetic energy of the air, and additional energy will be used for the bird's metabolism and will be lost as heat from the chemical reactions and mechanical systems involved in converting fuel into mechanical work. The parameter which the bird must strive to minimize is the metabolic power. This is related to the aerodynamic power by an efficiency factor, generally assumed, in common with other vertebrate locomotor systems, to be about 25%. There are a limited number of measurements of metabolic power consumption by birds flying free (see review in Greenewalt, 1975; also Hails, 1979) and by birds and bats flying in wind tunnels (Bernstein, Thomas & Schmidt-Nielsen, 1973; Thomas, 1975; Torre-Bueno & Larochelle, 1978; Tucker, 1968, 1972, 1973), but the difficulties involved in obtaining a meaningful sample of estimates in this way are formidable. In each case a U-shaped power against velocity curve is obtained, but the details of the curve are often at variance with the theoretical predictions if the efficiency is assumed to be constant. The minimum is substantially more shallow and appears to fall at a high flight velocity. Torre-Bueno & Larochelle (1978) found very little variation in flight power against velocity in starlings, and similar results were obtained for crows (Bernstein *et al.*, 1973) and gulls (Tucker, 1972). Torre-Bueno & Larochelle (1978) suggest that shortcomings in the Pennycuick power model (and later "improved" versions of it by Tucker (1973) and Greenewalt (1975)) could explain this, but the improvements they suggest were included in my own model (Rayner, 1979a), and only serve to heighten the discrepancy. Although it is possible to explain the discrepancy either by errors in the aerodynamic theories or by experimental errors in the metabolic measurements and the ensuing calculations, I prefer to point to the efficiency as the reason for the anomaly. The two values of power consumption can only be compared if the efficiency is known, and there is every reason to believe that it is dependent upon flight velocity and gait; any variations may have a material effect on the shape of the metabolic power curve as derived from aerodynamic theories.

Flight efficiency represents the proportion of energy input which is converted to mechanical work. (It is not a true efficiency, in that it is not a measure of energy wasted, for in flapping flight, and indeed in any natural locomotion, it is not realistic to consider any portion of the energy as "wasted".) It is likely to vary between different species owing to the different anatomical and biochemical composition of the flight muscles, and since gait and muscle contraction are related it can be expected to vary with flight velocity. I know of no means of estimating efficiency independent of both aerodynamic and metabolic values, which is clearly what is needed, but I believe detailed studies of flight muscle anatomy, physiology and biochemistry may indicate a pattern correlated with different animals and their power consumption. Comparisons and criticisms of the estimates of metabolic and aerodynamic powers cannot be encouraged until the efficiency is much better understood.

Birds and Bats Compared

There are many features common to both birds and bats, and many elements of their flight biology are remarkably convergent. The most striking differences between the two species are dictated by the different origins of the two groups. Most birds are diurnal, and few have attempted to colonize nocturnal niches; on the other hand, most bats are strictly nocturnal, and only fly by day when raptorial bird predators are absent. This nocturnality of the bats is probably a response to competition with diurnal birds, and is only possible in a fast-flying and manoeuvrable creature if some advanced sensory system is present, and the majority of bats benefit from echolocation, which they may have "inherited" from their gliding ancestors (Simmons & Stein, 1980; Scholey, in preparation a). The bat wing design probably leads to greater agility but higher fuel consumption compared to birds, but is ideally suited to the insectivorous habits of the majority of species; many bats have adopted other ecological niches and other food sources, but virtually all species rely on flight for feeding. Bats have been extraordinarily successful creatures, and are the most widespread (and possibly numerous) of mammalian orders; this is explained largely by their successful flight adaptation and their colonization of the aerial nocturnal niche where they are free of competition and predation. The only major habitats not occupied by bats are the polar regions (where food is scarce and thermoregulation probably impossible) and the oceans (again where food is hard to obtain, take-off from water impossible and daytime roosts absent). Apart from the ecological factors common to birds and bats, there are other features which relate them: the wing beat kinematics and the aerodynamic principles which govern lift and propulsion are common to both, although the structure of the wing permits the upstroke to be active in bats.

The morphometrics of bat wings show the same general trends of size variation with mass as found in birds (Greenewalt, 1962), but in general bat wing loadings are lower (Fig. 2) as a response to the greater agility needed for feeding. Bat wings are not particularly broad, but compared to birds the aspect ratio is low and therefore the surface area is rather larger. Flight morphology is of course very different in bats than birds owing to their quadrupedal ancestry. The hind feet support the rear edge of the wing membrane, and in many species a further membrane connects the hind feet and tail; this is used in insect capture and as an additional flight surface. The membrane itself is a thin layer of strong tissue spread between the extended fingers and the body, and requires highly specialized design features to enable it to keep its shape and efficacy in flight (Norberg, 1972). There are further differences between bats and birds in the anatomy of the pectoral girdle and of the wing muscles, and of course metabolism in bats, although highly specialized (Thomas, 1975), is that of a mammal.

The kinematics of bat wing beats have been described by Vaughan (1970), Norberg (1976a,b) and Altenbach (1979), and are generally similar to those of birds. The upstroke does not show a pronounced flick, and the wing tips are not as near to the body as in birds; the wing is highly twisted and appears to be aerodynamically active. Detailed analysis of the forces generated (Norberg, 1976a) shows that in forward flight of *Plecotus auritus* (body mass 9 g), which is typical of small microchiropteran bats, weight support is provided by the downstroke and thrust almost entirely by the upstroke. If this is representative there is a significant difference in this respect from small birds where the upstroke is passive. The reasons appear to be two-fold. First, the design of the wing membrane permits a negative angle of incidence during the upstroke without the excessive drag penalty which would occur in birds, and therefore thrust generation is possible during the upstroke. Secondly, the inclusion of the hind feet in the wing membrane assists in the control of wing camber and twisting. Since the bat upstroke is active, in some conditions at least, the vortex ring model developed for the bird wake will not be applicable to bats, for it assumes that no vorticity is shed by the upstroke; it is unlikely, however, that such calculations would greatly misestimate the flight power, since the aerodynamic principles which dictate the shape of the power curve are common to all flapping fliers.

One feature of bats as compared to birds which has not been explored is the explanation for the difference in range of body masses. No bat weighs more than about 1.5 kg, while birds weigh up to 10 or 12 kg (although sustained flapping flight is rare in birds over about 4 kg); the smallest passerine bird is the goldcrest (4 g), while the smallest bat known is about 1.5 g (probably also the smallest of all mammals). Hummingbirds fall between 2 and 20 g, but their kinematics are more similar to insects than to other birds. Like bats they use the upstroke extensively, but the wing is flipped over, and the gaits are not comparable. There are many reasons for this disparity in size, which together illustrate well the constraints within which flying animals must be designed and which must be considered in any study of flight biology.

To consider the upper size limit first, probably the most significant factor in limiting the maximum mass of bats is related to their nocturnality: virtually all large birds use gliding and soaring for much of their activity, and many smaller birds soar when conditions are suitable. Bats cannot do this simply because convective air currents are absent at night. Some bats undoubtedly do glide and slope-soar, but thermal soaring is impossible (dynamic soaring is only practical at sea). In addition, many birds possess separated primary feathers, which are used extensively in gliding as well as during the upstroke. Bats have a different strategy for force generation during the upstroke which avoids the need for drag reduction by feather separation, but the adaptation would be useful in gliding where its probable function is to spread the vortex wake behind the wing, thereby reducing the glide angle with little change in the glide speed and permitting a bird to have all the performance advantages of a long wing with none of the energetic disadvantages. Without either a very long wing (perhaps impractical for a bat), or separated primary feathers, the glide performance of a bat of large mass would be too poor for the adaptation to be successful, in much the same way that there is a maximum size for gliding mammals (see p. 155). There may also be a physiological factor involved: mammalian metabolism may not be able to provide sufficient power for a large bat to fly, or may make thermoregulation impossible; all the larger megachiropteran bats are found in the tropics, where thermoregulation is no problem. Biomechanical constraints may limit the tensile forces sustained by the legs during roosting without unacceptably heavy bone structure. Finally, the explanation may be ecological: large birds are either aquatic (impossible to imagine in a bat with membranes attached all around the body) or are carnivorous predators. Some bats are carnivorous, and some actually prey on other bats, but the morphological adaptations needed to acquire talons on the hindlimbs may not be feasible if these limbs are an integral part of the flight membrane. Hence, all the adaptations used by birds to obtain large size are morphologically impossible in bats. The design of the bat wing is highly successful for the ecological niches occupied, but it is in evolutionary terms unsuccessful since further radiation of the design to encompass other habitats and adaptations is unlikely to take place.

At the lower end of the size range, bats can be significantly smaller than birds. The use of the upstroke for lift generation is probably valuable, and if in addition the bat wing generates relatively little aerodynamic drag one of the main constraints on small birds is removed. Small birds with broad wings use bounding flight extensively to reduce profile drag; as far as I know, no bat uses this style of flight. Bats fly slower than birds, for their wing loadings are lower: agility rather than speed is helpful for catching flying insects, and therefore profile drag will be relatively less important. Also, if the upstroke is used for lift generation the wing beat frequency needed to generate sufficient lift at slow speeds without the wings stalling is lower, and so a smaller size can be reached before the biomechanical constraints on wing movements are significant. On the whole, I think the most important factor permitting small size in bats is the use of the upstroke, for the smallest bats and the hummingbirds both feed on nectar, which is very rich in energy, and hence energetic limitations are avoided; both use the upstroke extensively in order to reduce size without encountering aerodynamic problems. The absolute lower limit for flying vertebrates is probably determined by the biomechanics of wings and muscles, and by the upper limit to muscle contraction rates and wing beat frequencies. It has also been suggested (Scholey, in preparation a) that echolocation enables bats to reduce the size of head because they do not require large eyes for vision as do birds, and thereby can further reduce size.

CONCLUSIONS

Arguments such as those given above to explain the disparity in size between birds and bats indicate the variety of features which must be taken into account in order to understand fully animal flight biology. Each flying organism has adapted in its own way to a viable life style within constraints set by many different, interacting features of the animal's biology. I have discussed how energetics holds the key to the understanding of many flight problems, but there remain many aspects where our knowledge is too small and where the habits of the animals concerned are too little known for us to progress far with their analysis. It is still by no means clear how the variety of adaptive strategies which we can distinguish have arisen, but to explain and compare them is one of the most challenging problems confronting the flight biologist, calling as it does for a synthesis of each of the many factors influencing the habits and behaviour of an individual animal.

ACKNOWLEDGEMENTS

I wish to thank many of my colleagues for their ideas, help and advice in the course of the preparation of this chapter, and in particular M. A. Linley, Dr C. J. Pennycuick, K. D. Scholey and G. R. Spedding. My colleagues and I are indebted to the Science Research Council for financial support and encouragement.

REFERENCES

Aleyev, Yu. G. (1977). *Nekton.* The Hague: W. Junk.

Altenbach, J. S. (1979). Locomotor morphology of the vampire bat, *Desmodus rotundus. Spec. Publs Am. Soc. Mammal.* No. 6: 1–137.

Berger, M. & Hart, J. S. (1974). Physiology and energetics of flight. *Avian Biol.* 4: 360–415.

Bernstein, M. H., Thomas, S. P. & Schmidt-Nielsen, K. (1973). Power input during flight of the fish crow, *Corvus ossifragus. J. exp. Biol.* 58: 401–410.

Bilo, D. & Nachtigall, W. (1977). Biophysics of bird flight: questions and results. *Fortschr. Zool.* 24: 217–233.

Bock, W. J. (1965). The role of adaptive mechanisms in the origin of higher levels of organization. *Syst. Zool.* 14: 272–287.

Bock, W. J. (1979). The synthetic explanation of macroevolutionary change – a reductionistic approach. *Bull. Carnegie Mus. nat. Hist.* 13: 20–69.

Borodulina, T. L. & Blagosklonov, K. N. (1951). [On the biology of flying squirrels (*Pteromys volans* L.).] *Bull. Soc. Nat. Mosc.* 56(6): 18–24. [In Russian.]

Bramwell, C. D. & Whitfield, G. R. (1974). Biomechanics of *Pteranodon. Phil. Trans. R. Soc.* (B) 267: 503–592.

Brown, R. H. J. (1963). The flight of birds. *Biol. Rev.* 38: 460–489.

Carroll, R. L. (1978). A gliding reptile from the Upper Permian of Madagascar. *Palaeont. afr.* 21: 143–159.

Colbert, E. H. (1967). Adaptations for gliding in the lizard *Draco. Am. Mus. Novit.* No. 2283: 1–20.

Colbert, E. H. (1970). The Triassic gliding reptile *Icarosaurus. Bull. Am. Mus. nat. Hist.* 143: 85–142.

Cott, H. B. (1926). Observations on the life-habits of some batrachians and reptiles from the lower Amazon and a note on some mammals from Marajo Island, *Proc. zool. Soc. Lond.* 1926: 1159–1178.

Csicsáky, M. J. (1977). Body-gliding in the zebra finch. *Fortschr. Zool.* 24: 275–286.

Dathe, H. H. & Oehme, H. (1978). Typen des Ruttelfluges der Vögel. *Biol. Zbl.* 97: 299–305.

De Beer, G. (1954). *Archaeopteryx lithographica: a study based upon the British Museum specimen.* London: British Museum (Natural History).

Desmond, A. J. (1975). *The hot-blooded dinosaurs.* London: Blond & Briggs.

Duellman, W. E. (1970). The hylid frogs of Middle America. *Monogr. Mus. nat. Hist. Univ. Kansas* 1 & 2.

Feduccia, A. & Tordoff, H. B. (1979). Feathers of *Archaeopteryx*: assymetric vanes indicate aerodynamic function. *Science, N.Y.* 203: 1021–1022.

Greenewalt, C. H. (1962). Dimensional relationships for flying animals. *Smithson. misc. Collns* 144(2): 1—46.

Greenewalt, C. H. (1975). The flight of birds. *Trans. Am. phil. Soc.* 65(4): 1—67.

Hails, C. J. (1979). A comparison of flight energetics in hirundines and other birds. *Comp. Biochem. Physiol.* A 63: 581—586.

Herzog, K. (1968). *Anatomie und Flugbiologie der Vögel.* Stuttgart: Gustav Fischer Verlag.

Johnson-Murray, J. L. (1977). Myology of the gliding membranes of some petauristine rodents (genera: *Glaucomys, Pteromys, Petinomys,* and *Petaurista*). *J. Mammal.* 58: 374—384.

Klausewitz, W. (1960). Fliegende Tiere des Wassers. In *Der Flug der Tiere*: 145—158. Schmidt, H. (Ed.). Frankfurt am Main: Verlag Waldemar Kramer.

Klingel, H. (1965). Über das Flugverhalten von *Draco volans* (Agamidae) und verwandten Arten. *Zool. Anz.* 175: 273—281.

Kokshaysky, N. V. (1979). Tracing the wake of a flying bird. *Nature, Lond.* 279: 146—148.

Lawson, D. A. (1975). A pterosaur from the latest Cretaceous of West Texas: discovery of the largest flying creature. *Science, N.Y.* 187: 947—948.

Lighthill, M. J. (1974). Aerodynamic aspects of animal flight. *Bull. Inst. Maths Appls* 10: 369—393.

Lighthill, M. J. (1977). Introduction to the scaling of aerial locomotion. In *Scale effects in animal locomotion*: 365—404. Pedley, T. J. (Ed.). London & New York: Academic Press.

Lorenz, K. (1933). Beobachtetes über das Fliegen der Vögel und uber die Beziehungen der Flügel- und Steuerform zur Art des Fluges. *J. Orn., Lpz.* 81: 107—236. (Reprinted as *Der Vögelflug.* Pfüllingen: G. Neske (1965).)

May, R. M. (1979). Flight formations in geese and other birds. *Nature, Lond.* 282: 778—780.

Mertens, R. (1960). Fallschirmspringer und Gleitflieger unter den Amphibien und Reptilien. In *Der Flug der Tiere*: 135—144. Schmidt, H. (Ed.) Frankfurt am Main: Verlag Waldemar Kramer.

Nachtigall, W. (1975). Vögelflugel und Gleitflug: Einführung in die aerodynamische Betrachtungsweise des Flügels. *J. Orn., Lpz.* 116: 1—38.

Nachtigall, W. (1979a). Gleitflug des Flugbeutlers *Petaurus breviceps papuanus.* II. Filmanalysen zur Einstellung von Gleitbahn und Rumpf sowie zur Steurung des Gleitflugs. *J. comp. Physiol.* A 133: 89—95.

Nachtigall, W. (1979b). Gleitflug des Flugbeutlers *Petaurus breviceps papuanus.* III. Modellmessungen zum Einfluss des Fellbesatzes auf Umströmung und Liftkrafterzeugung. *J. comp. Physiol.* A 133: 339—349.

Nachtigall, W., Grosch, R. & Schultze-Westrum, T. (1974). Gleitflug des Flugbeutlers *Petaurus breviceps papuanus* (Thomas). Flugverhalten und Flugsteuerung. *J. comp. Physiol.* A 92: 105—115.

Norberg, U. M. (1972). Bat wing structures important for aerodynamics and rigidity. *Z. Morph. Tiere* 73: 45—61.

Norberg, U. M. (1975). Hovering flight in the pied flycatcher (*Ficedula hypoleuca*). In *Swimming and flying in nature* 2: 869—881. Wu, T. Y.-T., Brokaw, C. J. & Brennen, C. (Eds). New York: Plenum Press.

Norberg, U. M. (1976a). Aerodynamics, kinematics and energetics of horizontal flight in the long-eared bat *Plecotus auritus. J. exp. Biol.* 65: 179—212.

Norberg, U. M. (1976b). Aerodynamics of hovering flight in the long-eared bat *Plecotus auritus. J. exp. Biol.* 65: 459—470.

Norberg, U. M. (1979). Morphology of the wings, legs and tail of three coniferous forest tits, the goldcrest and the treecreeper in relation to locomotor pattern and feeding station. *Phil. Trans. R. Soc.* B 287: 131—165.

Oehme, H., Dathe, H. H. & Kitzler, U. (1977). Research on biophysics and physiology of bird flight. IV. Flight energetics in birds. *Fortschr. Zool.* 24: 257—273.

Oehme, H. & Kitzler, U. (1974). Untersuchungen zur Flugbiophysik und Flugphysiologie der Vögel. I. Über die Kinematik des Flügelschlages beim unbeschleunigter Horizontalflug. *Zool. Jb.* (Physiol.) 78: 461—512.

Oehme, H. & Kitzler, U. (1975a). Untersuchungen zur Flugbiophysik und Flugphysiologie der Vögel. II. Zur Geometrie des Vögelflügels. *Zool. Jb.* (Physiol.) 79: 402—424. (Translated as *NASA TT-F*-16901 (1976).)

Oehme, H. & Kitzler, U. (1975b). Untersuchungen zur Flugbiophysik und Flugphysiologie der Vögel. III. Die Bestimmung der Muskelleistung beim Kraftflug der Vögel aus kinematischen und morphologischen Daten. *Zool. Jb.* (Physiol.) 79: 425—458. (Translated as *NASA TT-F*-16902 (1976).)

Oliver, J. A. (1951). "Gliding" in amphibians and reptiles, with a remark on an arboreal adaptation in the lizard *Anolis carolinensis c.* Voigt. *Am. Nat.* 85: 171—176.

Olson, S. L. & Feduccia, A. (1979). Flight capability and the pectoral girdle of *Archaeopteryx*. *Nature, Lond.* 278: 247—248.

Ostrom, J. H. (1974). *Archaeopteryx* and the origin of flight. *Q. Rev. Biol.* 49: 27—47.

Ostrom, J. H. (1979). Bird flight: how did it begin? *Am. Scient.* 67: 46—56.

Pennycuick, C. J. (1968). Power requirements for horizontal flight in the pigeon *Columba livia. J. exp. Biol.* 49: 527—555.

Pennycuick, C. J. (1969). The mechanics of bird migration. *Ibis* 111: 525—556.

Pennycuick, C. J. (1972a). *Animal flight.* London: Edward Arnold.

Pennycuick, C. J. (1972b). Soaring behaviour of East African birds, observed from a motor glider. *Ibis* 114: 178—218.

Pennycuick, C. J. (1975). Mechanics of flight. *Avian Biol.* 5: 1—73.

Pennycuick, C. J. (1978). Fifteen testable predictions about bird flight. *Oikos* 30: 165—176.

Pennycuick, C. J. (1979). Energy costs of locomotion and the concept of foraging radius. In *Serengeti, dynamics of an ecosystem*: 164—184. Sinclair, A. R. E. & Norton-Griffiths, M. (Eds). Chicago: University of Chicago Press.

Pettigrew, T. H. (1979). A gliding reptile from the Upper Permian of North East England. *Nature, Lond.* 281: 297—298.

Polyakova, R. S. & Sokolov, A. S. (1965). [Structure of locomotor organs in the flying squirrel *Pteromys volans* L. in relation to its plane flight.] *Zool. Zh.* 44: 902—916. [In Russian.]

Rayner, J. M. V. (1977). The intermittent flight of birds. In *Scale effects in animal locomotion*: 437—443. Pedley, T. J. (Ed.). London & New York: Academic Press.

Rayner, J. M. V. (1979a). A new approach to animal flight mechanics. *J. exp. Biol.* 80: 17—54.

Rayner, J. M. V. (1979b). A vortex theory of animal flight. Part 2. The forward flight of birds. *J. Fluid Mech.* 91: 731—763.

Rayner, J. M. V. (1980). Vorticity and animal flight. *Seminar Ser. Soc. exp. Biol.* 5: 177—199.

Rayner, J. M. V. & Scholey, K. D. (In prep.). *Ecology and energetics of gliding flight in mammals*.

Regal, P. J. (1975). The evolutionary origin of feathers. *Q. Rev. Biol.* 50: 35—60.

Russell, A. P. (1979). The origin of parachuting locomotion in gekkonid lizards (Reptilia: Gekkonidae). *Zool. J. Linn. Soc.* 65: 233—249.

Scholey, K. D. (In prep. a). *Flight evolution in bats and its consequences for bat sensory systems*.

Scholey, K. D. (In prep. b). *Gliding flight of* Petaurista petaurista *(Rodentia: Sciuridae) and its ecological consequences*.

Scholey, K. D. (In prep. c). *Gliding flight in the flying lizard* Draco *(Reptilia: Agamidae)*.

Scott, N. J. & Starrett, A. (1974). An unusual breeding aggregation of frogs, with notes on the ecology of *Agalychnis spurrelli* (Anura: Hylidae). *Bull. S. Calif. Acad. Sci.* 73: 86—94.

Shelford, R. (1906). A note on flying snakes. *Proc. zool. Soc. Lond.* 1906: 227—230.

Siedlecki, M. (1908). Zur Kenntnis des Javanischen Flugfrosches. *Biol. Zbl.* 29: 704—737.

Simmons, J. A. & Stein, R. A. (1980). Acoustic imaging in bat sonar: echolocation signals and the evolution of echolocation. *J. comp. Physiol.* A 135: 61—84.

Smith, J. D. (1976). Comments on flight and the evolution of bats. In *Major patterns in vertebrate evolution*: 427—437. Hecht, M. K., Goody, P. C. & Hecht, B. M. (Eds). New York: Plenum Press.

Thomas, S. P. (1975). Metabolism during flight in two species of bats, *Phyllostomus hastatus* and *Pteropus gouldii*. *J. exp. Biol.* 63: 273—293.

Torre-Bueno, J. R. & Larochelle, J. (1978). The metabolic cost of flight in unrestrained birds. *J. exp. Biol.* 75: 223—229.

Tucker, V. A. (1968). Respiratory exchange and evaporative water loss in the flying budgerigar. *J. exp. Biol.* 48: 67—87.

Tucker, V. A. (1972). Metabolism during flight in the laughing gull *Larus atricilla*. *Am. J. Physiol.* 222: 237—245.

Tucker, V. A. (1973). Bird metabolism during flight: evolution of a theory. *J. exp. Biol.* 58: 689—709.

Vaughan, T. A. (1970). Flight patterns and aerodynamics. In *Biology of bats* 1: 195—216. Wimsatt, W. A. (Ed.). London & New York: Academic Press.

Wallace, A. R. (1869). *The Malay Archipelago*. London: Macmillan.

Weihs, D. (1974). Energetic advantages of burst swimming in fish. *J. theor. Biol.* 48: 215—229.

Weis-Fogh, T. (1976). Energetics and aerodynamics of flapping flight: a synthesis. *Symp. R. ent. Soc.* 7: 48—72.

Yalden, D. W. (1971). The flying ability of *Archaeopteryx*. *Ibis* 113: 349—356.

Symp. zool. Soc. Lond. (1981) No. 48, 173—197

Flight, Morphology and the Ecological Niche in some Birds and Bats

ULLA M. NORBERG

Department of Zoology, University of Göteborg,
Box 250 59, 400 31 Göteborg, Sweden

SYNOPSIS

The type of habitat a flying animal chooses to live in, as well as its way of exploiting the habitat (its ecological niche), are closely related to its body size, wing form, flight style and flight energetics. Natural selection is assumed to result in some near-optimal combination among these variables.

Optimal foraging theory, based on cost—benefit considerations, aims at explaining the ways animals forage. There should be mutual interrelationships among the optimal foraging behaviour, flight characteristics and function of ecologically important structures (ecological morphology). Such aspects are discussed and some studies on ecological morphology are reviewed.

The optimal flight speed varies with the type and abundance of food. Selection is likely to act towards a wing structure that minimizes the power required to fly at the speed and style optimal for the bird.

The greater the temporal heterogeneity of the environment, the more generalized a species must be both behaviourally and structurally. This is because the optimal foraging behaviour is likely to shift in a seasonal environment. The goldcrest, for example, changes its foraging behaviour drastically over winter.

Consideration of the energetics of hovering in hummingbirds suggests that the wing should be longer in birds at higher elevations than at lower, and also that their ecological behaviour should differ; this is also supported by observations.

The morphology of the wings of five bird species (three tits, the goldcrest and the treecreeper), who gather into mixed-species flocks in northern coniferous forests in winter, shows small but distinct differences; these can successfully be related to specific differences in choice of foraging site and behaviour.

The low wing loading in vultures enables the birds to start soaring early in the morning when thermals are weak and narrow. This gives the bird a chance to reach a carcass before mammalian scavengers do. Different species of vultures and birds of prey start soaring in the morning in the order of increasing wing loading. The vultures' low-aspect-ratio wings are an advantage at take-off from ground and trees.

Certain diving birds use their wings for flight in air as well as in water (in pursuit of prey). It seems an interesting problem to find out how the size and shape of their wings are related to the strongly conflicting demands associated with flight in two media with such different densities.

INTRODUCTION

Most birds and all bats can fly, and like other animals they occupy different niches to avoid competition. Different species therefore often have to fly in different manners, and this requires differing wing morphologies. Some birds and bats have long narrow wings and others have short broad wings, some have low wing loading and some have high and so on (Fig. 1). Some are adapted to soar over large areas in search of food, while others are adapted to hover in front of flowers while drinking nectar. Albatrosses and vultures can soar for hours with a minimum of energy cost, while hummingbirds must hover when foraging; hovering being the most energy-demanding type of locomotion. Widely differing wing morphologies are required for such different flight activities (cf. Fig. 8).

Bats span a size range from about 3 to 1380 g and birds one from about 2 to 12,000 g, and there are many size-dependent factors that also affect the wing morphology.

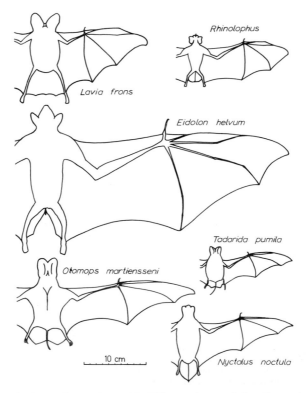

FIG. 1. Wing planform of some bats with different shapes of the wings and tail membrane. All are drawn to the same scale.

Adaptation is defined by Clutton-Brock & Harvey (1979) as a difference between two phenotypic traits (or complexes of traits) which increases the inclusive fitness of their carriers. Most adaptive explanations for such traits are founded initially on comparison. But, as the authors point out, the logical structure of many adaptive arguments is frequently weak, since they are initially *post-hoc*: a behavioural or morphological difference is described and then an explanation is constructed to fit the facts. Such explanations can often be misleading. Relatively few adaptive explanations originate as predictions from models that are then tested against real data.

As for the bird and bat wing, the function of various configurations is explicable in engineering terms, and, therefore, a *post-hoc* explanation of adaptations to different flight habits may not be very misleading. However, when ecological aspects are taken into account one should be more careful with *post-hoc* adaptive explanations. This is because of the large number of more or less well-known ecological factors that may cause selection pressures in various directions.

Since it is sometimes not at all obvious what should be maximized or minimized or what compromise should be made (optimization criterion), a *post-hoc* explanation may involve a subjective choice of arguments so as to fit the observed facts. The logic of optimization theory in biology has recently been discussed critically by Lewontin (1978), Maynard Smith (1978), and Oster & Wilson (1978).

In evolutionary explanations of behaviour, morphology and other traits it is usually assumed that the observed trait represents a near-optimal solution, meaning the solution that maximizes the inclusive fitness of the individual. Since fitness is not readily measurable, one usually singles out some other factor instead, which is related to fitness, but is more accessible for study. As an example, although the aim of optimal foraging theory is to predict the foraging behaviour that maximizes fitness, the models are often built to maximize net energy intake (energy maximizers), or to minimize foraging time (energy minimizers; Schoener, 1971). The main problem with such models is to define a proper optimization criterion, i.e. one that applies to the animal in its natural habitat.

Part of the theory of optimal foraging is highly relevant for an understanding of the morphology of ecologically important structures (ecological morphology). The structure and function of loco-motor organs are closely related to the manner (niche) in which a species exploits its environment.

This chapter will discuss some relations among flight style, morphology and ecology of birds and bats and give some examples of comparative studies on ecological morphology of the flight apparatus.

The theories of animal flight developed by Pennycuick (1968,

1969, 1972, 1975) and Rayner (1979a,b,c) will be used and readers are referred to these papers for detailed information on the aerodynamics of animal flight.

FLIGHT SPEED AS RELATED TO THE ECOLOGY OF THE SPECIES

The optimal flight speed varies among species and depends among other things on habitat structure, choice of food, foraging behaviour, density of food etc.

The power-versus-speed curve for a bird in powered flight (Fig. 2) was calculated by Pennycuick (1968, 1969, 1975) and Rayner (1979c). It is the sum of the induced power (power needed to support the weight), which forms the overwhelming part in hovering and slow flight, the parasite power (power needed to overcome the profile drag of the body), which is a major power drain at high flight speeds, and the profile power (power needed to overcome the profile

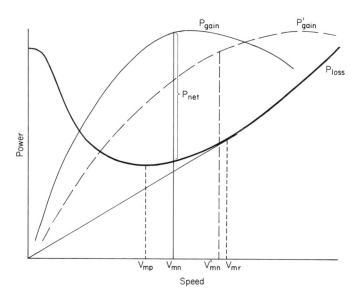

FIG. 2. Hypothetical power-versus-speed diagram for a bird catching insects during flight. The power required to fly, P_{loss}, follows a U-shaped curve (based on Pennycuick, 1968, 1969, 1975, and Rayner, 1979c), whereas the curve for rate of energy gain, P_{gain}, due to insect catching, increases with speed up to a maximum and then remains constant, or decreases again, as speed is further increased. This curve varies with the kind and density of food (exemplified by P_{gain} and P'_{gain}). P_{net} is the rate of the net energy gain. If a species tries to maximize its rate of net energy gain it should fly at speed V_{mn} (maximum net energy gain speed). This choice should be most beneficial when young are reared; the larger the amount of excess food the more young can be reared. In other circumstances there may be other goals with other conceivable speed optima. One is the minimum power speed, V_{mp}, that involves the lowest rate of working. Another is the maximum range speed, V_{mr}, at which a bird can cover the most distance on a given amount of energy (see Table I).

TABLE I

Optimal flight speeds for various goals

Goal	Occasion	Optimal flight speed	Source
Maximization of flight time on a given amount of energy	E.g. during night flights in swifts	Minimum power speed, V_{mp}	Pennycuick (1969, 1978)
Maximization of distance covered on a given amount of energy	During migration	Maximum range speed, V_{mr}	Pennycuick (1969)
Maximization of rate of net energy gain	During flights between some central place and foraging areas, with no foraging in flight	Faster than V_{mr}	R. Å. Norberg (in press)
Maximization of rate of net energy gain	During foraging in flight (e.g. insect catching)	Speed for maximum rate of net energy gain, V_{mn}	R. Å. Norberg (in press)

drag of the wings). Different theories have the profile power to be from constant value that is independent of speed (Pennycuick, 1968, 1969, 1975), to a variable increasing with speed raised to the power of three (Rayner, 1979c). The sum curve typically has a shallow U-shape. The bottom point on the curve defines the minimum power speed (V_{mp} in Fig. 2). The maximum range speed (V_{mr} in Fig. 2) at which the power/speed- ratio reaches its minimum could be found by drawing a tangent to the curve from the origin. This speed is higher than V_{mp}. One of these speeds is often considered to be the optimal speed. The maximum range speed should certainly be chosen during migration so that the birds could cover most distance on a given amount of energy (Pennycuick, 1969).

But if the optimization criterion is maximization of rate of net energy gain, animals should fly faster than V_{mr} (Table I) as shown by a graphical model on optimal speed during flights between some central place and foraging areas, and with no foraging in flight (R. Å. Norberg, in press). This model also predicts that the optimal speed increases with increasing food availability. If the bird were to fly slower than V_{mr} it would not only expend more energy in flying a given distance but would also require more time for travel and hence have less time for foraging. At speeds higher than V_{mr} the energy cost of flying a given distance becomes higher again, but the bird saves time and so gets more time for foraging.

For birds and bats catching insects (foraging) in flight, there will be another optimal speed. Since the volume of air searched for insects per unit time increases with increasing flight speed, the energy gained per unit time (P_{gain} in Fig. 2) should increase with the speed up to some limiting speed above which the yield should decrease again (R. Å. Norberg, in press). The peak occurs at different speeds for different species depending among other things on the size of the predator, size of prey and wing morphology. The net energy gain per unit time (P_{net} in Fig. 2) is the power gained minus the power required to fly (P_{loss} in Fig. 2),

$$P_{net} = P_{gain} - P_{loss}, \qquad (7.1)$$

which should be maximized. Selection forces may work towards a wing shape that minimizes the power required to fly at the speed optimal for the animal.

FLIGHT MODE AS RELATED TO THE ECOLOGY OF THE SPECIES

Optimal foraging theory was introduced by Emlen (1966) and MacArthur & Pianka (1966) and several authors have since then tried to predict animals' foraging behaviour by means of mathematical models. Reviews are given by Pyke, Pulliam & Charnov (1977) and Krebs (1978). They summarize the theory and experimental results with respect to optimal diet, patch choice, allocation of time to patches, and the foraging path.

In almost all foraging studies to date the basic hypothesis has been that animals try to maximize their net rate of energy intake.

The structure of the environment and the manner in which a species utilizes it are important for the wing design (Fig. 3). Different species utilize different food types and foraging habitats and hence different types of locomotion. Some species exploit a narrow range of foods and habitats and may be specialized, both behaviourally and morphologically. For other species it is better to be generalized, i.e. the optimal flight mode, and hence wing shape, are determined by diverse feeding behaviours due to, for instance, spatial or temporal heterogeneity of the environment (Fig. 3). Heterogeneous ("patchy") environments consist of patchworks of different resources, whereas homogeneous (or uniform) environments are those with similar or well mixed resources.

A productive *homogeneous* environment should be used in a more specialized way than a less productive one (e.g. Pianka, 1974). A model for utilization of patchy environments based upon optimization of the animal's time budget was made by MacArthur & Pianka (1966) and considered again by Pianka (1974). The model

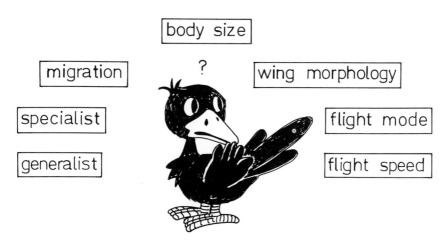

FIG. 3. What combination of ecological, behavioural and morphological factors a bird should have depends on the type of habitat it lives in as well as its way of exploiting the habitat.

predicts that animals which spend little effort searching for their prey should be more specialized than those with higher ratios of time spent in search to that spent on pursuit (pursuit, capture, and eating). When food is scarce, animals are unlikely to bypass potential prey, whereas when food is abundant, individuals may be more selective and specialize on the best food types. Thus, a high mean search time per item demands generalization, whereas a short mean search time per item (high expectation of locating prey items) allows some degree of specialization.

Pianka (1974) also considered a more *heterogeneous* patchy environment, in which the patches can be ranked in order of decreasing expectation of yield. The animal's time budget is divided into "hunting time", per item captured, spent within suitable patches, and "travelling time", per item caught, spent travelling between suitable patches. Travelling time decreases as an animal expands the number of different kinds of patches that it exploits, while the average hunting time within patches increases as more low-ranking patches are incorporated.

The optimal use of a patchy environment therefore depends upon the rate of decrease in travelling time (per prey item) associated with expanding the number of different patch types exploited. If the travelling time is long, selection for a wing form that decreases travelling expenses would be important.

If food density increases in all patches, both travelling time and hunting time become reduced. Animals that expend greater amounts of energy on search (searchers) will have their hunting time reduced more than those which spend relatively more energy on pursuit (pursuers). Under high food densities, pursuers should thus restrict the variety of patches more than searchers, and thus be more specialized.

In a temporally heterogeneous (fluctuating) environment different behaviours and different morphologies may be optimal at different seasons. As winter approaches and food supply becomes scarce, many bird species leave one area for another, especially if conditions differ greatly between winter and summer. These species may be allowed to be specialized and escape through migration. But migratory habits favour long wings which may not be compatible with the foraging behaviour. Some compromise may then have to be adopted. Many bird species stay during winter, and these should be generalized (Emlen, 1973).

R. Å. Norberg (1977) developed a model for calculation of a minimum energy budget and the corresponding time budget for foraging. An assumption is that the more energy-consuming a search method is, the more efficient it is in terms of area or space searched for food in unit time; consider running versus walking, flying versus sitting etc.

In a temporally homogeneous (stable) environment a species may specialize on the appropriate foraging technique and evolve behavioural and morphological adaptations accordingly. But since the optimal foraging technique may change in a temporally fluctuating environment an animal must be less specialized to be able to cope with different situations.

The model shows that the most energy-consuming search methods should be employed by a predator at the highest prey density. When prey density decreases (as at food bottle-neck periods) a predator should shift to progressively less energy-consuming methods although these are less efficient. Furthermore, the smaller the predator and the bigger the prey, the more energy-consuming (and efficient) search methods the predator should employ. When similar foods are eaten, small species, rather than big ones, should exploit the habitat patches demanding the most energy-consuming search methods, at least when food is scarce. When food is scarce, interspecific competition should lead to such habitat partitioning.

An example of seasonal shift in foraging behaviour in accordance with the above predictions has been found in the goldcrest (*Regulus regulus*). It is a common warbler in northern coniferous forests. Although the bird is insectivorous, large proportions of the populations remain over winter in their northern areas. There they have to eat an insect every few seconds throughout the entire day in winter in order to balance their energy budget. The selection pressure must therefore be strong for the bird to forage in efficient ways. In autumn, when food is abundant, it uses hovering flight to a large extent both in search for and capture of prey. As food becomes scarce later in winter it abandons hovering and shifts to less energy-consuming (and less efficient) types of locomotion for foraging (R. Å. Norberg, pers. comm.). Although these are associated with lower rates of gross energy intake per unit foraging time they are likely to give a higher rate of *net* energy gain per unit foraging time.

In conclusion, because of the seasonal shifts in optimal foraging behaviour the goldcrest can probably not develop profound morphological adaptations to either of the foraging techniques, since this would certainly make it less adapted to the other.

BODY SIZE AND HUNTING SUCCESS IN PREDATORS

For hunting success in raptors and owls aerial agility is important.

The aerial agility increases with decreasing size of the predator (Andersson & Norberg, 1981).

In most raptors and owls the sexes have different roles, the female incubating and guarding the nest and young against predators while the male forages for the whole family. Selection for aerial agility and, hence, small size relative to the size of the prey, should therefore be stronger in the male than in the female. This may be one reason for the reversed sexual size dimorphism in most raptors and owls, the male being smaller than the female (the reverse of the usual pattern among birds and mammals).

The most important aerial performances for success in prey capture are linear acceleration, horizontal speed, diving speed, rate of climb and manoeuvrability — the latter depending on angular acceleration and turning radius. Andersson & Norberg (1981) found that the only one of these flight functions which improves with increased predator mass is the terminal speed in diving, but its dependence on mass is weak. In all other five functions considered a small bird does better than a large one. Therefore, the size difference between a predator and its (flying) prey must not be too large.

Andersson & Norberg (1981) based their arguments on the assumption of geometric similarity. Certainly a species, or one sex of a species, could greatly enhance its hunting capacity by particular adaptations involving departures from geometric similarity. Many

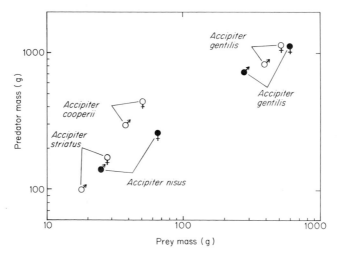

FIG. 4. Relations between predator mass and mean prey mass in three North American (open circles) and two European (solid circles) hawk species. The size differences among the species and the (reversed) sexual size dimorphism are associated with pronounced diet partitioning. Data from Newton (1979: table 1 p. 325; based on Storer, 1966, and Opdam, 1975).

alternative routes are conceivable and this should be an interesting area of research.

Among predators on flying birds there is a strong correlation between predator size and prey size (Newton, 1979), and this is probably caused partly by the need of predators to match (or surpass) the flight ability of their prey. Figure 4 shows the relation between predator mass and prey mass in three American and two European hawk species. Reversed sexual size dimorphism is also associated with pronounced diet partitioning.

STUDIES ON ECOLOGICAL MORPHOLOGY

Scaling

Various wing parameters (span, area, loading, aspect ratio) scale in different ways with body mass in different groups of bats and birds. One reason may be that birds and bats of different size use different types of flight. For instance, soaring flight occurs mainly in larger birds, bounding flight in smaller birds, and intermittent flight in certain intermediate-sized birds (Lighthill, 1977). The regression equations provide "norms" for the respective group. By comparing a particular species with these norms any morphological specialization can easily be identified with reference to the norm.

Figure 5 shows the wing loading and aspect ratio versus body mass in some bats and birds. Hummingbirds deviate from all other species by having a constant wing loading and a negative slope for aspect ratio. Ducks and members of the shorebird group have higher wing loadings than any other group. Frugivorous microbats have a slope similar to that of the frugivorous megabats, while the insectivorous microchiropteran families Molossidae and Vespertilionidae have similar, but steeper slopes for both wing loading and aspect ratio. Megabats and frugivorous microbats have rather similar slopes also for wing span, wing area, and forearm length, but they differ from all other groups (mainly insectivorous) in this respect in wing area, wing loading, and aspect ratio (U. M. Norberg, 1981). Megabats and frugivorous microbats thus show close convergence in wing form. These two groups have similar diet and foraging behaviour.

The slope of the regression lines for a particular measurement can, however, be different among groups owing to non-adaptive differences. Clutton-Brock & Harvey (1979) state that it appears to be a common biological phenomenon that size-dependent variables show progressively shallower slopes, when regressed on body size, the

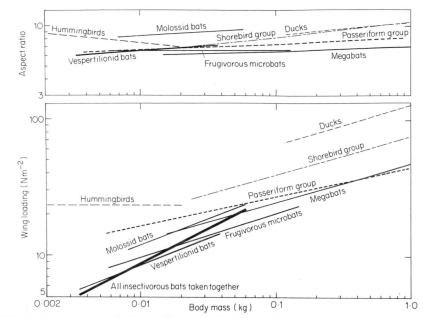

FIG. 5. Aspect ratio and wing loading versus body mass in some bats (from U. M. Norberg, 1981) and birds (recalculated from Greenewalt, 1975). The allometric equations were fitted by least squares regressions.

lower the taxonomic level that is considered. As a result, smaller species within any family would tend to show positive deviations from a regression line based on a higher taxonomic group (including the family), and large species negative deviations. This holds for several mammals. Also in the bats the slopes of regression lines for different families are different from those for higher taxonomic groups. But for the bats the slopes for different families for various direct measurements (span, area, forearm length) are instead steeper than for microbats or all insectivores taken together; i.e. than of groups of higher taxonomic status (see for example the regression line for wing loading for all insectivores taken together in Fig. 5, which is much shallower than those for separate insectivorous families; U. M. Norberg, 1981). As for regression lines it is thus important to compare groups at similar taxonomic levels.

At first sight one thus gets the impression that there is a large difference between insectivorous and frugivorous microbats. But the insectivorous microbats belong to several different families, each having its own regression line, while the frugivorous microbats belong to one family only. Nonetheless, there are differences in wing area, wing loading, and aspect ratio among frugivores and different insectivorous families, though of a lesser degree.

In many regression analyses in birds and mammals of a particular measurement versus body mass, the equations are conventionally fitted by least squares regression after transformation to logarithms ($Y = \alpha X^\beta$, where Y is the variable in question, X is body mass, and α and β are constants fitted to the data (e.g. Greenewalt, 1962, 1975; Lawlor, 1973; Alexander, 1977a,b). This is according to *Model I regression* in Sokal & Rohlf (1969). However, because the dependent as well as the independent variable is subject to error in such cases, Sokal & Rohlf (1969) suggest another method of curve fitting (*Model II regression*). A method suitable for relations discussed here is Bartlett's three-group method for Model II regression, described in detail by the authors. However, regression lines obtained by the conventional method usually do not differ to any greater extent from those obtained by the latter when the range along the X-axis is large relative to the error of measurement.

In conclusion, regression analysis is a valuable tool in biometric work. By comparing a particular species with the norms provided by the fitted equations, any morphological specialization can easily be identified with reference to the norm. These morphological data may be related to ecological information to elucidate adaptations to a particular niche.

Morphology and Foraging Energetics of Hummingbirds at Different Altitudes

Hummingbirds are the most excellent hoverers among birds. They are the most specialized nectar-feeding birds and exhibit a wide array of foraging behaviours. Feinsinger, Colwell, Terborgh & Chaplin (1979) studied the effects of elevation on morphology, flight energetics, and competition among hummingbirds, and their work will be reviewed here.

Among hummingbirds there are species (or sexes) which are highly territorial with interference competition, species which are non-territorial with exploitation competiton, and species which are intermediate. Feinsinger & Chaplin (1975) found that territorial species (or sexes) have shorter wing span than non-territorial species, and thus a higher wing disc loading and higher power output in hovering. A short span is better for manoeuvrability than a long one, and this is important for success in territory conflicts, but short wings increase the power required for hovering. However, ready access to abundant nectar supplies permits the territorial hummingbirds to refuel quickly, so saving energy by decreasing the hovering time. Instead they use more time for territory defence; but this activity is probably not as energy-consuming as hovering. Non-territorial hummingbirds are reduced to foraging among dispersed or nectar-poor flowers (Colwell, 1973), and they spend much time in foraging flight (hovering), and hardly any time in defence of food. They apparently compete through exploitation (Wolf, Stiles & Hainsworth, 1976). They may therefore be subjected to stronger selection for long wings than are territorial hummingbirds, since long wings reduce the wing disc loading and thus power for hovering.

Air density decreases with increased elevation and therefore the power required for hovering becomes larger for a given bird the higher the elevation. But, as Feinsinger *et al.* (1979) state, there must be an optimum wing-length that minimizes the total energetic cost for hovering, and therefore the wing-length should increase with the elevation at which the bird lives. The mean wing disc loading should then decrease with increased elevation, and the mean power for hovering remain the same irrespective of elevation. The authors state that, with increasing elevation, the overall pattern would be one of sequential replacement of non-territorial species by others with even lower wing disc loading, while territorial species would be continuously replaced, as they could no longer obtain sufficient energy to produce the required power for hovering, by other territorial species with somewhat lower wing disc loading (Fig. 6).

Feinsinger *et al.* (1979) tested the following four related predictions:

(1) A given hummingbird species will become increasingly territorial at higher elevations.

(2) The limits on power for hovering should remain substantially constant with elevation.

(3) Mean wing disc loading for hummingbirds as a group will decrease with increased elevation.

(4) Mean power for hovering for hummingbirds as a group will not vary with elevation.

The authors found that *Colibri thalassinus* is non-territorial at an elevation of 1400 m, but territorial at 3100 m. However, as the authors state, the first prediction was suggested by these differences and cannot be used to support it. Further data are not available to test this prediction. Field data verify the last three predictions, however. There are thus interrelationships among behaviour, morphology and energetics among hummingbirds. The authors considered only broad trends and population means, and they suggest further investigation on the effects of elevation on morphology, flight energetics and competition among other energy-limited flying animals.

Intraspecific and intersexual variation in foraging behaviour occur

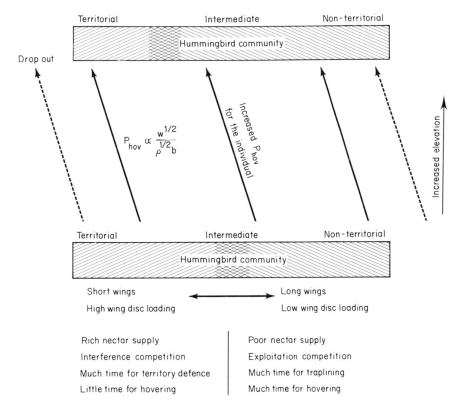

Mean P_{hov} constant for the community.

Extremes of P_{hov} constant for the community.

Mean wing disc loading for the community decreases with elevation.

FIG. 6. Hummingbird communities at different altitudes. The model is based on Feinsinger *et al.* (1979). The power for hovering, P_{hov}, increases with elevation for a given bird. Therefore this bird is supposed to decrease its hovering time and become more territorial to get at flowers with rich nectar supply. The mean wing span b for the entire community increases with increasing elevation, and hence the wing disc loading decreases correspondingly. Therefore, the mean power for hovering for the community remains about the same at different altitudes.

in many hummingbird species; Feinsinger *et al.* (1979) suggest that there may be a link between variation in wing disc loading and variation in foraging strategies among individuals in a population.

Ecological Morphology Among Species in a Pariform Guild

Apart from hummingbirds, several other bird species, and also some bats, hover, though not as excellently as do hummingbirds and not with the same kinematics. Hummingbirds move their wings in a horizontal figure of eight, maintain the wings extended and generate lift during the entire stroke ("normal hovering" in the terminology of Weis-Fogh (1973) and Rayner (1979c)). Other hovering birds and bats beat their wings forward—downward and then flex them to various degrees during the upstroke ("avian hovering" in the terminology of Rayner, 1979c), which contributes but small lift forces. An example of a bird with avian hovering is the goldcrest (*Regulus regulus*), which often gathers in mixed-species foraging flocks during the non-breeding season. These flocks often contain also the coal tit (*Parus ater*), the willow tit (*P. montanus*), the crested tit (*P. cristatus*), and the treecreeper (*Certhia familiaris*). These species thus are sympatric, occur in the same habitat, and partly overlap in choice of feeding station as well as food. Since they belong to the same foraging guild they probably influence each other via interspecific competition in an ecological, short time-scale, and, therefore, also in the evolutionary perspective. Their co-adaptation may be expected to have involved some divergence in feeding behaviour and adaptive morphology. The exploitation of different feeding stations should lead to different structural adaptations of wings and legs, and the question is how large are the differences that have evolved.

A way to identify adaptations to different niches is to seek answers to the following questions:

(1) Which are the quantitative, absolute, morphological differences in the locomotor apparatus among the species in question?

(2) Which are the relative morphological differences among these species, i.e. differences that are not due to scaling effects?

(3) What correlations can be found among differences in absolute and relative sizes and locomotor patterns?

(4) What correlations can be found between feeding station selection and locomotor pattern during foraging?

Applying these questions, U. M. Norberg (1979) compared the morphology of the wings, legs, and tail of the five bird species mentioned above.

Figure 7 illustrates the feeding station selection in spruce by these birds. The goldcrest forages on the needled outer parts of the branches. It is very good at manoeuvring and often hovers in front of, or underneath, the branches. It moves about within the mesh-work of sub-branches, has a very agile food-searching behaviour, and uses the wings much more than the other species. About 6% of its foraging time in autumn is spent hovering (R. Å. Norberg, personal communication). The coal tit also uses the outer parts of the branches. Both the goldcrest and the coal tit are more acrobatic than the other species. The coal tit often hangs under the branches and on the cones.

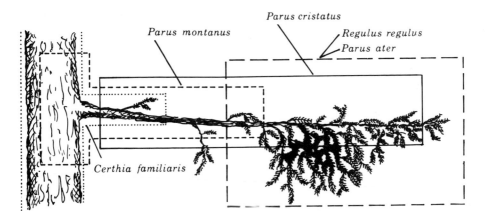

FIG. 7. Schematic illustration of the feeding station selection in spruce by three tits, the goldcrest and the treecreeper. From U. M. Norberg (1979; modified from Haftorn, 1956).

The crested tit mostly moves about with hops on branches and on the ground. It prefers the main and secondary branches. The willow tit also moves with hops on the branches, hangs under branches and cones, but also often clings to the trunk. The treecreeper climbs on the trunk in a vertical, headup position, but sometimes also moves about on the main branches.

The hovering ability and high manoeuvrability in the goldcrest enable the bird to find and collect food in places which the other species cannot utilize as efficiently and economically. It is mainly the adaptations for hovering and slow flight that will now be considered.

Low weight and low wing loading are advantageous especially for slow flight and hovering, and for manoeuvrability. Low wing loading enables the bird to produce enough lift in slow flight and in hovering, without having to use excessively high wing-beat frequencies that would result in large moments of inertia and hence an unnecessary waste of energy for production of inertial power.

The main power drain in hovering is the induced power (Pennycuick, 1968, 1969, 1972, 1975; Rayner, 1979c) while the other components (profile power, parasite power, inertial power) should be of minor importance. To minimize induced power, the wing disc loading should be low (low weight and proportionately long span: Pennycuick, 1968, 1972; Rayner, 1979c) and the stroke amplitude should be large (Rayner, 1979c).

Wing-beat frequency usually increases with decreasing flight speed as a compensation for the reduction of flight speed. The higher the wing-beat frequency the larger the inertial loads on the wing skeleton, and the larger also the inertial power (power needed to oscillate the wings). Since angular accelerations are larger in hovering than in forward flight, the inertial loads on the wing skeleton are particularly high in hovering. For this reason the supporting structures of the arm wing must be strong, and hence relatively heavy, in a hovering animal if a certain load safety factor against breakage is to be maintained. Therefore, it is important for a hovering animal to keep the wing's moment of inertia low by having a proximal location of the wing's mass. This may be achieved by reduced lengths of humerus and ulna and/or reduced angle between humerus and ulna. Thus, the shorter the arm wing is in relation to the total length of the wing, the less the inertial loads on the wing skeleton will be and also the less the inertial power (U. M. Norberg, 1979). The hand wing consists mainly of the primary feathers and thus has a low mass even if long.

Short arm wings in relation to the total wing length, as in hummingbirds, are associated with hovering and slow flapping flight, while long arm wings as in albatrosses, are associated with gliding (Fig. 8). The above argument may be one explanation for the short arm wings in hovering birds.

We will now see how the wing morphology of the five birds is correlated to hovering and slow flight. The differences in wing morphology are small but statistically significant in several respects.

The goldcrest and the treecreeper both have shorter arm wings in relation to the total length of the wings than any of the three tits (U. M. Norberg, 1979; Table II). In the former two species the

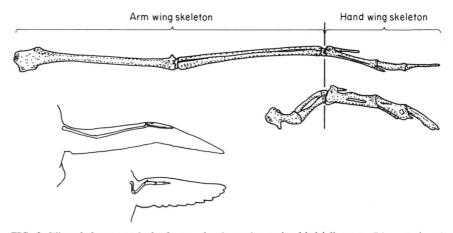

FIG. 8. Wing skeletons and planforms of a dynamic soaring bird (albatross, *Diomedea*) and a hovering bird (hummingbird, *Lampornis*) with the hand skeletons drawn to the same length. Modified from Hildebrand (1974; skeletons) and Herzog (1968; planforms).

TABLE II

Body mass, wing characters of importance for hovering flight, and power and specific power for hovering in five coniferous forest passerines. From U. M. Norberg (1979)

	Regulus regulus	*Certhia familiaris*	*Parus ater*	*Parus montanus*	*Parus cristatus*
Body mass (g)	5.8 ±0.11	9.1 ±0.22	9.1 ±0.08	10.9 ±0.08	11.5 ±0.37
Wing loading (Nm^{-2})	12.9 ±0.30	12.4 ±0.60	14.2 ±0.41	14.4 ±0.22	16.2 ±0.26
Wing disc loading (Nm^{-2})	3.4 ±0.07	3.5 ±0.19	3.7 ±0.10	4.0 ±0.07	4.3 ±0.15
Length of hand wing/length of arm wing	2.9 ±0.09	3.0 ±0.10	2.6 ±0.10	2.3 ±0.05	2.5 ±0.11
Induced power P_i (W)	0.07	0.10	0.11	0.14	0.15
Specific induced power P_i/M ($W\,kg^{-1}$)	1.19	1.13	1.21	1.28	1.31

increase in hand wing length compensates for the shortened arm wing, but does not add much to the total mass of the wing.

The goldcrest is the lightest of the species, has low wing loading and the lowest wing disc loading. Although the wing span is not especially long, the induced power (absolute value) during hovering is the lowest among the five species. Why is not the span longer so as to reduce the power for hovering still more? We have seen that the induced power decreases with increased wing span. On the other hand, the longer the wings the larger the inertial loads on the wing skeleton, and the less the manoeuvrability. Furthermore, long wings are not practical for flapping flight in dense vegetation. Therefore, the actual wing length in the goldcrest probably is a near-optimal compromise between the following opposing tendencies: minimization of induced power (tending towards long span) against minimization of inertial loads on the wing skeleton, and improvement of manoeuvrability and practicability among branches (all tending towards short span).

Turning now to the treecreeper we see that it would be rather good at hovering since it has very low wing and wing disc loadings. It is a treetrunk forager adapted to vertical climbing (long tail, long toes, long curved claws, proportionately short tibia; U. M. Norberg, 1979). It climbs upwards and then flies downwards to a lower level in the next tree and climbs upwards again. A long span should be of little hindrance to a treecreeper when foraging on tree trunks (the others fly about among branches) and it has no need of especially manoeuvrable flight. Furthermore, the treecreeper (as well as the goldcrest and the coal tit) is partly migratory and therefore should benefit by long wings (the power required to fly at the maximum range speed being proportional to $(span)^{-3/2}$; Pennycuick, 1969). These arguments may explain why opposing selection pressures balance at a relatively long span in the tree creeper.

The coal tit has relatively low weight and relatively long wings and, therefore, relatively low wing and wing disc loadings and rather low induced power output for hovering. But it does not have especially short arm wings. It is adapted to slow flight but the relatively long arm wings do not make it particularly adapted to hovering. The willow tit has relatively large body mass, rather short span and high wing disc loading, and the longest arm wing in relation to the total wing length. It is therefore not especially adapted to hovering or to manoeuvrable flight. The crested tit has the highest wing and wing disc loadings of the five species, and hence is the least adapted to hovering and manoeuvrable flight. It often hops about on the branches or on the ground when searching for food instead of flying

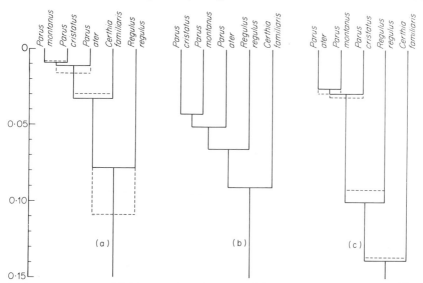

FIG. 9. Phenogram on (a) wing skeleton characters, (b) wing form characters, and (c) leg characters of five coniferous forest bird species, based on the weighted pair-group method with arithmetic averages. These birds often gather in mixed-species foraging flocks in winter and then exploit partly different sites in the trees (cf. Fig. 7) and use partly different loco-motory patterns. Solid lines show the result when deviations from geometric similarity are compared among the species, whereas dashed lines show the result with reference to elastic similarity. The higher up the branching point, the greater the phenetic similarity. The co-efficients along the vertical axis indicate degree of dissimilarity, where each unit ($k = 0.01$) means 1% dissimilarity. From U. M. Norberg (1979).

about among the branches as the other tits and, in particular, the goldcrest more often do. Its large weight instead makes it better able to survive cold winter nights than the other species.

A clustering process can be used to illustrate the phenetic resemblances among species regarding the morphology of the locomotor apparatuses. This was done for the present five species.

Figure 9 shows that there are more differences in leg morphology than in wing morphology among the five species in the guild. These differences can easily be correlated with the birds' different types of locomotion, though the differences in many cases are not large. The three tits, which are congeneric, are much alike as regards the wing and leg skeletons. The wing form, however, is not dependent on the wing skeleton, but primarily on the lengths of the feathers. The wing form is, in fact, rather diverse among the tits. Thus, as regards aerial locomotion in these three tits, the adaptations to different niches, and hence to different flight patterns during foraging, have led to divergent evolution of the wing feathers rather than of the wing skeleton.

The niche differentiation among the five species is associated with clear structural differentiation of the locomotor organs but to rather different degrees for various structures.

Soaring and Ecology

Many large birds, and even some insects and small birds, use soaring when searching for food and when they migrate (Pennycuick, 1972). Soaring costs a minimum of energy as compared with other types of flight. Only power for corrective manoeuvres and for holding the wings down in the horizontal position is needed. During soaring energy is extracted from the atmosphere.

Some aspects of Pennycuick's (1971) work on thermal soaring in vultures will be reviewed here. Birds which use thermal soaring as their main means of locomotion, such as vultures, storks, and eagles, typically have wings of very low aspect ratio ($\leqslant 7$: Pennycuick, 1971). Their wings are broad, giving large wing areas and, hence, low wing loadings. Birds with low wing loadings can glide slowly with low sinking speed and are good at exploiting weak and narrow thermals, while birds with high wing loadings can glide fast without excessive steepening of the gliding angle, however with a larger radius of turn. This follows from the following arguments. During balanced gliding the minimum gliding speed is proportional to the square root of wing loading ($V_{stall} \propto (W/S)^{1/2}$), and the sinking speed equals the minimum gliding speed times the inverse of the lift to drag ratio ($V_s = V_{stall} (D/L)$). The radius of a banked turn is proportional to wing loading ($r \propto W/S$).

Albatrosses have high aspect ratios and are excellent at dynamic soaring (extracting energy from horizontal wind shear). Pennycuick (1971) asked why vultures do not have longer wings. He made performance estimates for an imaginary "albatross-shaped vulture", which had the proportions of the wandering albatross (*Diomedea exulans*, body mass 8.5 kg) scaled down to the white-backed vulture's (*Gyps africanus*) mass of 5.38 kg.

Pennycuick calculated the wing area of the albatross-shaped vulture to be two-thirds of that in the white-backed vulture. Its wing loading and turning radii would therefore be 1.5 times as great. The albatross-shaped vulture should climb better than the real white-backed vulture and have far higher average speeds when travelling across country. But it would be inferior in one respect, the ability to remain airborne in thermals with radii between 14 and 17 m. In short, an albatross-shaped vulture, once airborne, would soar superbly in thermals, and would far outdistance storks and eagles on migration.

Why then do vultures, storks, and eagles not have albatross-shaped (high aspect ratio) wings? Pennycuick suggested that it may be of great importance for vultures to be able to soar in very small thermals and to be able to start patrolling early in the morning when the thermals are weak and narrow; this would not be possible with a high wing loading.

Hankin (1913) observed that different species of vultures and birds of prey start soaring in the morning in the order of increasing wing loading. The earlier in the day that the vultures can start searching for remnants of animals killed by large carnivores in the night the better their chances of reaching a kill before mammalian scavengers get there. This is particularly important since the latter are usually dominant over the vultures and keep them away from the carcasses. On the other hand an advantage of high wing loading is a high cross-country speed which would enable the vulture to search larger areas and to transport food to the nest over large distances in a day.

Pennycuick further suggested that it is perhaps more probable that the low-aspect-ratio wing is adapted to the requirements of take-off from the ground or trees which may call for a large wing area. For this purpose long wings might be a hindrance.

Flight in Air and Water

Some birds use their wings for flight in air as well as in water (in pursuit of prey). Examples are the diving petrels (family Pelecanoididae of the order Procellariiformes) in the southern oceans, the alcids (family Alcidae of the order Charadriiformes) in the northern oceans, and the dippers (e.g. *Cinclus cinclus*, family Cinclidae of the order Passeriformes) in running water.

The optimal span and area of wings used for flight in air as well as for "flight" in water should be quite different from those used for flight in air only, or water only (e.g. penguins), because of the enormous difference in density between the two media. It seems an interesting problem to find out how the size and shape of such wings are related to the strongly conflicting demands associated with flight in these two media. To some extent, some of these birds can obviously adjust their wings behaviourally to the higher density of water by maintaining the wings only partly open during the wing beat; this has been described in, for example, the dipper (Goodge, 1959).

REFERENCES

Alexander, R. McN. (1977a). Allometry of the limbs of antelopes (Bovidae). *J. Zool., Lond.* **183**: 125–146.

Alexander, R. McN. (1977b). Mechanics and scaling of terrestrial locomotion. In *Scale effects in animal locomotion*: 93–110. Pedley, T. J. (Ed.). London & New York: Academic Press.

Andersson, M. & Norberg, R. Å. (1981). Evolution of reversed sexual size dimorphism and role partitioning among predatory birds, with a size scaling of flight performance. *Biol. J. Linn. Soc.* 15: 105–130.

Clutton-Brock, T. H. & Harvey, P. H. (1979). Comparison and adaptation. *Proc. R. Soc. Lond.* (B) 205: 547–565.

Colwell, R. K. (1973). Competition and coexistence in a simple tropical community. *Am. Nat.* 107: 737–760.

Emlen, J. M. (1966). The role of time and energy in food preference. *Am. Nat.* 100: 611–617.

Emlen, J. M. (1973). *Ecology: an evolutionary approach*. Reading, Massachusetts: Addison-Wesley.

Feinsinger, P. & Chaplin, S. B. (1975). On the relationship between wing disc loading and foraging strategy in hummingbirds. *Am. Nat.* 109: 217–224.

Feinsinger, P., Colwell, R. K., Terborgh, J. & Chaplin, S. B. (1979). Elevation and the morphology, flight energetics, and foraging ecology of tropical hummingbirds. *Am. Nat.* 113: 481–497.

Goodge, W. R. (1959). Locomotion and other behaviour of the dipper. *Condor* 61: 4–17.

Greenewalt, C. H. (1962). Dimensional relationships for flying animals. *Smithson. misc. Collns* 144(2): 1–46.

Greenewalt, C. H. (1975). The flight of birds. The significant dimensions, their departure from the requirements for dimensional similarity, and the effect on flight aerodynamics of that departure. *Trans. Am. phil. Soc.* 65(4): 1–67.

Haftorn, S. (1956). Contribution to the food biology of tits especially about storing of surplus food. IV. A comparative analysis of *Parus atricapillus* L., *Parus cristatus* L. and *Parus ater* L. *K. norske Vidensk. Selsk. Skr.* 1956(4): 1–54.

Hankin, E. H. (1913). *Animal flight: a record of observation*. London: Iliffe.

Herzog, K. (1968). *Anatomie und Flugbiologie der Vögel*. Stuttgart: Gustav Fischer Verlag.

Hildebrand, M. (1974). *Analysis of vertebrate structure*. New York: John Wiley & Sons.

Krebs, J. R. (1978). Optimal foraging: decision rules for predators. In *Behavioural ecology: an evolutionary approach*: 23–64. Krebs, J. R. & Davis, N. B. (Eds). Oxford: Blackwell Scientific Publ.

Lawlor, T. E. (1973). Aerodynamic characteristics of some neotropical bats. *J. Mamm.* 54: 71–78.

Lewontin, R. C. (1978). Fitness, survival, and optimality. In *Analysis of ecological systems*: 3–21. Horn, D. H., Mitchell, R. & Stairs, G. R. (Eds). Columbus, Ohio: Ohio State Univ. Press.

Lighthill, M. J. (1977). Introduction to the scaling of aerial locomotion. In *Scale effects in animal locomotion*: 365–484. Pedley, T. J. (Ed.). London & New York: Academic Press.

MacArthur, R. H. & Pianka, E. R. (1966). On optimal use of a patchy environment. *Am. Nat.* 100: 603–609.

Maynard Smith, J. (1978). Optimization theory in evolution. *A. Rev. Ecol. Syst.* 9: 31–56.

Newton, I. (1979). *Population ecology of raptors.* Berkhamsted: T. & A. D. Poyser.

Norberg, R. Å. (1977). An ecological theory on foraging time and energetics and choice of optimal food-searching method. *J. Anim. Ecol.* 46: 511–529.

Norberg, R. Å. (In press). Optimal flight speed in birds when feeding young. *J. Anim. Ecol.*

Norberg, U. M. (1979). Morphology of the wings, legs and tail of three coniferous forest tits, the goldcrest, and the treecreeper in relation to locomotor pattern and feeding station selection. *Phil. Trans. R. Soc. Lond.* (B) 287: 131–165.

Norberg, U. M. (1981). Allometry of bat wings and legs and comparison with bird wings. *Phil. Trans. R. Soc. Lond.* (B) 292: 359–298.

Opdam, P. (1975). Inter- and intraspecific differentiation with respect to feeding ecology in two sympatric species of the genus *Accipiter. Ardea* 63: 30–54.

Oster, G. F. & Wilson, E. O. (1978). *Caste and ecology in the social insects.* (Monographs in Population Biology No. 12.) Princeton: Univ. Press.

Pennycuick, C. J. (1968). Power requirements for horizontal flight in the pigeon. *J. exp. Biol.* 49: 527–555.

Pennycuick, C. J. (1969). The mechanics of bird migration. *Ibis* 111: 525–556.

Pennycuick, C. J. (1971). Gliding flight of the white-backed vulture *Gyps africanus. J. exp. Biol.* 55: 13–38.

Pennycuick, C. J. (1972). *Animal flight.* London: Arnold.

Pennycuick, C. J. (1975). Mechanics of flight. In *Avian biology* 5: 1–75. King, J. R. & Farner, D. S. (Eds). London & New York: Academic Press.

Pennycuick, C. J. (1978). Fifteen testable predictions about bird flight. *Oikos* 30: 165–176.

Pianka, E. R. (1974). *Evolutionary ecology.* New York: Harper & Row.

Pyke, G. H., Pulliam, H. R. & Charnov, E. L. (1977). Optimal foraging: a selective review of theory and tests. *Q. Rev. Biol.* 52: 137–154.

Rayner, J. M. V. (1979a). A vortex theory of animal flight. Part 1. The vortex intake of a hovering animal. *J. Fluid Mech.* 91: 697–730.

Rayner, J. M. V. (1979b). A vortex theory of animal flight. Part 2. The forward flight of birds. *J. Fluid Mech.* 91: 731–763.

Rayner, J. M. V. (1979c). A new approach to animal flight mechanics. *J. exp. Biol.* 80: 17–54.

Schoener, T. W. (1971). Theory of feeding strategies. *A. Rev. Ecol. Syst.* 2: 369–404.

Sokal, R. R. & Rohlf, F. J. (1969). *Biometry.* The principles and practice of statistics in biological research. San Francisco: W. H. Freeman & Co.

Storer, R. W. (1966). Sexual dimorphism and food habits in three North American accipiters. *Auk* 83: 423–436.

Weis-Fogh, T. (1973). Quick estimates of flight fitness in hovering animals, including novel mechanisms for lift production. *J. exp. Biol.* 59: 169–230.

Wolf, L. L., Stiles, F. G. & Hainsworth, F. R. (1976). Ecological organization of a tropical, highland hummingbird community. *J. Anim. Ecol.* 45: 349–379.

Symp. zool. Soc. Lond. (1981) No. 48, 199–218

Echolocation for Flight Guidance and a Radar Technique Applicable to Flight Analysis

J. D. PYE

Department of Zoology and Comparative Physiology,
Queen Mary College, Mile End Road, London E1, England

SYNOPSIS

Flight involves movement in a three-dimensional medium and requires an external frame of sensory reference for orientation. Vision and echolocation can both provide this as well as serving for orientation to specific objects such as obstacles, food and water.

Any echo system such as radar can measure either target distance (range) or relative velocity (range rate). The first needs signals of broad bandwidth, either very brief impulses or pulses with a frequency sweep, while the second requires signals of long duration and ideally of constant frequency. Bats and echolocating birds may collectively or individually use any of these signal types and show other close analogies with various man-made systems.

Working in reverse from a fixed reference point, echoes can also be used to study echolocation and flight. Ultrasound has been used on a small scale and microwave Doppler radar has been developed for field observations. In conjunction with ultrasound recording, this allows the frequency emitted by a flying bat to be measured regardless of acoustic Doppler shifts due to its motion. The radar trace also gives interesting information on flight performance of bats and birds. Flight speed is readily measured despite geometrical distortion due to the "one-dimensional" display of Doppler radar. Wing movements can be registered from the 3.5 beats per second of herons to 57 beats per second in hummingbirds; the rate may vary with conditions, at least in bats. Especially interesting traces are obtained from the leaping flight of small passerines.

Various components of the echo can be enhanced in tame or captive animals by attaching minute resonant dipole reflectors to different parts of the body. This technique can therefore be used to analyse the "radar signature" of flying animals and so to examine the normal flight behaviour of identified species.

AIR-BORNE ECHO SYSTEMS

Flight is a form of movement that occurs in a three-dimensional medium and this raises special problems of orientation and co-ordination. Vision of surrounding, stationary objects gives an external reference frame by which flight can be controlled but special problems arise when vision is restricted. Pioneer aviators, with limited instrumentation, described severe disorientation in thick cloud and even emerged upside-down to their surprise. Apparently the linear and angular accelerometers in the vestibular labyrinth of the ear are unable to cope with the strong and rapidly changing g-forces that are experienced, although the same principles are used effectively in the ballistic navigation instruments of atomic submarines and jumbo-jets.

The labyrinth may serve underwater, where it was first developed, because there is a strong vertical pressure gradient that can provide a reference direction; this can be detected by an air-filled swim-bladder and the clupeids, which regularly migrate vertically, have developed a system that combines the swim-bladder and the labyrinth in a fascinating way (Denton & Blaxter, 1976). This vertical pressure gradient is very much weaker in air and may be rejected as a normal source of flight information.

Where vision is weak, as in most microchiropteran bats, or ineffective as in cavernicolous birds or fruit-bats, an alternative sensory mechanism is echolocation. One tends to think of echolocation as a means of avoiding obstacles and of intercepting flying prey, but it can also be considered as a means of referring to external objects for flight control. Griffin (1971) has discussed the problems of high-flying bats in obtaining an adequate ground echo: because of the great attenuation of high-frequency sounds by the atmosphere, bats flying high above the ground must emit intense cries of relatively low frequency. Such sounds can often be detected from the ground.

ALTERNATIVE APPROACHES

The mechanism of, and the information provided by, any echolocation system are determined by the nature of the signal emitted. Theory developed for man-made radar and sonar systems shows that there are two principal ways to operate: by wide-band signals to measure the range of targets and by long duration signals to measure the relative velocity or first derivative of range. Both methods are used by bats and may be considered equivalent for the present purposes; nevertheless it is instructive to examine the differences briefly.

Wide bandwidth is a property of very short impulsive sounds. It is clear that the time-delay due to propagation to a target and back to the receiver can be measured accurately with such signals. Echoes from two or more targets at slightly different ranges will tend not to overlap and so in principle can be detected separately if the impulse is short enough. Such impulses are represented by the click-like sounds of odontocete Cetacea underwater or of rousettine fruit-bats and echolocating birds in air. The situation is a little more complicated in the latter cases, as grouping of impulses within each unitary click can give the possibility of some velocity information. Thus in *Rousettus* and in several species of the swiftlets *Aerodramus*, each click is composed of two discrete impulses that are not resolved by human ears, while in the oilbird *Steatornis* and at least one species of *Aerodramus* it consists of a brief burst of impulses (Medway & Pye, 1977).

Broad bandwidth can alternatively be obtained by using a tone-pulse whose frequency is swept rapidly over a wide range. This has several advantages that are exploited in frequency modulated (f.m.) "chirp" radar: the pulse may now be considerably longer than a click and so can have a much greater energy, giving increased sensitivity. Range measuring accuracy is maintained because each part of the pulse is labelled by its characteristic frequency, while overlapping echoes from closely spaced targets will have slightly different frequencies at any instant during their reception. F.m. sweep pulses of this kind are widely used by microchiropteran bats, either with a single "carrier" wave swept over an octave or more, or with a shallower sweep whose bandwidth is extended by harmonics.

The measurement of relative velocity depends on the fact that an echo is not only delayed by propagation to the target and back but also returns at a different frequency if the target is moving relative to the radar. This doppler shift is produced by a change in echo duration because the range, and therefore the propagation delay, changes while the signal is actually being reflected. This change is difficult to detect with short, broad-band signals but can be measured with longer signals. Bandwidth is here ideally reduced to a minimum in order to increase echo detection in the presence of background noise. Signals of long duration with most or all of their energy at a constant frequency are regularly emitted by a number of microchiropteran bats. Others, such as the pipistrelle, emit this kind of signal under certain conditions but change to shorter f.m. signals at other times (Pye, 1978).

It is difficult to believe that bats would use two kinds of optimal signal or even change from one to the other unless they are able to exploit the corresponding advantages of each. That they may do so has now been demonstrated by clear behavioural tests on different species. Simmons (1970) has shown that the ability of f.m.-sweep bats to discriminate between the ranges of two targets in different directions closely matches the theoretical predictions for their signals.

At about the same time Schnitzler (1968, 1970; Gustafson & Schnitzler, 1979) showed that constant frequency bats of at least three families are sensitive to the doppler shift. On detecting an echo of higher frequency from an approaching target, they accurately lower their emitted frequency so that subsequent echoes return at the original frequency. This doppler compensation behaviour thus keeps important echoes exactly tuned to the extremely narrow band of sensitive hearing of these bats. While the polar co-ordinates of direction and range provide a direct representation of spatial relationships, the corresponding co-ordinates of direction and relative velocity may at first sight seem to be less useful. It will be shown later that this is far from true if a target is followed along its track.

GROUND-BASED ECHO SYSTEMS

Airborne radar conveys information to the pilot but ground-based radar is equally useful to defence systems and to air-traffic controllers whose job is to follow the movements of aircraft. In just the same way echo systems can be used to study flight or even echolocation itself. One is then operating from the fixed reference frame to observe the behaviour of the moving organism.

In an early experiment ultrasonic doppler echolocation was used to record the rapid ear movements of certain constant-frequency bats (Pye, Flinn & Pye, 1962; Pye & Roberts, 1970). More recently a microwave doppler radar technique has been developed to measure the frequency of ultrasound emitted by bats flying in the field. It is hoped that this will enable us to resolve whether doppler compensation is used by species, such as the pipistrelle, that emit long, constant frequency signals only when flying in the open, well clear of the ground. Ultrasonic signals recorded from these bats show fluctuations in frequency that are, in part at least, due to doppler shifts from their movements relative to the microphone. Frequency changes of similar magnitude in the bat's actual emissions can only be distinguished if flight velocity relative to the microphone is known for each pulse detected.

The equipment used for this purpose has been developed with generous assistance from Marconi's Low-power Systems Radar Laboratory and has been described elsewhere (Halls, 1978; Pye, 1978). A continuous microwave radio beam of either 9 GHz or 13.5 GHz (wavelengths 3.3 cm and 2.2 cm respectively) is trained on the flying bat. Doppler shifts in the returning radar echo are presented

FIG. 1. Stroboscopic images of a pipistrelle bat, *Pipistrellus pipistrellus*, flying indoors. The interfemoral membrane is alternately extended backwards and lowered almost to the vertical, apparently to prevent stalling at low speed. In outdoor flight one wingbeat would carry the bat about half a metre.

as a difference tone or doppler note whose frequency represents the radial velocity of the target. As this frequency is quite low, it can be recorded on magnetic tape as a permanent record and is also presented to the operator through headphones to help him to keep the beam of the radar trained on the target.

These signals can reveal a great deal about the flight of the bat. We soon discovered that a pipistrelle flying round quite a large indoor laboratory moves at only half the speed of the same species cruising in the open, yet it uses a wingbeat rate that is 50% higher. Flight at low speed is known to be hard work and the need to avoid stalling may explain why the legs and tail lower the interfemoral membrane on each stroke when flying indoors (Fig. 1). This may also explain why high-speed bats, such as noctules and molossids, fly only clumsily if at all indoors.

ANALYSIS OF DOPPLER RADAR RECORDINGS

The use of doppler radar as an observing instrument has also given us an insight into what might be achieved by bats working in the velocity mode. Unlike the bat, our radar has only one receiver and is thus restricted to observation in only one dimension: it can only measure radial velocity of the target along the microwave beam. Although the beam may of course be turned to follow the target as it moves, angular movements of the radar are not recorded nor can the position of the target within the beam be determined at any instant. Nevertheless considerable information is available if the velocity trace is examined over a period of time.

The doppler frequency (f_d) available from the radar can be converted to the observed velocity (v') by the simple relationship:

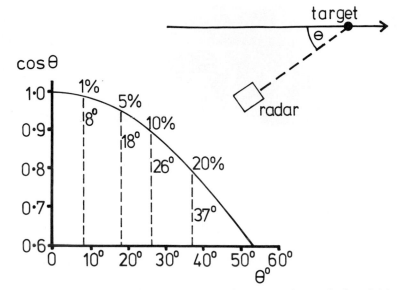

FIG. 2. When a doppler radar observes a target moving at an angle θ to the line of sight, the observed velocity is the true velocity times $\cos \theta$. The graph shows $\cos \theta$ for $\theta < 53°$ and the angles that produce some specific percentage errors.

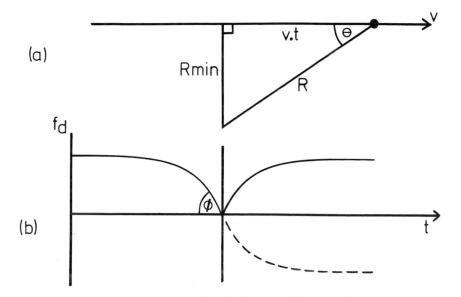

FIG. 3. If a target moves at constant velocity along a straight path (a), the output frequency of a doppler radar follows a curve that is reflected at the baseline (b). The greatest slope of frequency against time ($\tan \phi$) allows minimum range to be measured and thus the ensuing track to be calculated.

$$v' = c \cdot \frac{f_\mathrm{d}}{2f_\mathrm{t}} \qquad (8.1)$$

where c = velocity of light, and f_t = transmitted radar frequency. It is often easier to think that one cycle of f_d is produced when the target moves nearer (or recedes) by half a wavelength of the radar beam. Thus:

$$v' = f_\mathrm{d} \cdot \frac{\lambda}{2} \qquad (8.2)$$

where λ = transmitted radar wavelength.

When the line-of-sight of the radar is at an angle θ to the track of the target, the radial velocity observed (v') is the true velocity (v) multiplied by $\cos \theta$ (Fig. 2). The true velocity is only "seen" when the radar "looks" directly along the track so that θ is zero. Fortunately the properties of the cosine mean that the errors only increase slowly at first as θ increases from zero. If measurements are made of short tracks at all angles within a plane containing the radar beam, then 8.1% of the measurements will be in error by 1% or less (Table I). If all possible tracks in three dimensions are included, the proportion becomes equal to the maximum error itself, i.e. 1% of brief

TABLE I

Proportion of short tracks for which the velocity observed with a doppler radar lies within various error limits of the true velocity

% error in velocity measurement	% of tracks randomly orientated in 2 dimensions	% of tracks randomly orientated in 3 dimensions
0.5	6.4	0.5
1	8.1	1
2	12.8	2
3	15.6	3
5	20.2	5
10	28.7	10

measurements will be within 1% of the true value (Table I). For flying animals vertical mobility will be less than horizontal mobility so that a realistic situation is somewhere between these two cases. If a complete trace is observed, until the target recedes beyond maximum range, a reasonably accurate measure of true velocity will be obtained every time (unless the target follows a spiral path!).

When the target is at its nearest point to the radar, θ becomes a right angle for an instant (Fig. 3a) and the observed velocity falls to zero (Fig. 3b). Furthermore, a simple doppler radar cannot distinguish between approach and recede velocities, both of which are represented with the same sign, the "negative" frequencies of the receding target being "reflected" above the zero axis (Fig. 3b). However, the slope of the trace as it passes through (or is reflected at) zero allows the distance of nearest approach to be calculated. This minimum range is given by:

$$R_{\min} = \frac{2 \cdot v^2}{\lambda \left(\dfrac{df_d}{dt}\right)_{\max}} \tag{8.3}$$

where v = true velocity, λ = wavelength of radar transmission, f_d = doppler frequency, t = time, $df_d/dt = \tan \phi$ (Fig. 3b); or with a *velocity* axis:

$$R_{\min} = v \cdot \tau$$
$$= v^2 \cdot \cot \phi \tag{8.4}$$

where $\tau = v \cdot \cot \phi$, and $\cot \phi$ = maximum slope of the velocity trace. As velocity, time and minimum range are all known, a straight track can now be mapped out: distance moved from nearest approach to maximum range = velocity × time elapsed, i.e. $v \cdot t$, so that:

$$\text{maximum range of observation} = \sqrt{\{(R_{\min})^2 + (v \cdot t)^2\}} \tag{8.5}$$

Also:

$$\text{total length of observed track} = \text{velocity} \times \text{total duration of trace}$$
$$= v \cdot t_{\text{total}} \tag{8.6}$$

These relationships assume a straight track. In practice many tracks are effectively straight over their observed section and others can be eliminated from spoken comments recorded on another track of the tape. Alternatively if speed is assumed to be constant, integration of the trace allows curvature of the track to be plotted.

SOME OBSERVATIONS OF BIRD FLIGHT

Birds are much easier to record on radar than bats because they are more abundant while good visibility facilitates identification of the subjects and training of the radar beam. They also show much greater variability of size and behaviour. The author's own observations of birds have been greatly extended in the course of a project undertaken by Mr B. W. McDaid. For these observations the radar doppler signal was recorded on one track of a battery-powered stereo tape-

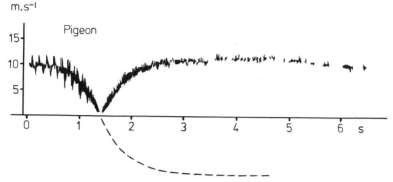

m.s⁻¹

FIG. 4. A straight overhead flight of a London pigeon, *Columba livia*, recorded by B. W. McDaid with the 9 GHz radar. For details of analysis see the text.

recorder and a voice commentary was recorded on the other track. Traces were later analysed on a sonagraph sound spectrum analyser and the frequency scale was converted to relative velocity as described above. Most records were made with the 9 GHz (3.3 cm) radar so that 1 m s^{-1} gave a doppler frequency of 60.6 Hz and normal range of the display extended to 16.5 m s^{-1} (37 m.p.h.) at 1 kHz top frequency.

Figure 4 shows a trace of a London pigeon that flew a straight track passing almost overhead. The general form of this trace may be compared with Fig. 3b. Flight speed was 11.5 m s^{-1} (26 m.p.h.); total track recorded was 75 m flown; nearest point was about 5.5 m and maximum line-of-sight range at the end of the trace was therefore 58 m. At this point θ was 5.4° so that $\cos \theta = 0.9956$, giving a velocity error of only 0.44%.

It is rarely possible to record the approach of a bird at maximum range. Indeed one gets an impression that approaching birds veer away when the radar is aimed at them. This may well be a subjective illusion since receding tracks are only (and always) obtained from a bird that has already made a rather close approach.

Figure 4 also shows clear wingbeats superimposed on the approaching section of the trace at a rate of 7.1 beats per second. The velocity amplitude of these movements with respect to the body motion is striking: values range up to 3.4 m s^{-1}. Eastwood (1967) stated that the wings contribute little to the *intensity* of the radar echo from birds; Edwards & Houghton (1959) estimated that outstretched wings contribute only 5% of the total echo while Eastwood (1967) attributed fluctuations of echo intensity at the wingbeat rate to changes in body shape due to muscular contraction. With doppler radar, however, the trace shows *velocities*. If we assume for simplicity

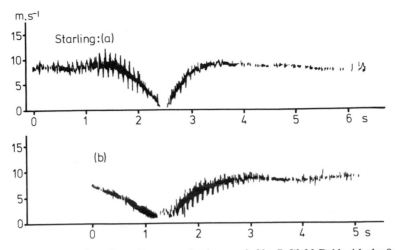

FIG. 5. Two traces of starlings, *Sturnus vulgaris*, recorded by B. W. McDaid with the 9 GHz radar. (a) Velocity $9.3\,\mathrm{m\,s^{-1}}$ (21 m.p.h.), total track flown 59 m, minimum range 6 m, maximum range 36 m, wing-beat rate $10.8\,\mathrm{s^{-1}}$. (b) Velocity $9.1\,\mathrm{m\,s^{-1}}$ (20.5 m.p.h.), total track flown 46 m, minimum range 5.6 m, maximum range 34 m, wing-beat rate $10.3\,\mathrm{s^{-1}}$.

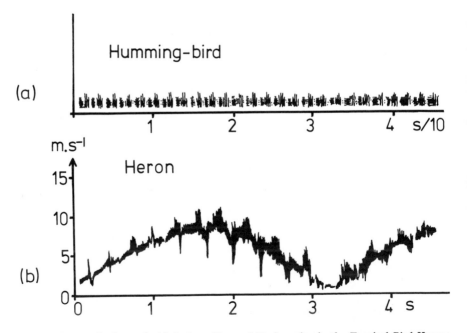

FIG. 6. Traces of a hummingbird, *Amazilia amabilis*, hovering in the Tropical Bird House, and of a heron, *Ardea cinerea*, flying in the Snowdon aviary. Recorded with B. W. McDaid and J. A. T. Halls, at the London Zoo with the 9 GHz radar. For details of analysis see the text.

that each wing beats in simple harmonic motion through 2 radians (114°), then the radius equals the peak displacement a, and

$$a_{max} = v_{max} \cdot \frac{1}{2\pi f} \tag{8.7}$$

$$= 7.6 \text{ cm}$$

where v_{max} = maximum velocity (3.4 m s^{-1}) and f = frequency (7.1 Hz). It is clear, then, that a considerable part of the wing is being observed at times.

Wing-beats are not observed on this trace as the pigeon recedes. It appears that the nature of the echo from a flying bird (its "signature") is aspect-sensitive. Figure 5 shows two traces of starlings. This is another species that often shows straight steady tracks although the asymmetry of trace (b) indicates that the bird changed course. Trace (a) shows wing-beats on approach but very much smaller wing traces on recede, while trace (b) shows just the opposite. This need not be surprising since the wing motions of birds are complex (see for instance tracings reviewed by Gray, 1968) so that the cycle of relative velocities seen will depend strongly on the angle of view. In Fig. 5a the bird flew almost overhead as in Fig. 4 and these traces are consistent with wing motions quoted by Gray (1968): 223, fig. 9—20D). The bird of Fig. 5b flew to one side of the observer, over some trees. In such a case the velocity profiles of the wings will be more complex and will differ from each other as seen by the radar.

Figure 6 shows the extremes of wing-beat rate that have been observed here. The hummingbird was hovering at its feeding-bottle in the Tropical Bird House at the London Zoo. Its wing-beat rate was then 55—57 beats per second. The heron was recorded in the Snowdon Aviary. It is remarkable that such a large bird could achieve a flight velocity of up to 8.7 m s^{-1} (19.5 m.p.h.) in a relatively confined space and the manoeuvres involved may account for rapid changes of wing-beat rate in the range 3.0—3.5 beats per second. Similar top speeds were recorded in the same aviary for red-billed teal (*Anas erythrorhynca*: 8.9 m s^{-1} — 20 m.p.h.), alpine chough (*Pyrrhocorax graculus*: 10.3 m s^{-1} — 23 m.p.h.) and sacred ibis (*Threskiornis aethiopicus*: 12.1 m s^{-1} — 27 m.p.h.).

LEAPING FLIGHT

Much the most interesting traces have been obtained for the leaping or bounding flight of small passerines in which short bursts of wing-flapping are interspersed with periods of passive "body-gliding" with wings closed. The resulting traces consist of straight sections with alternating slopes, representing constant accelerations and decelerations. These have to be interpreted with care and would justify a much more thorough analysis than has yet been possible.

Rayner (1977) assumed that in leaping flight the active, flapping phase follows a circular arc (concave side upwards) while the gliding phase is a parabolic trajectory in which lift and drag are ignored. Csicsáky (1977) produced evidence from wind-tunnel observations that body-gliding generates lift so that the trajectory is flatter than a ballistic parabola. The radar traces would appear to have much to offer here and raw data are easily obtained under natural conditions, but at present no quantitative statement can be offered. What is clear from the traces is that the bird experiences high accelerations with frequent reversal, giving a very "rough ride".

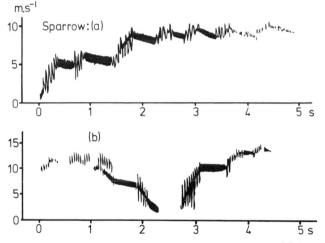

FIG. 7. Traces of the flight of house sparrows, *Passer domesticus*. (a) One bird receding from the nearest point. Recorded with M. Flinn with the 13.5 GHz radar. Mean velocity 10.0 m s^{-1} (22.4 m.p.h.), total track flown 49 m, wing-beat rate 19–21 s^{-1}. (b) Recorded while approaching and receding by B. W. McDaid with the 9 GHz radar. Alternating active and passive phases show inversion of their acceleration slopes after the nearest point. Mean velocity 13.9 m s^{-1} (31 m.p.h.), total track flown 62 m.

From the initial trace obtained (Fig. 7a) the bird was only followed as it receded from the nearest point. At first it was assumed that the active, flapping phase represented horizontal acceleration while the passive, gliding phase showed deceleration due to drag. But later traces that include the bird's approach (Fig. 7b) show that the two phases of acceleration become inverted on the trace when the bird passes its nearest point. Active phases are decelerating towards the

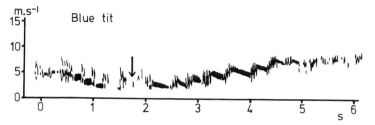

FIG. 8. A trace of a blue tit, *Parus caeruleus*, in leisurely undulating flight recorded by B. W. McDaid with the 9 GHz radar. Mean velocity 7.9 m s⁻¹ (17.7 m.p.h.), total track flown 50 m.

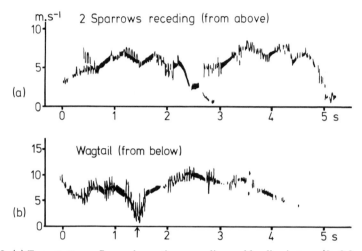

FIG. 9. (a) Two sparrows, *Passer domesticus*, receding and landing in turn (0–2.9 s and 2.7–5.2 s) recorded from above (the flat roof of a five-storey building) by J. A. T. Halls with the 13.5 GHz radar. Acceleration slopes of the active and passive phases are inverted from those seen when viewed from below. (b) Trace of a wagtail, *Motacilla* sp., approaching and receding, recorded by B. W. McDaid with the 9 GHz radar. Here the slopes of active and passive phases do not invert after the nearest point.

radar during approach while passive gliding represents an apparent acceleration. This means that the acceleration must be vertical: downwards on the passive phase and upwards on the active phase.

Wingbeat rates observed for this species are about 20 per second, considerably higher than Meinertzhagen's value quoted by Greenewalt (1962) but close to the figure found by Vaughn (1974). The number of wing-beats in each active phase and the durations of the gliding phases both vary considerably. Figure 8 shows a trace of a blue tit with a large number of short flap-glidge cycles in slow flight; again the slopes reverse when the bird passes the nearest point (arrowed).

The predominantly vertical component of the accelerations is confirmed in the traces of Fig. 9a in which two sparrows were observed from above as they flew away and landed. Here the acceleration slopes are again inverted compared with similar flights observed from below (compare Fig. 7a). Figure 9b, however, represents something of a puzzle. This wagtail trace clearly shows leaping flight but the acceleration slopes do not invert as the bird passes the nearest point. The voice commentary gives no clue except that the bird was recorded as it approached and flew overhead. Possibly in this unique case the bird was first observed flying close to the ground (i.e. from above) but rose as it passed the observer.

INTERPRETIVE MODELS

The traces of leaping flight suggested a number of physical observations that have been useful in assessing the sensitivity of the techniques and interpreting the results obtained. Figure 10a shows the trace recorded when a "superball" was dropped from a first-floor window onto a concrete path and observed by the vertically directed radar. Except for the moment of impact and reversal, the ball is constantly accelerating downwards at 1 g. The dotted calibration line shows that air resistance can be ignored at these velocities. The traces are V-shaped, however, owing to the radar's inability to distinguish between approach and recede velocities. The true 1 g slopes are shown in Fig. 10b where negative velocities are transposed to their true, unreflected positions.

On the first fall, the trace shows that the ball reached a maximum velocity of $8.9 \mathrm{~m~s}^{-1}$, indicating a fall of 4.01 m. This is in gratifying agreement with the measured fall of 4.0 m. On the second fall (first bounce) maximum velocity was $8.3 \mathrm{~m~s}^{-1}$ indicating a peak height of 3.54 m or a conservation of 88% of the initial energy after one impact. Later bounces show progressively less efficiency (86% and 83%). This may be due in part to the ball acquiring some spin (the concrete was not very flat) but the decrease is largely apparent, owing to the trajectory moving slightly to one side and introducing a small $\cos \theta$ error. In the third and fourth bounces of Fig. 10a, the trace can be seen to divide as the radar catches sight of the ball's image reflected in the concrete; before this the trace shows signs of beats between the two echoes. The window from which the ball was dropped projected from an arch, otherwise a vertical wall to ground level would have produced two further images, one of the real ball and one of its underground image.

Figure 11 shows the trace obtained as the "superball" bounced past the radar which was held by an operator standing on the ground.

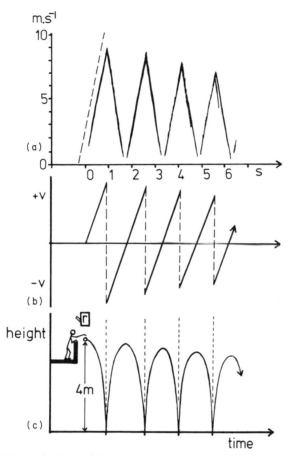

FIG. 10. (a) Trace obtained with the 13.5 GHz radar looking vertically down on a "super-ball" dropped 4.0 m onto a concrete path. The dotted line shows a 1 g acceleration slope for comparison. (b) The theoretical velocity trace of the ball with continuous 1 g acceleration downwards except at the moment of impact and reversal (dotted lines). (c) Time course of bounces along the vertical trajectory.

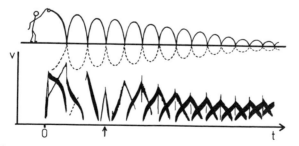

FIG. 11. Below: the trace obtained as a "superball" was bounced past the 13.5 GHz radar which was held about 1.5 m above ground level. The arrow indicates the time of nearest approach. The upper diagram gives an idea of the trajectory of the real ball and (dotted) that of its radar image below ground level.

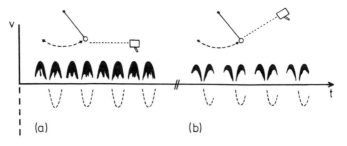

FIG. 12. (a) Trace obtained when the radar observed a pendulum from a horizontal view-point; dotted sections below the baseline are traced from alternate sections above to show the sinusoidal variation of real velocity. (b) The same when the radar was elevated to look slightly down at the pendulum; the peaks of maximum relative velocity move closer to one end of the swing.

Again the ball shows a continuous 1 g downward acceleration but its horizontal velocity here lifts the complete trace above zero as a saw-tooth line. The complementary negative 1 g slopes represent the image of the ball bouncing "beneath the ground". An interesting feature is that the "real ball" traces are slightly sigmoid whereas the image traces are much more linear. This is presumably because the trajectory of the real ball is nearer to the radar so that its angle to the line-of-sight varies more than that of the underground image.

This interpretation was checked in an experiment, conceived by Justin Halls, in which a pendulum was recorded from different view-points. Viewed horizontally the velocity should be a sinusoidal func-tion, giving a "full-wave rectified" trace (Fig. 12a). When the radar was raised to look slightly downwards, the trace became "distorted" in time with the velocity peaks moving nearer to one end of the swing (Fig. 12b).

The next stage is to apply the insights of these simple models to the more complicated traces obtained from birds in leaping flight and to the interpretation of wing movements as part of the radar "signa-tures" of birds.

DISCUSSION AND CONCLUSIONS

The observations reported here have interesting implications both for the study of flight mechanisms and for echolocation by constant frequency bats. For flight studies the technique offers a simple means of acquiring data on the performance of named species under natural conditions. It can thus form a valuable link between labora-tory or wind-tunnel measurements, where conditions are far from normal and performance may necessarily be stereotyped, and studies made with long-range radar where identity of the subject is often uncertain (Eastwood, 1967).

This method is not entirely new although earlier studies failed to reveal the fine structure that is available: Schnell (1965) recorded only flight speed from a meter indicator while Martinson (1973), with a similar technique to the present one, was only able to resolve wing-beat rates up to 10 per second. Green & Balsley (1974) used a much higher-powered and less portable doppler system while most other radar studies have used high-powered pulse radars. No previous observations of leaping flight by doppler radar are known.

Identification of birds by radar is of considerable interest both for ornithology and for the avoidance of bird-strikes by aircraft (Houghton, Blackwell & Wilmot, 1973). Unfortunately little is known about the interaction between a flying bird and a radar beam so that identifications by radar alone tend to be very uncertain (Eastwood, 1967; Able, 1977). A bird undoubtedly forms a complex and changing target but a new technique is now in hand by which the echo of a flying bird can be analysed so that its radar "signature" can be interpreted. This relies on enhancing various parts of the echo by attaching resonant reflectors, tuned to the radar frequency, to different parts of the body in turn.

A half-wave dipole seen broad-side on has a radar cross-section area of $0.88 \lambda^2$. This gives 9.6 cm^2 at 9 GHz or 4.3 cm^2 at 13.5 GHz. These figures have been checked by comparison with a range of steel balls and are comparable to the maximum, broad-side cross-section of a small bird. Inclination of the dipole with respect to the beam or to the plane of polarization gives similar reductions in echoing strength that are close to a cosine-squared low, with half amplitude at 45°. The average cross-section for random orientation is $0.18 \lambda^2$, but several dipoles can be added together to boost the return. Now modern chaff dipoles for microwave frequencies are very small and light. The best material for our purpose is silvered monofilament nylon which at 0.44λ (the optimum length) or 9.7 mm at 13.5 GHz weighs only 125 μg. Such a dipole can be attached to a single feather by a smear of silicone grease.

Selective echo enhancement in this way can be studied throughout the wing-beat cycle or from various viewpoints (as a function of aspect) in wind-tunnel observations. It is also hoped to try the effects of several dipoles on trained falcons or domestic pigeons in outdoor flight. In this way we hope to derive further clues to the traces obtained from wild birds and eventually, perhaps, to provide data useful to the operators of pulsed air-traffic control radar.

It has already been stated that the use of these doppler radars has given us an insight into the capabilities of doppler bats. When listening to moving targets through the headphones one gains a vivid impression of the world as experienced by such velocity-sensitive animals although most of them benefit from shorter wavelengths. The interpretations of Figs 4–12 give some idea of what can be achieved. Nevertheless the bats are potentially capable of at least three further kinds of doppler echo processing that we have not yet used with the radar.

First, it is not necessary to wait until a linearly moving target reaches its nearest point before its track can be established. Any two measures of doppler frequency from an approaching target that differ from the maximum value give two measures of cos θ. Together with the distance flown by the target in the intervening time (v.t) this allows the track to be constructed and the nearest point to be predicted if required.

Secondly, by observing the intensity of the echo received, the bat can obtain range information even if the target track is non-linear (Pye, 1967). Echo intensity for small targets as a function of distance follows the radar inverse fourth law. Thus for two different ranges the intensities have the ratio:

$$\frac{\text{intensity}_1}{\text{intensity}_2} = \left[\frac{\text{range}_2}{\text{range}_1}\right]^4 \tag{8.8}$$

or

$$\frac{r_2}{r_1} = \left[\frac{\text{intensity}_1}{\text{intensity}_2}\right]^{1/4} \tag{8.9}$$

But the difference in range is easily derived from a doppler system:

$$r_1 - r_2 = f_d \cdot t \cdot \frac{\lambda}{2} \tag{8.10}$$

By solving these simultaneous equations the ranges can be obtained. For example an increase of intensity by 16 times (or + 12 dB) indicates that range has been halved; if the target has approached by 1 m by any route in the meantime, the remaining range must be 1 m. The pronounced ear-movements of doppler bats, that are correlated with emitted pulses, may thus serve to scan the target area and so ensure that the received echo intensity is not obscured by the directionality of the ear itself.

Thirdly, the bat operates a double receiver, the two apertures of which will have slightly different cos θ unless the target is moving within the median plane of the bat's head. At close ranges this difference may well be detectable (Pye, unpublished calculations) and the sharp null in the forward direction could be very useful to a bat that is intercepting its prey in flight.

As a final conclusion it may be said that, although the polar co-ordinates of direction and range may seem to be the most "natural" ones for any echolocating system, a doppler velocity-measuring system can also provide a great deal of information if observations are made over a short period of time. The total information available in a sequence of doppler echoes can be used to resolve many of the uncertainties that may at first sight seem to be present.

ACKNOWLEDGMENTS

The author is greatly indebted to Marconi's Low Power Systems Radar Laboratory, Chelmsford, for help in developing the two radars used here, to Mr Justin Halls for help in maintaining the equipment and in developing techniques for its use, to Mr Brian McDaid for energetically employing it on birds and to Mr Peter Olney, Curator of Birds, Zoological Society of London, for allowing us and helping us to record bird flights in the Society's aviaries.

REFERENCES

Able, K. P. (1977). The flight behaviour of individual passerine nocturnal migrants: a tracking radar study. *Anim. Behav.* 25: 924–935.

Csicsáky, M. J. (1977). Body-gliding in the zebra finch. In *Physiology of movement: biomechanics*: 275–286. Nachtigall, W. (Ed.). Stuttgart, N.Y.: Fisher.

Denton, E. J. & Blaxter, J. H. S. (1976). The mechanical relationships between the clupeid swimbladder, inner ear and lateral line. *J. mar. biol. Ass. U.K.* 56: 787–807.

Eastwood, E. (1967). *Radar ornithology*. London: Methuen.

Edwards, J. & Houghton, E. W. (1959). Radar echoing area polar diagrams of birds. *Nature, Lond.* 184: 1059.

Gray, J. (1968). *Animal locomotion*. London: Weidenfeld and Nicolson.

Green, J. L. & Balsley, B. B. (1974). Identification of flying birds using a doppler radar. *Conf. on biol. aspects of the bird-aircraft collision problem*: 491–508. Gauthreaux, S. A. (Ed.). Clemson Univ. prepared for AFOSR.

Greenewalt, C. H. (1962). Dimensional relationships for flying animals. *Smiths. misc. Collns* 144(2): 1–46.

Griffin, D. R. (1971). The importance of atmospheric attenuation for the echo-location of bats (Chiroptera). *Anim. Behav.* 19: 55–61.

Gustafson, Y. & Schnitzler, H. U. (1979). Echolocation and obstacle avoidance in the hipposiderid bat *Asellia tridens*. *J. comp. Physiol.*, 131: 161–167.

Halls, J. A. T. (1978). Radar studies of bat sonar. In *Proc. 4th Int. Bat Res. Conf.*: 137—143. Olembo, R. J., Castelino, J. B. & Mutere, F. A. (Eds). Kenya Literature Bureau.

Houghton, E. W., Blackwell, F. & Wilmot, T. A. (1973). Bird strike and the radar properties of birds. In *Radar: present and future*: 257—262. London: I.E.E.

Martinson, L. W. (1973). *A preliminary investigation of bird classification by Doppler radar*. RCA Govt. and Commer. Systems, Missile and Surface Radar Divn., Moorestown, N.J. 08057. Prepared for NASA, Wallops Station, Wallops Island, VA 02337.

Medway, Lord & Pye, J. D. (1977). Echolocation and the systematics of swiftlets. In *Evolutionary ecology*: 225—238. Stonehouse, B. & Perrins, C. M. (Eds). London and Basingstoke: Macmillan Press Ltd.

Pye, J. D. (1967). Theories of sonar systems and their application to biological organisms: discussion. In *Animal sonar systems, biology and bionics*: 1121—1136. Busnel, R. G. (Ed.). Jouy-en-Josas, France: Laboratoire de Physiologie Acoustique.

Pye, J. D. (1978). Some preliminary observations on flexible echolocation systems. In *Proc. 4th Int. Bat Res. Conf.*: 127—136. Olembo, R. J., Castelino, J. B. & Mutere, F. A. (Eds). Kenya Literature Bureau.

Pye, J. D., Flinn, M. & Pye, A. (1962). Correlated orientation sounds and ear movements of horseshoe bats. *Nature, Lond,* 196: 1186—1188.

Pye, J. D. & Roberts, L. H. (1970). Ear movements in a hipposiderid bat. *Nature, Lond.* 225: 285—286.

Rayner, J. (1977). The intermittent flight of birds. In *Scale effects in animal locomotion*: 437—443. Pedley, T. J. (Ed.). London & New York: Academic Press.

Schnell, G. D. (1965). Recording the flight-speed of birds by doppler radar. *Living Bird* 4: 79—87.

Schnitzler, H. U. (1968). Die Ultraschall—Ortungslaute der Hufeisen-Fledermäuse (Chiroptera-Rhinolophidae) in verschiedenen-Orientierungssituationen. *Z. vergl. Physiol.* 57: 376—408.

Schnitzler, H. U. (1970). Echoortung bei der Fledermaus, *Chilonycteris rubiginosa*. *Z. vergl. Physiol.* 68: 25—38.

Simmons, J. A. (1970). Distance perception by echolocation: the nature of echo signal processing in the bat. *Bijd. Dierk.* 40: 87—90.

Vaughn, C. R. (1974). Intraspecific wingbeat rate variability and species identification using tracking radar. *Conf. on biol. aspects of the bird-aircraft collision problem*: 443—476. Gauthreaux, S. A. (Ed.). Clemson Univ. prepared for AFOSR.

Symp. zool. Soc. Lond. (1981) No. 48, 219–238

The Use of Muscles During Flying, Swimming, and Running from the Point of View of Energy Saving

G. GOLDSPINK

Muscle Research Unit, Department of Zoology, University of Hull, Hull, England

SYNOPSIS

During evolution energy conservation has been a very important factor. This chapter attempts to explain how the musculature has been adapted to produce locomotory movements with minimum expenditure of energy.

Electromyographic studies have revealed that gliding flight requires the recruitment of only a few motor units. The majority of muscle fibres in bird flight muscle are of the fast contracting oxidative type. However, in the herring gull, approximately 8% of the fibres are of the slow type and it is suggested that these are the ones that are recruited for gliding. The majority of muscle fibres in the pectoralis have an intrinsic rate of shortening that is matched to the flapping frequency of the bird and this is one reason, along with aerodynamic reasons, for the U-shaped power: velocity curve.

Fish swimming differs from bird flight in that both frequency and amplitude of beat are altered. Changes in the frequency without loss of efficiency necessitate that myotomal muscle is made up of several types of muscle fibres. EMG studies have shown that only the red, slow contracting fibres that are usually situated near the lateral line are used for slow cruising movements. The fast contracting white fibres, which make up 80% of the musculature, are recruited only for fast swimming speeds, whilst the pink or intermediate fibres are used for intermediate cruising speeds.

Tendons play an important part in terrestrial locomotion, particularly running and galloping. The role of the muscles seems to be to tension the tendons and they do it in such a way that they are contracting isometrically. The energy turnover in isometric contraction is considerably less than during isotonic contraction. Therefore, this represents an important energy saving. An even greater energy saving is obtained when the muscles are doing negative work, and this provides the explanation as to why very little effort is needed for terrestrial animals to descend a hill. Throughout an attempt is made to explain the energy requirements, based on what is known about the crossbridge mechanism of contraction.

INTRODUCTION

Energy conservation is often a topic of conversation these days. However, as far as animals are concerned, this has been one of the main factors that have shaped the course of evolution. The species that has been able to move to a suitable food supply and use the energy in the food efficiently has, in general, been the successful species. In very large herbivores this has, perhaps, been the most important criterion. However, for carnivorous animals and small herbivores speed of movement has also been an important factor. Some animals rely on armour or camouflage for protection against predators and these are invariably slow-moving creatures. It is also apparent that, although most aquatic and terrestrial animals are capable of swift movements, they very rarely move rapidly. Most species of fish can swim at up to 10 body lengths s^{-1} but most of the time they swim at less than 0.3 body length s^{-1}. From an energetic point of view there seems, therefore, to be some merit in moving slowly. Hence the musculature in most animals has been adapted for producing slow movements most of the time but nevertheless it is capable of producing fast powerful movements when the need arises. This has led to a division of labour between muscle fibres. Indeed, from this chapter and some of the others presented in this book, it will be apparent that most vertebrate muscles contain different types of muscle fibres which are adapted for carrying out the different functions effectively and economically.

The situation regarding bird flight is rather different because birds have to fly fast enough to generate sufficient lift in order to remain airborne. Hence, the metabolic power requirements for flying, particularly for small birds, are very high.

The metabolic cost of locomotion is to a large extent indirectly determined by the rate of energy consumption of the muscles. It is therefore important for us to understand what is happening at the muscle level and even at the molecular level. Unfortunately, we have to rely mainly on information obtained from isolated muscle preparations, but some meaningful predictions can be made. In this chapter I shall attempt to explain the energy cost of the three main types of vertebrate locomotion in terms of muscle physiology and muscle energetics.

Some of the factors that determine the rate of energy turnover in muscle during locomotion can perhaps be listed at this point.

(1) The number of muscle fibres (motor units) recruited.
(2) The type of muscle fibres (motor units) recruited.
(3) The kind of contraction involved.

(4) The duration of contraction.

(5) The velocity and extent of shortening or lengthening of the muscle.

These could be dealt with in an exhaustive way but as the topic of the symposium is not muscle energetics, perhaps a more interesting approach is to discuss these factors in the context of each type of locomotion.

BIRD FLIGHT

There are several types of bird flight, including flapping flight, undulating flight, gliding and hovering and the energy cost for these different modes differs widely (Table I).

Electromyographic studies have been carried out on the herring gull (*Larus argentatus*) to see the extent of muscle fibre recruitment during both gliding and flapping flight (Goldspink, Mills & Schmidt-Nielsen, 1978). In this work three fine wire electrodes were inserted into the pectoralis major of gulls which were flown in a wind tunnel. The wire electrodes were connected to a small preamplifier that was carried on the back of the bird. This was, in turn, connected via a flexible lightweight screened lead to the recording apparatus. The gulls that were used had been previously trained to fly in the wind tunnel and they could be induced to glide by changing the angle of the tunnel to approximately 7°.

As will be seen from Fig. 1 and Table II, electrical activity of the pectoral muscles was present during gliding but it was very low, indicating that very few muscle fibres (motor units) were being recruited. As the action potentials were low, we felt that it would be interesting to load the gulls with a weight equivalent to a heavy meal. As will be seen from Fig. 1, this resulted in a considerable increase in electrical activity, indicating that the postural problem of preventing the wings from flipping up over the body is much greater, even though the weight of the bird was increased by only 13%. Hence it would seem that the payload is very critical in the design of the bird for flight, particularly for gliding flight. For aerodynamic reasons only large and medium-sized birds glide. However, there are some species of large birds, e.g. swans and geese, which do not glide and it is probable that their relative payload is too high.

From Table I it will be seen that normal gliding represents a very considerable energy saving. Indeed, the metabolic rate is only doubled during gliding whilst it is increased by six or seven times during flapping flight (Tucker, 1972; Baudinette & Schmidt-Nielsen, 1974). If

TABLE I

Approximate power requirements for different types of flight

Bird	Type of flight	Velocity	Power (W kg^{-1})	Reference
Hummingbird	Hovering	0	240	Lasiewski (1963)
Budgerigar	Level flapping flight	10 ms^{-1}	120	Tucker (1968)
Budgerigar	Undulating flight	7 & 10 ms^{-1}	125	Data from Tucker (1968)
Fish crow	Flapping flight	10 ms^{-1}	80	Tucker (1972)
Laughing gull	Flapping flight	10 ms^{-1}	60	Bernstein, Thomas & Schmidt-Nielsen (1973)
Herring gull	Gliding	—	15	Baudinette & Schmidt-Nielsen (1974)

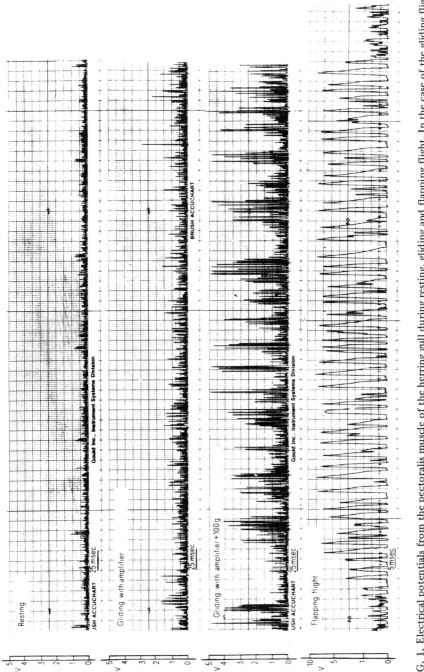
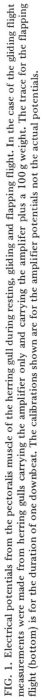

FIG. 1. Electrical potentials from the pectoralis muscle of the herring gull during resting, gliding and flapping flight. In the case of the gliding flight measurements were made from herring gulls carrying the amplifier only and carrying the amplifier plus a 100 g weight. The trace for the flapping flight (bottom) is for the duration of one downbeat. The calibrations shown are for the amplifier potentials not the actual potentials.

TABLE II

Mean integrated values in arbitrary units (dimensions $mV\ s^{-1}$) of recorded potentials from the pectoral muscle of herring gulls

Recording	Resting	Gliding Amplifier only (23 g)	Gliding Amplifier + weight (123 g)
1	34 ±9 (7)	100 ±34 (16)	241 ±38 (20)
2	36 ±3 (15)	83 ±25 (31)	185 ±43 (19)
3	22 ±2 (14)	72 ±18 (25)	206 ±51 (30)
Mean	30	89	211

Values given ± the SD (Student's t-test). The number of times the traces were sampled are given in parentheses.

one looks at the electromyographic traces for flapping flight, it is
apparent that this does involve the recruitment of many fibres but
these are, of course, activated in a rhythmical way. One of the main
factors in energy saving during gliding is the much smaller number of
motor units that are involved. One question that should be asked is,
are the fibres (motor units) that are involved in gliding different to
those used for flapping flight? Talesara & Goldspink (1978) found
that 8% of the fibres in the pectoralis major of the herring gull
showed the characteristics of slow contracting oxidative fibres.
Although in the majority of species that fly, all the pectoral fibres
were of the fast oxidative type, the pigeon is interesting in that it
possesses two types of fast fibres; a small oxidative type and a large
less oxidative type. The pigeon is a strong flyer that has the ability to
take off almost vertically and it is suggested that the large fast
glycolytic fibres may be used for rapid take-off but not for sustained
flapping flight.

There seem to be two important aspects of gliding as a means of
energy saving.

(i) Only a few muscle fibres (motor units) are recruited.

(ii) It involves isometric or semi-isometric contraction.

The energy saving by involving only a few motor units is perhaps
obvious and requires no further comment, except to point out again
that in some species there may be a different kind of motor unit to
the others in the pectoral muscle. Figure 4a shows the energy turnover in different types of fibres during isometric contraction (zero
velocity) and during shortening at differing velocities. From this plot
it will be seen that isometric contraction is less expensive than isotonic contraction. Also it will be noted that the energy turnover in a
slow contracting fibre is lower than in a fast contracting fibre. In
slow fibres the myosin crossbridges, which are the force generators,
have a longer cycle time and therefore do not hydrolyse ATP as
rapidly as fast contracting fibres. No energy is required whilst the
crossbridges are engaged with the actin filaments (Fig. 2). ATP is
only required to reprime the crossbridges at the end of each force
generating cycle. In isometric contraction the engagement time for
the crossbridges is longer than if the sarcomeres are shortening and
the actin and myosin filaments are moving over each other. Hence
isometric contraction is less costly than isotonic contraction. The
muscle action during gliding cannot be absolutely isometric because
the bird has to make slight adjustments to allow for changes in the
air currents; this the bird does in a very precise way so that the
muscle force balances the torque on the wings. It could be argued
that the need for control is the only reason why muscle has to be

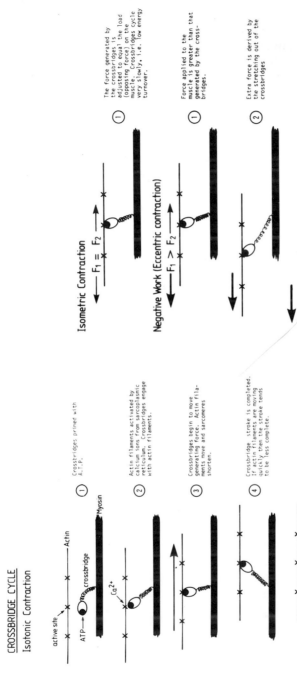

FIG. 2. A diagrammatic representation of action of the crossbridges during (a) isotonic, and (b) isometric and negative work contractions. The energy used by muscles is directly relatable to the rate of turnover of ATP by the crossbridge. One molecule of ATP is believed to be hydrolysed each time a crossbridge completes its cycle of force generation. The cycle time depends on the type of muscle fibre, i.e. whether it is a fast or a slow fibre, and on the type of contraction.

involved, otherwise a "catch mechanism" could be employed to hold the wings in the horizontal position[*]. Nevertheless, as the muscle action is primarily isometric, and involves only a few motor units, a considerable energy saving is made during gliding. From Table I, it will be seen that flapping flight is expensive. Indeed the power produced by hummingbird muscles probably represents the highest sustainable power that muscle tissue can develop. In spite of this, flapping flight, with the exception of hovering, is more efficient than a terrestrial locomotion as a means of covering distance; that is to say the net cost of transport in $J\ kg^{-1}\ m^{-1}$ is less than for terrestrial locomotion. In small birds, such as the budgerigar, a "U"-shaped curve is obtained for the metabolic power requirements for level flying at different velocities, $10\ ms^{-1}$ being the velocity at which the power requirements are the lowest for this species. In general, bird flight muscles are homogeneous in that they are composed of fast contracting oxidative, glycolytic fibres only. The power requirements for flight are very high and the pectoral muscles account for about one-third of the weight of the bird. Therefore the bird cannot afford the luxury of carrying populations of different types of muscle fibres. The term "fast" is relative because for each species the intrinsic rate of contraction is matched to the flapping frequency of the bird. Hence the muscle fibres in the pectoral muscles of a large stork have a much lower intrinsic rate of shortening than do those of a small wren. However, as the pectoral muscle fibres within a given bird all possess the same optimum velocity of shortening, this means that the flapping frequency is virtually fixed. Hence birds tend to alter the amplitude of the wing beat rather than the frequency. If they do flap faster or slower than the optimum frequency as determined by the muscle characteristics then the metabolic power requirement is increased considerably. This is one of the reasons, along with aerodynamic reasons, why the metabolic power—velocity curve is "U"-shaped (Tucker, 1968). Let us now contrast this situation with fish swimming, as fish are capable of altering both the frequency and the amplitude of tail beat.

[*] It was pointed out by Rayner at this symposium that Pennycuick has recently examined the wing attachment of the albatross and there is evidence of a catch mechanism in this bird. The albatross glides for many months at a time and it is probably the species that is best adapted for gliding; indeed, it very rarely indulges in flapping flight.

FISH SWIMMING

Unlike bird flight muscle, fish myotomal muscle is very definitely compartmentalized and each compartment contains a different type of muscle fibre. As shown by Johnston (this volume), the myotomal musculature consists of three basic types of muscle fibre. The red fibres have a low specific ATPase activity and possess many mitochondria. Therefore, they are presumably slow contracting and have an oxidative metabolism (SO fibres). These constitute only about 5% of the musculature and tend to be found in a thin line on the outside of the myotomes near to the lateral line (Fig. 3). The bulk of the musculature is made up of white fibres which have a myosin with a high specific ATP activity. These fibres are therefore fast contracting. They contain virtually no mitochondria; they rely, therefore, mainly on glycolysis for their energy supply (FG fibres). Using histochemical methods it is possible to identify a pink or intermediate type of fibre. These fibres have a myosin with a fairly high ATPase activity and have high levels of both oxidative and glycolytic enzymes. Hence they are reasonably fast contracting and are capable of deriving energy from both oxidative and glycolytic pathways (FOG fibres).

Measurements of metabolites (Johnston & Goldspink, 1973a, b) and EMG studies (Bone, 1966; Hudson, 1972; Davison, Goldspink &

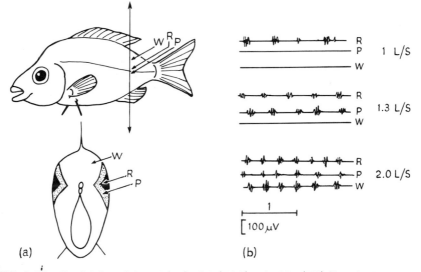

(a)   (b)

FIG. 3. The distribution of the red (SO), pink (FOG) and white (FG) fibres in the carp. The recruitment of the different types of fibres at different swimming speeds is shown by the EMG traces. At low cruising speeds (1 body length s^{-1}) the slow contracting fibres were used, at slightly faster speeds (1.3 body lengths s^{-1}) the pink (FOG) fibres were also recruited. At 2.0 body lengths s^{-1} some of the white fibres were also involved. (Data from Davison et al., 1976.)

Johnston, 1976) have indicated that there is a hierarchy of recruitment of the fibre types. It seems that only the red (SO) are recruited
at low swimming speeds. However, as the speed increases, the pink
(FOG) fibres begin to be recruited. The white fibres, on the other
hand, are recruited only for rapid swimming movements (see Fig. 3).
It may seem strange that 80% of the musculature is reserved for rapid
movements and therefore recruited very infrequently. However, the
fish is in a buoyant medium and therefore its payload is not as critical
as that of the bird. The other point is that drag is proportional to the
cube of the velocity and therefore high speed swimming requires a
lot of power and hence a lot of muscle.

The hierarchy of fibre recruitment, which has also been described
for mammalian muscle (Henneman & Olson, 1975) is very interesting
from an energetics point of view. Muscle action during swimming is
isotonic although rather surprisingly the muscle fibres shorten by
only about 5% of their length. If we look at the efficiency of doing
work (work done per joule of available energy) as shown in Fig. 4b,
we see that the SO fibres have the highest efficiency peak, providing
they are contracting slowly. Therefore, it is desirable to use the SO
fibres for slow movements, such as slow cruising. The efficiency of
the slow fibres, however, drops very sharply as the velocity of shortening increases, but it is at this point that the FOG fibres are recruited

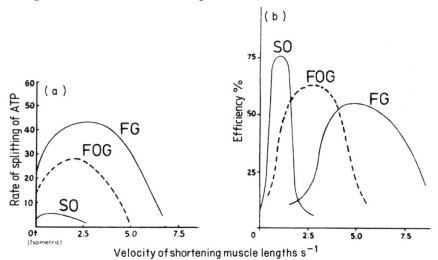

Velocity of shortening muscle lengths s^{-1}

FIG. 4 a and b. The energy turnover (a) and efficiency of converting high energy phosphate
into work (b) for the different types of muscle fibres. These data were obtained by measuring
the breakdown of creative phosphate in muscles that are predominantly fast (FG fibres)
and predominantly slow (SO fibres). The position of the FOG fibres is speculative. The
muscles were first treated with iodoacetate and nitrogen or cyanide to prevent glycolysis
and oxidative phosphorylation and hence the replenishment of ATP and CP. The efficiency
values are based on a free energy value of 468.4 KJ for each mole of creative phosphate split.

and, therefore, the work output and the efficiency of the musculature is maintained. For the higher swimming speeds there is no alternative but to use the white (FG) fibres as the power requirements are high. Although the fibres split ATP at a very rapid rate and therefore the energy turnover is very high (Fig. 4a), they do have a reasonable efficiency for working at the higher shortening velocities. The problem is that the fibres cannot possibly replenish the ATP as quickly as it is used. Therefore the fish fatigues very quickly when swimming at high speed and the immediate energy supplies can only be replenished after the exercise bout is over.

TERRESTRIAL LOCOMOTION

After viewing the American astronauts moving around on the surface of the moon, we can appreciate that much of the energy requirements for terrestrial animals are related to overcoming the effects of gravity. However, recent work (Taylor, Shkolnik et al., 1974; Alexander, 1974, 1977; Cavagna, 1977; McMahon, 1977) has indicated that many terrestrial animals use gravity to good effect in that they propel themselves by bouncing along on their tendons. The analogy of a pogo stick has been used, although this action can be fairly readily appreciated if one watches a gazelle running or a kangaroo hopping. Work by Taylor, Shkolnik et al. (1974) indicated that, from an energetics point of view, the conversion of energy stored in the tendons as elastic strain energy into kinetic energy is the main action. These workers found that a goat, a gazelle and cheetah of about the same weight consumed almost identical amounts of energy when running at the same velocity (Fig. 5). At first sight this is surprising as the configuration of the limbs of these animals is rather different. However, if one considers the elastic recoil of the tendons to be the major action, rather than the swinging of the legs backwards and forwards, then the thickness of the limbs is not a major consideration. The kangaroo has perhaps exploited this system more than any other animal. It will be seen from Fig. 5 that the power requirements for locomotion actually decrease as the velocity of locomotion increases (Taylor, Shkolnik et al., 1974). The kangaroo has large Achilles tendons and the transference of elastic strain energy into kinetic energy increases as the animal takes longer and longer hops.

What does this storage of energy in the tendons mean as far as muscle action and energetics are concerned? It has been pointed out by Alexander (1977) and McMahon (1977) that the changes in muscle length with this system are much less than one might other-

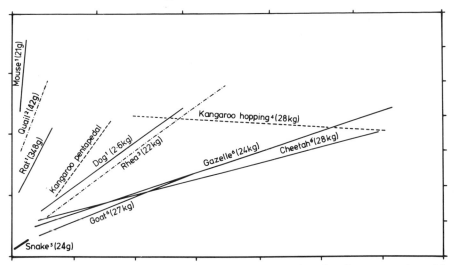

FIG. 5. The metabolic requirements for terrestrial locomotion. Crawling ——; quadrupedal running ——; bipedal running -·-·; hopping - - -, at different velocities. (1) Taylor, Schmidt-Nielsen & Raab (1970). (2) Fedak, Pinshow & Schmidt-Nielsen (1974). (3) Taylor, Oniel, Fedak & Schmidt-Nielsen (1971). (4) Dawson & Taylor (1973). (5) Chodrow & Taylor (1973). (6) Taylor, Shkolnik *et al.* (1974). Taken from Goldspink (1977b).

wise expect. Indeed, it seems that the muscles develop force at about the same rate at which the tendons recoil. It also means that they tension the tendons whilst the limb is off the ground and whilst the load on the muscles is low. Therefore the contraction is essentially an isometric one. The energy turnover for isometric contraction is considerably lower than in isotonic contraction. This represents a considerable energy saving. Of course, even greater energy saving would be made if muscles were not involved at all, as the animal could still bounce on its tendons, but there would be no means of control and, to some extent, the muscles are also used as a braking system. This brings us to a consideration of negative work, a situation in which a muscle that is activated is stretched. The result is that a greater total amount of force is developed by the system. The extra force is believed to be derived from the stretching of the crossbridges whilst they are engaged with the actin filaments (Fig. 2). It has already been pointed out that energy is used only at the detachment phase and not during the engagement phase. Therefore, the force derived from the compliance of the crossbridges is free. Indeed Curtin & Davies (1975) have shown that in an isolated frog muscle the energy turnover was only 25% of the isometric turnover rate when the muscle was being lengthened at 0.13 muscle length s^{-1} (Table III). Positive work at about 0.9 muscle lengths s^{-1} was almost twice as costly as isometric force and therefore about eight times more costly than

TABLE III

ATP breakdown during stretching and shortening. From Curtin & Davies (1975)

	Velocity $1_0 s^{-1}$	n	Mean rate of total chemical change $\mu mol\ g^{-1}\ s^{-1}$
(A)	Isometric contraction		
	0.00	7	$+0.72 \pm 0.09$
(B)	Stretching		
	0.13	11	$+0.18 \pm 0.042$
	0.18	10	$+0.18 \pm 0.042$
	0.33	7	$+0.22 \pm 0.143$
	0.66	18	$+0.49 \pm 0.163$
	1.33	10	$+0.45 \pm 0.173$
	2.00	18	$+0.84 \pm 0.351$
(C)	Shortening		
	0.20	10	$+0.69 \pm 0.035$
	0.33	13	$+0.75 \pm 0.072$
	0.66	16	$+1.08 \pm 0.105$
	0.91	9	$+1.41 \pm 0.099$
	1.33	9	$+0.73 \pm 0.065$
	2.10	10	$+0.94 \pm 0.149$

negative work. As the crossbridges operate only over a short distance there must be slippage, that is to say the crossbridges must be pulled off the actin filaments. However, they probably engage with the next active site, without necessarily completing their cycle. Indeed, the evidence would suggest that the crossbridge cycle is longer in negative work situations, otherwise the energy turnover would not be so low. The fact that negative work requires less total energy than positive work was demonstrated to the Royal Society by A. V. Hill, who arranged two bicycle ergometers back to back and, using this system, showed that the efforts of a large powerful male athlete doing positive work could be easily resisted by a slightly built female doing negative work. In a less dramatic but perhaps more practical way, Margaria (1968) demonstrated the importance of negative work in human locomotion. To do this he measured the energy consumption in walking uphill and walking downhill. Walking is a complex action involving many muscles. Therefore it is unlikely that the energy saving due to the negative work component is as high as the eight times obtained for the isolated frog muscle even when running downhill. For walking or running on the level no overall saving is obtained from negative work because the negative work has to be paid for by positive work at another point in the cycle. However, it does mean that we can come down the stairs or come down a rope with less effort than it takes to go up the stairs or the rope, although muscles still have to develop approximately the same force to keep the body erect.

Walking and running also involve quite a lot of isotonic contraction and, in this context, it is interesting to note that most muscles in terrestrial vertebrates are mixed, in that fast contracting and slow contracting fibres are found side by side within the same anatomical muscle. As with fish, the fast contracting fibres can be divided into two types: fast glycolytic and fast oxidative glycolytic.

MAMMALIAN MUSCLE FIBRE CHARACTERISTICS

Tonic Muscle Fibres

These fibres are very slow contracting and do not usually show a pro-pagated muscle action potential, hence when stimulated with a single stimulus they do not produce a significant response. They are multiply innervated (en grappe motor end-plates) and have a graded response to stimulation of different frequencies. Tonic fibres usually have a very low specific myosin ATPase activity and this is no doubt why they are capable of developing and maintaining isometric tension very economically (Goldspink, Larson & Davies, 1970; Matsumoto, Hoekman & Abbott, 1973). A good example of this type of muscle is the anterior latissimus dorsi muscle in the bird which holds the wings against the body. The fibres of this muscle like other tonic fibres are able to remain contracted for long periods of time with very little utilization of energy.

Slow Phasic Fibres (SOs or Type I)

These fibres are slow contracting but they are not usually as slow as the tonic muscle fibres. They have a propagated action potential and are hence referred to as twitch fibres. Slow phasic fibres are respon-sible for both maintaining posture and carrying out slow repetitive movements and are economical and efficient in carrying out these functions (Goldspink, 1975). Since they usually contain many mito-chondria and since their myofibrils hydrolyse ATP only very slowly this type of fibre is very fatigue-resistant.

Fast, Phasic, Glycolytic Fibres (FG, FF or Type IIb)

These fibres are adapted for a high power output and have a reason-able thermodynamic efficiency for producing work. They have a high or very high intrinsic speed of shortening and hence a high myosin

ATPase specific activity. Fast phasic fibres are usually used only when very rapid movement is required. They usually possess very few mitochondria because they cannot possibly replenish ATP as fast as it is used. They therefore fatigue very rapidly and most of the replenishment of their energy supplies has to take place whilst the fibres are inactive. During contraction the energy that is supplied comes mainly from glycolysis and the immediate energy stores (ATP and phosphoryl creatine).

Fast, Phasic, Oxidative Fibres (FOG, FR or Type IIa)

These fibres are essentially the same as the fast glycolytic fibres except that they contain more mitochondria. In some cases they may have a slightly lower intrinsic rate of shortening than the fast glycolytic fibres. These fast oxidative fibres are apparently adapted for fast movements of a repetitive nature and are probably recruited next after the slow twitch fibres. Because of the larger numbers of mitochondria they are less subject to fatigue and are able to recover relatively quickly after exercise.

DISCUSSION

It is argued by the author that the main reason for the presence of different fibre types within the same muscle is one of energetics. From the mechanical point of view, it might be thought that the slow contracting fibres are needed to produce slow movements. However, fast fibres will produce slow movements if they are subjected to a high load, in other words if only a few fast motor units are recruited to produce the movement. If we again look at Fig. 4b we see that each fibre type has its own efficiency maximum. There is good evidence from neurophysiological studies (Henneman & Olson, 1975) and from histochemical studies involving glycogen depletion (Gollnick, Karlsson, Piehl & Saltin, 1974; Gollnick, Piehl & Saltin, 1974). Moreover, mammalian muscle fibres show the same sequence of recruitment as in fish (Armstrong, this volume): that is to say, the SOs first, the FOGs next and the FGs last. By bringing in the different fibre types at different velocities of movement the muscle is able to maintain its thermodynamic efficiency reasonably well. Indeed whole body efficiency measurements indicate that muscle efficiency is fairly constant and also independent of the velocity of the movement or the duration of the exercise (Bannister & Jackson, 1967; Pugh, 1975; Gaesser & Brooks, 1975). The question of energy turnover during long-term exercise is an interesting one. Although the slow oxidating fibres are fairly fatigue-resistant, they will nevertheless eventually fatigue and this will necessitate the recruitment of fast fibres if the force development is to be maintained. The fast fibres have a lower maximum efficiency than the slow fibres and, of course, they fatigue much more rapidly. However, it is interesting to note that fatigue is associated with a slowing of the crossbridge cycle and hence there is a lower turnover of energy (Awan & Goldspink, 1972; Edwards, Hill & Jones, 1975). This reduction in energy turnover with duration of contraction is not fully understood as it is not simply a lack of ATP in the fibres. However, it does involve a slowing of the muscle and therefore it results in conservation of energy.

As stated above, most muscles in terrestrial vertebrates are mixed. However, the proportion of the fibre types within an individual muscle differs very considerably, according to the function of the muscle, the type and size of the animal and its mode of life. Some animals, such as dogs and horses, are adapted for running relatively quickly for quite long periods of time and they tend to have a high percentage of SO and FOG fibres. Cats, on the other hand, in particular lions, rely on a short burst of speed to catch their prey and hence they have a high percentage of FG fibres (see Armstrong, this volume). It is well known that the net cost of transport for larger animals is less than for smaller animals. As previously pointed out (Goldspink, 1977a) there are at least two reasons for this. One reason is that the stride is longer for larger animals. This means that the muscle will be activated fewer times per minute. A certain amount of energy is required for the activation and relaxation processes (moving Na^+, K^+ and Ca^{++} ions) and therefore in larger animals this activation – relaxation component should be less. However, the main factor is probably the fact that the larger animals possess a higher percentage of slow fibres and all the fibres, even the FGs, tend to have a relatively low intrinsic rate of contraction. As pointed out, slow (SO) fibres not only are more economical for developing isometric force but they have a greater efficiency for doing work. Therefore these slow fibres will have a lower energy turnover irrespective of whether they are involved in tensioning the tendons (isometric contraction) or in swinging the legs backwards and forwards. This and the fact that they have a lower resting metabolic rate means that, from an energy point of view, the larger animals "pound for pound" are more efficient than smaller animals.

Much of what has been said in this chapter has relied on the extrapolation of energetic measurements that have been carried out on isolated muscle, to what is believed to occur in muscles during locomotion. What we need now are methods that enable us to measure the energy turnover in individual fibres during locomotion. Only in this way will we be able to get a full understanding of the use of the different kinds of muscle fibres during locomotion.

ACKNOWLEDGEMENTS

The work on muscle fibre recruitment during bird flight was supported by an S.R.C. Research Grant to the author and by an N.I.H. Research Grant HL-02228 to Professor Schmidt-Nielsen. The work on muscle energetics was also supported by the S.R.C. The work on recruitment of muscle fibres during fish swimming was supported by N.E.R.C.

REFERENCES

Alexander, R. McN. (1974). The mechanics of jumping by a dog. *J. Zool., Lond.* **173**: 549–573.

Alexander, R. McN. (1977). Terrestrial locomotion. In *Mechanics and energetics of animal locomotion*: 168–203. Alexander, R. McN. & Goldspink, G. (Eds). London: Chapman and Hall.

Awan, M. Z. & Goldspink, G. (1972). Energetics of the development and maintenance of isometric tension by mammalian fast and slow muscles. *J. Mechanochem. Cell Motility* **1**: 97–108.

Bannister, E. W. & Jackson, R. C. (1967). The effect of speed and load changes on oxygen intake for equivalent power outputs during bicycle ergometry. *Int. Z. angew. Physiol.* **24**: 284–290.

Baudinette, R. V. & Schmidt-Nielsen, K. (1974). Energy cost of gliding flight in herring gulls. *Nature, Lond.* **248**: 83–84.

Bernstein, M. H., Thomas, S. P. & Schmidt-Nielsen, K. (1973). Power input during flight of the fish crow, *Corvus ossi fragus*. *J. exp. Biol.* **50**: 401–410.

Bone, Q. (1966). On the function of the two types of myotomal muscle fibres in elasmobranch fish. *J. mar. biol. Ass. U.K.* **46**: 321–349.

Cavagna, G. A. (1977). Aspects of efficiency and inefficiency of terrestrial locomotion. *Biomechanics* **6A**: 3–26.

Chodrow, R. E. & Taylor, C. R. (1973). Energy cost of limbless locomotion in snakes. *Fedn. Proc. Fedn Am. Socs exp. Biol.* **32**: Abs: 1128.

Curtin, N. A. & Davies, R. E. (1975). Very high tension with little ATP breakdown by active skeletal muscle. *J. Mechanochem. Cell Motility* **3**: 147–154.

Davison, W., Goldspink, G. & Johnston, I. A. (1976). The division of labour between fish myotomal muscles during swimming. *J. Physiol., Lond.* **263**: 185–186.

Dawson, T. J. & Taylor, C. R. (1973). Energetic cost of locomotion in kangaroos. *Nature, Lond.* **246**: 313–314.

Edwards, R. H. T., Hill, D. K. & Jones, D. A. (1975). Heat production and chemical changes during isometric contractions of the human quadriceps muscle. *J. Physiol., Lond.* **251**: 303–315.

Fedak, M. A., Pinshow, B. & Schmidt-Nielsen, K. (1974). Energy cost of bipedal running. *Am. J. Physiol.* **227**: 1036–1044.

Gaesser, G. A. & Brooks, G. A. (1975). Muscular efficiency during steady-state exercise in effects of speed and work rate. *J. appl. Physiol.* **38**: 1132–1139.

Goldspink, G. (1975). Biochemical energetics for fast and slow muscle. In *Comparative physiology. Functional aspects of structural materials*: 173–185. Botis, C., Maddrell, S. H. P. & Schmidt-Nielsen, K. (Eds). North Holland. Elsevier.

Goldspink, G. (1977a). Mechanics and energetics of muscle in animals of different sizes with particular reference to the muscle fibre composition. In *Scale effects in animal locomotion*: 37–55. Pedley, T. J. (Ed.). London & New York: Academic Press.

Goldspink, G. (1977b). Energy cost of locomotion. In *Mechanics and energetics of animal locomotion*: 153–167. Alexander, R. McN. & Goldspink, G. (Eds). London: Chapman and Hall.

Goldspink, G., Larson, R. E. & Davies, R. E. (1970). Thermodynamic efficiency and physiological characteristics of the chick anterior latissimus dorsi muscle. *Z. vergl. Physiol.* **227**: 848–850.

Goldspink, G., Mills, C. & Schmidt-Nielsen, K. (1978). Electrical activity of the pectoral muscles during gliding and flapping flight in the herring gull, *Larus argentatus. Experientia* **34**: 862–864.

Gollnick, P. D., Karlsson, J., Piehl, K. & Saltin, B. (1974). Selective glycogen depletion in skeletal muscle fibres of man following sustained contractions. *J. Physiol., Lond.* **241**: 59–67.

Gollnick, P. D., Piehl, K. & Saltin, B. (1974). Selective glycogen depletion pattern in human muscle fibres after exercise of varying intensity and at varying pedalling rates. *J. Physiol., Lond.* **241**: 45–57.

Henneman, E. & Olson, C. B. (1975). Relations between structure and function in the design of skeletal muscle. *J. Neurophysiol.* **28**: 560–580.

Hudson, R. C. L. (1972). On the function of the white muscles in teleosts at intermediate swimming speeds. *J. exp. Biol.* **58**: 509–522.

Johnston, I. A. & Goldspink, G. (1973a). Quantitative studies of muscle glycogen utilisation during sustained swimming in crucian carp. (*Carassius carassius* L.) *J. exp. Biol.* **59**: 607–615.

Johnston, I. A. & Goldspink, G. (1973b). A study of glycogen and lactate in the myotomal muscles and liver of the coalfish *Gadus virens* L during sustained swimming. *J. mar. biol. Ass. UK.* **53**: 17–26.

Lasiewski, R. C. (1963). Oxygen consumption of torpid, resting, active and flying hummingbirds. *Physiol. Zool.* **36**: 122–140.

Margaria, R. (1968). Positive and negative work performances and their efficiencies and human locomotion. *Int. Z. angew. Physiol.* **25**: 339–351.

Matsumoto, Y., Hoekman, T. & Abbott, B. C. (1973). Heat measurements associated with isometric contraction in fast and slow muscles of the chicken. *Biochem. Physiol.* **46A**: 785–797.

McMahon, A. (1977). Scaling quadrupedal galloping: frequency stresses and joint angles. In *Scale effects of animal locomotion:* 143–152. Pedley, T. J. (Ed.). London & New York: Academic Press.

　　　　　　　　　　G. Goldspink

Pugh, L. G. C. E. (1975). The relation of oxygen intake and speed in competition cycling and comparative observations on the bicycle ergometer. *J. Physiol., Lond.* **241**: 795—808.

Talesara, G. L. & Goldspink, G. (1978). A combined histochemical or biochemical study of myofibrillar ATPase in pectoral, leg and cardiac muscle of several species of bird. *Histochem. J.* **10**: 695—710.

Taylor, C. R., Oniel, R., Fedak, M. & Schmidt-Nielsen, K. (1971). Energetic cost of running and heat balance in a large bird, the rhea. *Am. J. Physiol.* **221**: 597—601.

Taylor, C. R., Schmidt-Nielsen, K. & Raab, J. L. (1970). Scaling of energetic cost of running to body size in mammals. *Am. J. Physiol.* **219**: 1104—1108.

Taylor, C. R., Shkolnik, A., Daniel, R., Baharar, D. & Borat, A. (1974). Running in cheetahs, gazelles and goats: energy cost and limb configuration. *Am. J. Physiol.* **227**: 848—850.

Tucker, V. A. (1968). Respiratory exchange and evaporative water loss in the flying budgerigar. *J. exp. Biol.* **48**: 67—68.

Tucker, V. A. (1972). Metabolism during flight in the laughing gull *Larus atricilla. Am. J. Physiol.* **222**: 237—245.

Symp. zool. Soc. Lond. (1981) No. 48, 239–267

Stance and Gait in Tetrapods: An Evolutionary Scenario

STEPHEN C. REWCASTLE

39, Christmaspie Avenue, Wanborough, Guildford, Surrey, England

SYNOPSIS

Two forms of stance and locomotion exist in modern tetrapods. Limb motion is parasagittal in the larger and cursorial mammals (erect limb posture), and non-parasagittal in amphibians, reptiles and smaller mammals (sprawling limb posture). The general characteristics of these two locomotor categories are discussed, with particular attention to hindlimb structure in lizards.

The sprawling stance is adaptive for small tetrapods because of its stability, and may be of special value in climbing. It does not appear to be adaptive for large or cursorial tetrapods because of problems of support, and vertical displacements of the centre of mass are not produced by limbs moving in a near-horizontal plane. Thus, the suspended phase in sprawlers is short, and a gallop is absent. In mammals, the ability to gallop is related to sagittal bending in the lumbar region of the vertebral column, a faculty apparently never developed in reptilian lineages. However, adoption of a bipedal stance when running promotes a long suspended phase.

Early reptiles possessed a suite of hindlimb features comparable to modern lizards, particularly asymmetrical feet and, in advanced forms, an opposable fifth digit. Such features were retained in lizards, and appear to represent basic climb-in adaptations. Loss of the fifth digit is characteristic of thecodonts, and correlates with the evolution of erect limb posture and possibly ground-dwelling cursorial habits.

INTRODUCTION

It has been recognized for some time that two basic types of stance and locomotion exist amongst modern tetrapods, broadly characterized as "sprawling", in amphibians and reptiles, and "erect" in mammals, and dinosaurs and their avian descendants. The sprawling stance is generally defined as one in which the limbs have a net lateral orientation, or more particularly, that excursion of the propodials is restricted to a horizontal or near horizontal plane. In the erect stance, in contrast, the limb axes are vertical, and limb excursion is largely confined to a parasagittal plane. The classic studies of

Gregory & Camp (1918), Romer (1922) and Schaeffer (1941) indicated that the sprawling stance was ancestral for the tetrapods, and that the erect stance arose independently in both "sauropsid" (i.e. reptilian) and "theropsid" (i.e. mammalian) lineages.

In the literature dealing with the evolution of tetrapod locomotor systems the development of erect stance is frequently viewed, retrospectively, as an "improvement" (e.g. Romer, 1956; Charig, 1972). It is generally assumed that the erect stance is more efficient in supporting the body, and that "fore-and-aft" limb motion enhances cursorial ability. The corollary of such observations, sometimes implied and sometimes unequivocally stated (Desmond, 1975), is that the sprawling stance is energetically and mechanically inefficient. Thus, the sprawling type of locomotion tends to be viewed as a "grade" of locomotor organization, subordinate to the erect "grade", rather than as a purely different type of locomotor adaptation. The somewhat chauvinistic attitude towards the "superiority" of the erect stance has been rather reinforced in recent years by suggestions concerning the possible physiological correlates of these types of locomotion. Ostrom (1969a) and Bakker (1971) have argued that the erect gait reflects possession of an endothermic and homeothermic metabolism, implying a causal relationship between gait and physiology although the polarity of this relationship is not well clarified. Bakker's (1971) argument seems to be that the erect gait enhances foraging ability (necessary to fuel an elevated metabolism) by improving the cursorial faculties of speed and endurance. Bakker also implies that support of a large body by sprawling limbs increases the energetic expense of body support to a critical level where the additional expense of endothermy cannot be borne. In defence of Bakker's thesis, Dodson (1974) has construed non-attainment of erect posture in reptiles as a "failure". Once again, the implication is that the adaptive scope of reptiles (excluding dinosaurs) has been attenuated by the sprawling limb posture, in this instance by the preclusion of endogenic thermal homeostasis.

Bennett & Dalzell (1973) have provided an elegant critique of Bakker's (1971) paper, and have suggested that, for the correlation between gait and physiology to hold, one might expect a pronounced difference in the energetic cost of locomotion between sprawling and erect forms; specifically, that the erect stance should be more expensive and thus liberate more heat. (One might argue that the erect stance should be *less* expensive given the energetic burden of nonmuscular thermogenesis, an important heat source in mammals.) But Bakker (1972) has shown that the energetic cost of locomotion in lizards is equal to or somewhat less than that for mammals of similar size running at similar speeds. The essential difference seems to be in endurance, as lizards typically tire quickly when running at speed, although this probably relates to a considerable dependence on anaerobic glycolysis for strenuous activity (Bennett & Dawson, 1976). Regal (1978) also suggests that the incomplete double circulation of lizards may also restrict strenuous exercise, as the separation of pulmonary and systemic circulations may break down with increasing heart beat rate. In summary, it would appear that there is no *a priori* energetic evidence for the sprawling gait being mechanically inferior to the erect.

In a comparative study of activity levels and behaviours of lizards and mammals, Regal (1978) has stressed the importance of viewing lizards and mammals as members of largely discontinuous "adaptive zones" (*sensu* Simpson, 1953; Van Valen, 1971). Without this perspective, comparisons between the two groups may be misleading since one is comparing inherently different suites of adaptations. The concept of "adaptive zone" is, like the "niche", rather nebulous in that the sum total of adaptations of a taxon define the adaptive zone. Nevertheless, the occurrence of "key innovations" (*sensu* Liem & Osse, 1975), serving as the focal points (and possibly determinants) of subsequent adaptive diversity or character, seem to play a major role in the origin and radiation of higher taxons, and such characters may thus specify adaptive zones (Van Valen, 1971; Russell, 1979).

Body size is clearly an important parameter of an adaptive zone, and size has a direct effect on locomotor capabilities (Coombs, 1978). Small size places energetic constraints on endurance and absolute speed (Schmidt-Nielsen, 1972; Taylor, Schmidt-Nielsen & Raab, 1970; Tucker, 1970), and so small size tends to preclude enhancement of "cursorial ability". Thus, animals deemed "good" cursors are generally moderate to large sized animals (Coombs, 1978). Furthermore, a given kind of terrain may have very different "qualities" for animals in different size classes, as pointed out by Robinson (1975), and small size to a certain extent demands a degree of scansorial ability. The conclusion that a given body size precludes some adaptations whilst potentiating others is inescapable.

By and large, lizards (the most numerous and varied assemblage of sprawling tetrapods) are restricted to the small-size class. Hard data are not available, but one might make an educated guess that the great majority of species mass less than a kilogram. Given this, one would not expect widespread cursorial adaptations within the group; at least, not in the form of cursorial adaptations seen in mammals (Gregory, 1912: Smith & Savage, 1955). Small size seems to correlate with the sprawling limb posture, in modern faunas at least.

Jenkins (1971) has shown that in small mammals, limb motions may be closely comparable to those seen in lizards; i.e. propodial motion is markedly non-parasagittal. Thus, parasagittal limb motion is not necessarily typically mammalian but it is typical of cursors.

This study will focus attention largely on features of limb structure and locomotion in sprawling tetrapods. The aim is to generate a broad comparison between these two types of locomotion in an attempt to identify their basic functional characteristics. It is hoped that this will contribute towards an understanding of the evolution and adaptive radiation of the tetrapod locomotor system.

MAJOR FEATURES OF STANCE AND GAIT

In reptiles and amphibians the humerus and femur tend to be horizontal and laterally directed, determined in large part by the terminal position of the proximal articular surfaces on the shafts, and the postero-lateral facing glenoid surface (Romer, 1922; Ostrom, 1969a; Bakker, 1971). In lizards, extreme adduction of the femur is also precluded by the expansion, proximo-ventrally, of the internal trochanter (Fig. 1) (Snyder, 1954; Charig, 1972). In both theropsids and advanced thecodonts, the shift to erect posture, with vertical orientation of the propodials, was effected chiefly by the development of a mesially inturned femoral head, and a ventral-facing glenoid.

The distinction between the sprawling and erect modes of locomotion is therefore primarily postural and concerns support. But there is also a kinematic distinction, the term "kinematic" referring to kinds of motion and the geometric effects of motion (MacConaill & Basmajian, 1969). The erect stance can be described as kinematically two-dimensional since motions of all limb segments are largely restricted to a single (parasagittal) plane. The sprawling stance is, in contrast, kinematically three-dimensional, the limb segments moving in different planes in space, and in some instances with differing senses of motion. Finally, there may be real differences in gait between sprawling and erect tetrapods. A "gait", as defined by Hildebrand (1974), is "a regularly repeating sequence and manner of moving the legs in walking or running", and specific gaits are usually defined on the basis of footfall patterns (Muybridge, 1887; Howell, 1944; W. F. Walker, 1972). Amphibians, reptiles and mammals may show similar footfall patterns during locomotion (Howell, 1944; Snyder, 1952; Sukhanov, 1968), but their gaits may differ in parameters of stride and step, and in the nature of the contribution by vertebral flexion (Slijper, 1946; Gray, 1968). These three aspects of locomotion (support, kinematics and gait) will be addressed in turn.

Posture and Support

The most frequently cited "deficiency" of the sprawling limb posture is that more muscular effort is required to support the body when slung between the epipodials than if the limb were vertically oriented. Thus, as analysed by Nauck (1924), a bending moment tending to collapse the limb exists about the knee since the knee is at some horizontal distance from the hip (Fig. 1). In contrast, the mammalian limb is viewed as a vertical compression member, and support of the body is presumed to be by the limb skeleton alone. But, as Bennett & Dalzell (1973) point out, in both mammals (excluding graviportal forms) and reptiles there is generally one flexed joint in each limb, and so in both there is a tendency for limb collapse. Jenkins' (1971) data for small mammals, demonstrating that both knee and elbow joints tend to be acutely flexed in both a static pose and during locomotion, suggest that the muscular effort required to prevent collapse of a flexed limb is likely to be of comparable magnitude in both reptiles and mammals of equivalent size. Romer (1956) and Bennett & Dalzell (1973) have argued, however, that the sprawling limb posture is non-adaptive for large animals because of the large shearing stresses exerted on a non-vertical limb element. Large tetrapods are posturally erect.

This static consideration of support is somewhat inadequate. In large mammals, joint-locking mechanisms prevent limb collapse whilst at rest, and in reptiles a resting posture (as when basking) is one without a limb support, the animal resting on its belly. During locomotion controlled "limb collapse" is advantageous in large animals since it prevents impact loading of limb bones. Furthermore, a mechanism is provided for the deformation of elastic tendons by which potential and kinetic energy may be stored as strain energy, released on extension of the limb joint (Alexander, Maloiy, Njau & Jayes, 1979; Alexander, Maloiy, Hunter, Jayes & Nturibi, 1979). Such mechanisms provide a means of energy saving during locomotion. In short, the "support problem" needs examining from a different perspective.

Along more positive lines, the sprawling posture provides a highly stable support since the body's centre of mass lies well within the support quadrilateral of the feet. The habitual symmetrical form of gait of sprawling tetrapods is also a stable one, especially at low speeds (Gray, 1944). Stability would appear to be of prime adaptive value in small animals since, being of low mass, they have low inertia and so are easily toppled by small forces. In climbing, stability is of

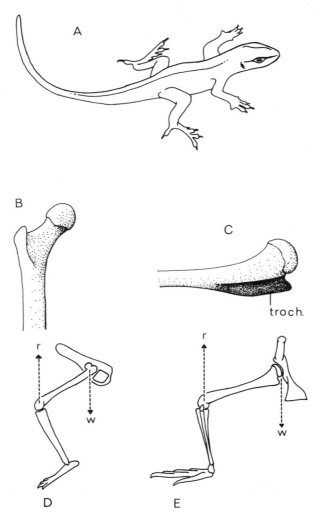

FIG. 1. (A) *Anolis grahami* in "resting" posture with limbs maximally sprawled (from photograph). (B) Posterior view of typical mammalian femur with mesially offset femoral head. (C) Posterior view of typical reptilian femur with terminal femoral head. (D) Mammalian limb, lateral view left, and (E) reptilian limb, posterior view, left. In both the reaction at the knee (r) acts at some horizontal distance from force imposed by body weight (w).

importance and, since the foot is placed on the substrate lateral to the body in a climbing lizard, limb usage can proceed even when the body is close to, or rests upon, the substrate. The stability of the sprawling posture may also be of value in "cursorial" forms with markedly elongate limbs — a small long-legged tetrapod with erect posture, and hence a narrow trackway, would appear to be highly unstable.

Osteokinematics

During parasagittal limb retraction the limb develops longitudinal thrust parallel to (but in the reverse sense of) the direction of progression. Because the distal end of the limb describes an arc about the acetabulum (or glenoid), the thrust developed also has a vertical component producing a vertical displacement of the centre of mass during one step (Alexander & Jayes, 1978a, b). Such displacements of the centre of mass promote a suspended phase between successive steps and so provide a mechanism for increasing stride length. Because all the limb segments move in one plane, "functional" limb length approximates to total limb length and so, in the absence of a suspended phase, stride length will be a reflection of total limb length. In addition, during retraction the orientational relationships of the axes of the major limb joints, relative to a parasagittal plane, do not alter, remaining perpendicular to that plane. So the flexion—extension arcs of all limb segments are longitudinal.

In sprawling tetrapods, and in small mammals (Jenkins, 1971), fore- and hindlimbs tend to remain acutely flexed at elbow and knee during retraction. Femoral (and humeral) retraction is in a horizontal or near-horizontal plane, whilst the crus is moving in a near-vertical plane. The sprawling gait would thus appear to differ from the erect gait in two important respects. First, femoral retraction produces a force with potential posterior and lateral components, since the knee is describing an arc about the acetabulum in a horizontal plane. Displacements of the centre of mass thus tend to be horizontal rather than vertical. Secondly, functional limb length would appear to be less than total limb length because of limb flexure at knee and elbow. Kinematically, the sprawling gait differs from the erect gait in that the orientation of the axes of the major limb joints, relative to a parasagittal plane, alters during retraction (Fig. 2). At the beginning of

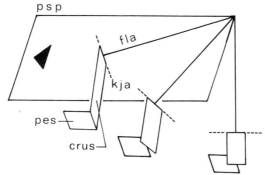

FIG. 2. Kinematics of horizontal femoral retraction. As the femur is retracted the axis of the knee joint (kja) tends to become parallel to a parasagittal plane (psp). fla — femoral long axis.

retraction, the femoral long axis is antero-laterally directed, making an angle of between 20° and 40° to a parasagittal plane (Sukhanov, 1968; Jenkins, 1971). In such a position the knee joint axis trends antero-mesial to postero-lateral. As retraction proceeds, the knee axis tends to become parallel to the body axis, and so the plane of limb flexure becomes transverse. This demands that rotation occur between knee and pes, since the knee is pivoting about some fixed point on the substrate. The rotation is expressed as a rotation of the crus about its long axis, the sense being anti-clockwise in the case of the left limb viewed proximally, or "lateral" rotation. The axial rotation of the crus can be further defined as a "conjunct" rotation, since it is an indissociable geometric effect of the motion of a separate element (Barnett, Davis & MacConaill, 1961; MacConaill & Basmajian, 1969). A conjunct rotation is distinguishable from an "adjunct" axial rotation, motion of an element about its long axis not induced by the separate motion of another element. Both types of rotation are marked in sprawling locomotion, and their consequences will be discussed later.

The above description of limb movement in sprawling tetrapods is highly stereotypic, but serves as a model with reference to which morphological analysis may proceed. In particular, the model suggests a number of mechanical "problems" which concern the operation of the distal limb (Rewcastle, 1980, in press). The "solutions" to such "problems" are discussed in the next section.

Gait, Axial Motion and Limb Proportions

The characteristic gait of both lizards and urodele amphibians is a symmetrical trot in which contralateral fore- and hindlimbs alternate in body support (Howell, 1944; Snyder, 1952). At very low speeds, however, a lateral sequence walk, in which ipsilateral feet move in pairs and three feet form a support, is seen (Gray, 1944; W. F. Walker, 1972; Edwards, 1977). Thus the gaits of both mammals and "lower" tetrapods are closely comparable, but as Sukhanov (1968) has noted, the trot in lizards is one in which the hindlimbs predominate. That is, the hind feet contact the substrate for a greater percentage of the total period of a locomotor cycle than do the forefeet, particularly with increasing speed (Urban, 1965).

Speed increase, in lizards, is associated primarily with an increase in step length[*] and some increase in limb retraction velocity (Snyder,

---

[*] Step length is used here to define the distance the animal moves during one period of foot/substrate contact. Stride length is the distance travelled between successive contacts of one foot with the substrate.

1952). Stride length may equal or slightly exceed snout-vent length (Sukhanov, 1968; Daan & Belterman, 1968) in fast runners, and although there are periods when only one foot is on the ground there is no tendency for a prolonged suspended phase. Thus step frequency may remain constant over the range of running speeds, or show only a slight decrease with increasing speed (Daan & Belterman, 1968). In mammals, speed increase is associated with an increase in the length of the suspended phase, and in particular the transition to an asymmetrical gallop which promotes an increase in stride length and a decrease in step frequency. The gallop has not been observed in lizards, but has been described by Cott (1961) in small Nile crocodiles, and fortuitously filmed in a small *Crocodylus porosus* by Zug (1974).

The ability to gallop may be closely related to the faculty for sagittal vertebral flexion—extension. In a galloping mammal the spine is maximally flexed as the hindfeet strike the ground, extending during retraction of the hindlimbs. Extension of the spine increases step length, stride length, and hindlimb retraction velocity (Hildebrand, 1959). A second functional requirement for gallopers concerns limb proportions (see below).

Sagittal vertebral bending is characteristic of the lumbar vertebrae in mammals, permitted by the vertical longitudinal orientation of the zygapophyseal facets (Slijper, 1946), most marked in carnivores (Savage, 1977). Truncal vertebrae of reptiles and amphibians differ from those of mammals not only in the lack of vertebral differentiation, but in the horizontal or inclined (radial) orientation of the zygapophyses (Romer, 1956; Hoffstetter & Gasc, 1969). Flexibility of the vertebral column in "lower" tetrapods is thus largely horizontal. Aside from the cervical vertebrae, only in crocodilians is there seen a tendency towards vertical zygapophyses, notably on the first caudal vertebra between both sacrum and second caudal (Hoffstetter & Gasc, 1969). Zug's (1974) records of crocodilian galloping indicate that maximum flexion of the body is in the sacral region, suggesting that the whole trunk is extended on the tail base during hindlimb retraction (forelimb protraction). By analogy with Snyder's (1949, 1954) arguments concerning the role of the tail as a cantilever in bipedal lizards, one might suggest that the tail acts as a cantilever during the crocodilian gallop. Ostrom (1969b) suggests a similar role for the highly specialized tail of the small theropod *Deinonychus*. Interestingly, the anterior caudal vertebrae in this animal (particularly the fifth as illustrated by Ostrom) show development of vertical zygapophyses.

A regionally well-differentiated vertebral column and lumbar

vertebrae with vertical zygapophyseal facets appear to be uniquely
mammalian features. Thecodonts (forms generally accepted as close
to the ancestry of the archosaurs) are closely comparable to modern
lizards in vertebral construction, and there seems no tendency towards regional differentiation of the column (Cruickshank, 1972;
Ewer, 1965; Hughes, 1963; Romer, 1972a, b, c, d). The zygapophyseal facets are typically inclined and never vertical. Advanced
thecodonts were posturally erect, indicated by the presence of a
mesially-inturned femoral head (as in *Lagerpeton* and *Lagosuchus*,
Romer, 1971, 1972d; *Lewisuchus*, Romer, 1972c), and crocodilians
may also have achieved erect stance (e.g. *Hallopus*, A. D. Walker,
1970). The above-mentioned genera were small, seemingly cursorial,
animals, yet their vertebral construction is not indicative of galloping
abilities; at least, sagittal vertebral flexion—extension would have
been very limited. Amongst both bipedal and quadrupedal dinosaurs,
vertebral construction usually reflects rigidity in the trunk region,
with a strong tendency for ossification of tendons of the spinal
musculature, horizontal zygapophyses and development of hypo-
sphene—hypantrum articulations. These considerations indicate (but
do not conclusively demonstrate) that the faculty for galloping did
not develop in the archosaur lineage. Why a mammalian-type
vertebral column did not appear in the thecodont—archosaur line is
not clear, and it may be that such an innovation in mammals was
serendipitous. It might be suggested that the initial trend towards
vertebral differentiation could be the localization of rib-bearing
vertebrae to the thoracic region, correlated with the appearance of
diaphragmatic breathing (Brink, 1954).

    A spine capable of sagittal flexion—extension may be advantageous
for galloping tetrapods (especially small- and medium-sized forms),
but it is not a mechanical essential. A galloping tetrapod can be
viewed as two bipeds hopping one behind another (Alexander &
Jayes, 1978b), with the forefeet moving half a cycle out of phase
with the hindfeet (i.e. with the forefeet at the end of retraction, the
hindfeet at the beginning of retraction). As long as parameters of
stride, step and force output are equal for fore- and hindlimbs, no
force is transmitted between fore- and hindlimbs (R. McN. Alexander,
personal communication), and so there is no mechanical requirement
for any form of flexible member linking fore- and hindlimbs. However, for a tetrapod to gallop in this way it is important that the
phase relationships of fore- and hindlimbs should be maintained.
Mammals, and particularly those that gallop with a relatively rigid
spine, may achieve this simply by possessing fore- and hindlimbs of
equal length so that forelimb step and hindlimb step are potentially

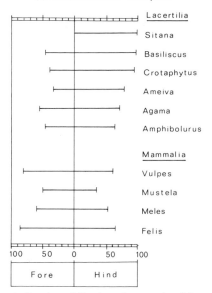

FIG. 3. Limb proportions in lizards and (for broad comparison) in mammals. Fore- and hind-limb length is expressed as a percentage of snout-vent length in lizards. Data for mammals from Savage (1977), the latter expressing limb length as a percentage of presacral column length. Lizard species — *Sitana ponticeriana; Crotaphytus collaris; Ameiva dorsalis; Agama ?agama; Amphibolurus barbatus.*

equal. The limb proportions of lizards, however, are very different. Typically, hindlimb length exceeds forelimb length to a marked extent, particularly in fast-running lizards (Russell & Rewcastle, in preparation) (Fig. 3, Table I), where the hindlimb may be between 40% and 50% longer than the forelimb. As both Snyder (1952) and Sukhanov (1968) have noted, the hindfoot strikes the ground before the contralateral forefoot and leaves the ground after it during a symmetrical trot. Hindlimb step length is therefore greater than fore-limb step, but the phase relationships of fore- and hindlimbs are maintained by delaying forelimb protraction. Thus the forefoot drags during the early stages of retraction of the homolateral hindfoot. Increase in speed is achieved primarily by increase in hindlimb step length, so the period during which the forefoot drags increases (Snyder, 1952; Urban, 1965; Daan & Belterman, 1968). That is, a fast-running lizard is, in the broadest sense, bipedal.

Considering the symmetrical pattern of gait in association with the sprawling limb posture, it is possible to suggest why, in lizards, there should be a disproportion in length between fore- and hindlimbs. At the beginning of hindlimb retraction the pes is placed on the sub-strate lateral to, or in advance of, the homolateral manus, the latter being at the end of retraction. (Note that hindlimb length may be more than 80% of snout-vent length, and often exceeds the distance between glenoid and acetabulum.) During protraction, the manus swings forwards as well as outwards (Renous & Gasc, 1977), since motions of the humerus are largely in a horizontal plane. Thus if the excursion arcs of fore- and hindlimbs overlapped to any great extent the hindlimb would interfere with protraction of the forelimb.

Long hindlimbs and short forelimbs would seem to be a functional requirement for sprawling cursorial tetrapods, and such limb proportions would seem to preclude the possibility of galloping. Small crocodiles may be able to gallop because the forequarters can be raised on the tail cantilever during hindlimb retraction. This compensates for the tendency of the simultaneous thrust of both hindlimbs to topple the hindquarters over the forequarters, which are unsupported during hindlimb retraction.

Most fast-running lizards are potential bipeds, at least in terms of limb proportions. This is reflected in the widespread occurrence of facultative bipedalism recorded amongst lizard genera (see Snyder, 1949, 1952, 1954; Sukhanov, 1968). Limb proportions of thecodonts were closely comparable to those of modern lizards (Table I), which has led to the view that a bipedal "trend" was well established in dinosaur ancestors. But such limb proportions are not necessarily indicative of bipedal abilities — as noted, disproportionately long hindlimbs may be a functional requirement for cursorial, quadrupedal sprawlers. Such limb proportions are, however, pre-adaptive for bipedal progression. Bipedal running may be advantageous for tetrapods which cannot gallop, or at least lack the spinal structure for efficient galloping. Clear data is not available, but the bipedal stance in lizards may increase the duration of the suspended phase; Snyder (1952) records a stride length of three times snout-vent length in bipedal lizards. It seems of some significance that the smaller and faster dinosaurs were bipedal genera (Coombs, 1978; Alexander, 1976). The choice between bipedal or quadrupedal habits for cursors might be determined by the presence or absence of a sagittally flexible vertebral column.

HINDLIMB STRUCTURE AND LOCOMOTION IN "LOWER" TETRAPODS

Hindlimb structure and function show great diversity amongst modern reptiles and amphibians. However, within the major taxa, specific types of locomotor adaptation are not as striking or as amenable to analysis as the locomotor adaptations of mammals. Within taxa one gains the impression that single key innovations have been the focus of subsequent locomotor developments, or have potentiated certain kinds of locomotor adaptations. Thus, the Anura are systematically diverse but (in gross detail) morphologically homogeneous, especially in the form of the locomotor system. The key innovation may be identified as a mobile ilio-sacral articulation, the development of which was foreshadowed (or potentiated) by the heritage character of one pair of sacral ribs in the Amphibia (Whiting, 1961). This innovation seems to be central to the structural complex facilitating both jumping and swimming. Similar arguments concerning the development and diversification of sub-digital pads in geckoes have been forwarded by Russell (1976, 1979).

In this section, two major kinds of limb structure and locomotion will be examined, as exemplified by urodele (caudate) amphibians, and lizards.

Urodeles

Schaeffer (1941) has suggested that the urodele tarsus, if not homologous on all points of detail to the tarsus of amphibians ancestral to reptiles, is at least structurally analogous to it. In both urodeles and ancestral amphibians the tarsus consists of a mosaic of small elements without a well-defined region of cruro-pedal articulation. This structure, with many discrete inter-tarsal, cruro-tarsal and tarso-metatarsal articulations, is inherently flexible but tends to be structurally unstable (Schaeffer, 1941).

In urodeles femoral excursion is largely in a horizontal plane, although at speed some adduction occurs, leading to a narrowing of the trackway and an increase in stride and step (Peabody, 1959). In horizontal femoral retraction, the angular change in orientation of the knee joint (relative to a parasagittal plane) is equal to the arc through which the femur is retracted, provided that the knee axis remains horizontal during retraction. This demands an equal amount of conjunct rotation of crus on pes. Conjunct rotation appears to occur via a torsion of crus and tarsus about the long axis of the epipodial, such that the tibia comes to cross obliquely in front of the fibula (Fig. 4). Such motions, essentially passive, are comparable to forearm rotations as seen in man (MacConaill & Basmajian, 1969) and lizards (Renous & Gasc, 1977).

Haines (1942) argued that rotation of the crus of "lower" tetrapods occurred chiefly as a result of independent (adjunct) axial rotations of tibia and fibula, demanding that rotation occur at the knee. However, Carleton (1941) had previously suggested that the knee in tetrapods is strictly a hinge joint with rotation as a part of a joint-locking mechanism in mammals. As Carleton (1941) points out,

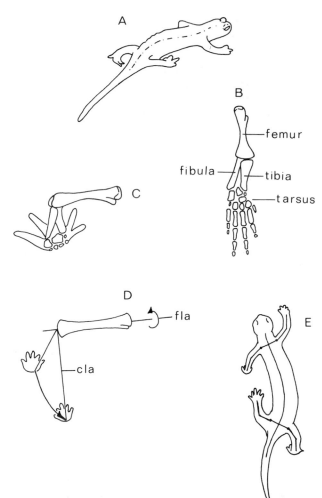

FIG. 4. Major features of urodele limbs and locomotion. (A) *Salamandra salamandra* (from photograph). (B) Right limb, *Salamandra*, dorsal view (drawn from alizarin transparency). (C) Left limb, dorsal view, with femur fully retracted. (D) The "double crank" system, with senses of motion of femur and crus indicated. fla, femoral long axis; cla, crus long axis. (E) role of lateral undulation in promoting limb retraction by girdle rotation.

urodeles (and lizards) lack extensively developed tibio-fibular muscles which might promote adjunct axial rotations of either element.

During limb retraction the foot remains stationary on the substrate, extending on the crus at the end of retraction. The long axis of the foot (usually between the third and fourth digits) is directed forwards during locomotion (Schaeffer, 1941; Peabody, 1959; Sukhanov, 1968), as seems to have been the case in ancestral amphibians (Romer & Byrne, 1931). Robinson (1975) describes the primary role of the urodele foot as one of gripping the substrate. The foot is symmetrical, with the tips of the digits lying on a semi-circle. A high degree of flexibility within the metatarsus is possible, permitting divergence of the first and fifth digits from the long axis of the pes. Upon flexion, these digits come into opposition, and the foot forms a symmetrical grip.

Locomotion is accomplished by three different kinds of motion of the axial and appendicular skeleton. Propodial retraction contributes between 56% and 62% of step length, but decreases in importance with increasing speed (Edwards, 1977). Propodial adjunct axial rotation (being a clockwise rotation in the case of the left limb viewed proximally) also contributes between 26% and 28% of step length (Fig. 4). This system, termed a "double-crank" by Barclay (1946) is analogous to a wheel (Gray, 1968) in which the propodial axial rotation is transferred into a conjunct motion of epipodial retraction.

Lateral undulation of the vertebral column is an important component of urodele locomotion, especially at moderate and high speeds of progression. Daan & Belterman (1968) show that the contribution of lateral undulation may be "direct", in which lateral undulations consist of travelling waves developing locomotor forces transferred to the limbs. Alternatively, the contribution may be "indirect", in which case lateral undulations produce standing waves promoting rotation of the limb girdles in a horizontal plane about a vertical sagittal axis. Girdle rotation provides a mechanism for increasing the retraction angles of humerus and femur, increasing step and stride length and increasing the force output of the limbs. Daan & Belterman (1968) calculate that girdle rotation contributes 43% and 33% of step length in fore- and hindlimbs respectively.

Edwards (1977) has shown that lateral undulation in salamanders has pronounced travelling wave characteristics, especially at highest speeds. This contrasts with Daan & Belterman's (1968) conclusions that, in lizards at least, standing waves predominate at high speeds. Edwards demonstrates that waves of undulation passing from pectoral to pelvic girdles can promote protraction—retraction of contralateral limbs. Edwards (1977) terms this gait the "travelling wave trot", and points out that it is only feasible in combination with the sprawling limb posture because of the stability of the posture and the horizontal orientation of the propodials. His conclusion is that the inheritance in early tetrapods of a lateral undulation system capable of promoting travelling waves may have resulted in selection pressures favouring retention of the sprawling limb postures.

Lizards

In lizards, locomotion may be either "limb-dependent", in which axial undulation is of relatively minor importance, or "axial-dependent", in which the limbs show marked reduction and axial undulations promote travelling waves (Daan & Belterman, 1968). Limb-dependent forms and especially "cursorial" species, are typically "short-coupled" (Peabody, 1959) with a relatively short inter-girdle region and relatively long hindlimbs (Snyder, 1954; Russell & Rewcastle, 1979, and in preparation). Presacral vertebral counts typically lie between 20 and 30 vertebrae, whilst in axial-dependent species presacral counts exceed 40 vertebrae (Hoffstetter & Gasc, 1969). Limb reduction and loss accompanying increase in presacral count has occurred several times in squamates (Essex, 1927; Lande, 1978), and Gans (1975) has indicated that this phenomenon correlates with either burrowing or detritus dwelling. It should be noted that reduction in limb length (relative to snout-vent length) is not equivalent to limb degeneration, and does not necessarily indicate axial dependence. The skink *Tiliqua* has remarkably short and apparently feeble limbs yet, as Daan & Belterman (1968) have shown, axial undulation is not marked during locomotion.

As a general statement it would appear that limbed lizards are capable of utilizing two different forms of limb configuration for different types of progression. In individuals the type of limb usage and locomotion may vary with both speed of progression and substrate type (Sukhanov, 1968; personal observations). In slow locomotion, in climbing or in locomotion on a shifting substrate (as in "sand-swimming") the hindlimb extends in a near-horizontal plane and the body is close to or rests upon the substrate. In fast locomotion on a level substrate, however, the femur becomes markedly adducted during retraction and the body is held well off the substrate. In the former, lateral undulation with travelling waves may predominate, but with increasing speed the amplitude of lateral undulation decreases, and standing waves predominate (Daan & Belterman, 1968). The key to understanding these two types of locomotion is in the morphology of the hip joint.

The proximal femoral condyle has a convex surface and, in proximal view, has a compressed oval outline (Fig. 5). The orientation of this oval, and the major axes of the hip, is most conveniently defined with reference to the femur horizontally oriented, perpendicular to a parasagittal plane and with the axis of the knee horizontal. The maximum extent of the articular surface, determining the sense of greatest motion at the hip, can be defined by the long axis of the oval of the condyle. This axis is inclined from antero-dorsal to

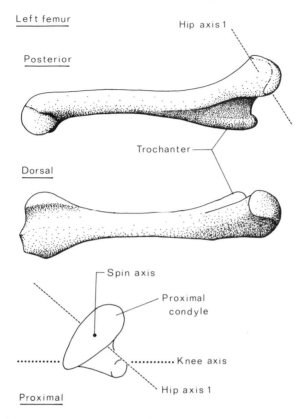

Left femur

Hip axis 1

Posterior

Trochanter

Dorsal

Spin axis

Proximal
condyle

Knee axis

Hip axis 1

Proximal

FIG. 5. Left femur, *Varanus*, in posterior (lateral), dorsal and proximal views. "Hip axis 1" is the principal axis of the hip, about which femoral retraction proceeds. The "spin axis" is the axis about which femoral long axis rotation proceeds.

postero-ventral. An axis perpendicular to this (and to the femoral long axis) determines the principal axis of the hip joint, i.e. the axis about which femoral retraction proceeds. Thus, the principal axis of the hip and the axis of the knee lie in different planes. In mammals exhibiting parasagittal limb motion these axes typically lie in the same plane, perpendicular to a parasagittal plane.

The principal motion of the femoral condyle within the acetabulum is a "slide" (*sensu* Barnett, Davis & MacConaill, 1961; Mac-Conaill & Basmajian, 1969), the external effect of which is a "swing" of the femur, as during retraction. But the femoral condyle can also "spin" within the acetabulum, the external effect of which is adjunct axial rotation of the femur. Both types of motion influence the orientation of the axis of the knee, and two kinds of femoral motion can be defined representing two kinematic and locomotory extremes. In the first, femoral retraction proceeds in a horizontal plane, and

represents the type of locomotion previously defined as slow or scansorial. In the second, femoral motion is in a non-horizontal, non-
parasagittal plane, and is characteristic of fast locomotion (Snyder,
1949, 1952; Urban, 1965).

In horizontal femoral retraction, medial axial rotation of the
femur (clockwise in the case of the left limb viewed proximally)
brings the principal axis of the hip into a vertical position. If the
femoral long axis is laterally directed, this motion constitutes retraction of the crus, and is equivalent to the "double-crank" system seen
in urodeles. The double-crank system is an important part of the
forelimb mechanism (Haines, 1952; Renous & Gasc, 1977), but may
have a rather different role in the hindlimb, especially in non-
horizontal retraction. Adjunct axial rotation of the femur brings
about inclination of the knee joint axis, from postero-dorsal to
antero-ventral (Fig. 6). Thus the crus slopes posteriorly and ventrally.
Retraction of the femur is accompanied by extension of the knee
and ankle, and so propulsive forces are developed in a near-horizontal
plane.

The major muscles involved in femoral retraction are the m. caudifemoralis longus and brevis (Snyder, 1954), originating on caudal
vertebrae, though developmentally differentiated from ventral
intrinsic limb muscles (Romer, 1942). These muscles insert together
via a broad tendon onto either side of the internal trochanter (Fig. 6).
Since this insertion is ventral to the femoral long axis the caudifemoralis is potentially a femoral long axis rotator. Also possibly involved in promoting axial rotation of the femur is part of the m.
pub-ischio-femoralis internus, originating within the girdle and passing out over its anterior rim. This slip passes back dorsal to the
femur and inserts adjacent to the insertion of the caudifemoralis.

Horizontal femoral retraction is particularly characteristic of
scansorial locomotion (Fig. 6). In climbing geckoes, however, limb
usage seems to be comparable to that of salamanders. Russell's
(1975) analysis of locomotion in *Gekko gecko* suggests that limb
retraction is largely as a result of rapid axial undulation, which may
indicate that the "travelling wave trot" is the form of gait used in
climbing.

In non-horizontal femoral retraction the axis of the knee is
horizontally oriented at the beginning of retraction. The sense of
motion of the femur, as defined by the sense of maximum extent of
the condylar surface, is downwards and backwards. Femoral adduction is, in part, a conjunct motion of retraction. As retraction proceeds, the axis of the knee suffers a change in orientation relative to
a parasagittal plane, but this is less than in the case of horizontal

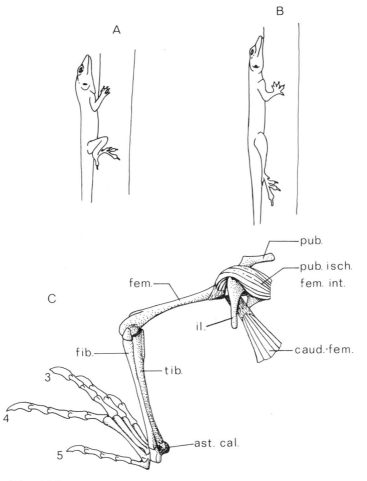

FIG. 6. (A) and (B) *Anolis garmani* climbing a vertical branch (from photographs). (C) Limb skeleton of a generalized lizard, left limb, dorsal view. Abbreviations — ast. cal., astragalo-calcaneum (proximal tarsal element); caud.-fem., m. caudifemoralis; fem., femur; fib., fibula; il., ilium; pub., pubis; pub. isch. fem. int., m. pubo-ischio-femoralis internus (in part); tib., tibula.

femoral retraction as described on pp. 245–246. That is, the angular change in orientation of the knee axis concomitant with retraction is less than the angle through which the femur is retracted. This kinmatic effect forms part of a general theorem of osteokinematics provided by MacConaill (1966), and the analysis of such motions is beyond the scope of this work. Of greatest relevance here is the observation that non-horizontal, non-parasagittal femoral retraction decreases the necessary amount of conjunct rotation between crus and pes. Concomitant with this, the tendency of the flexion—extension

plane of the crus to become transverse during femoral retraction is
also reduced.

Conjunct rotation of the crus, and re-orientation of the knee joint
axis, can theoretically be eliminated by an adjunct axial rotation of
the femur in a reverse sense to that of conjunct rotation. It has
already been noted that conjunct axial rotation of the crus has a
lateral sense (i.e. anti-clockwise in the case of the left limb viewed
proximally) and that adjunct axial rotation of the femur is medial in
sense. That is, adjunct axial rotation of the femur occurring when the
femur is inclined below the horizontal maintains a near-perpendicular
relationship of the axis of the knee to a parasagittal plane (Fig. 7).
The corollary is that the flexion—extension arc of the crus is essentially longitudinal, as in the case of parasagittal limb motion.

Analysis of additional features of the distal limb, in particular the
knee and mesotarsal joints and the structure and function of the pes,

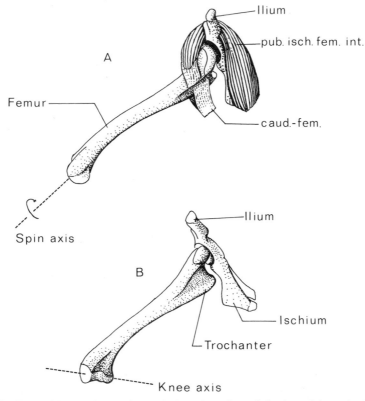

FIG. 7. Femoral long axis rotation and the orientation of the knee joint axis. Posterior
views, left limb and half-girdle (*Iguana*). (A) Femur depressed below horizontal, without
axial rotation having occurred. (B) Femoral long axis rotation re-orients the knee axis which
becomes sub-perpendicular to a parasagittal plane. Abbreviations — caud.-fem., m. caudifemoralis; pub. isch. fem. int., m. pubo-ischio-femoralis internus (in part).

suggest ways in which the operation of the distal limb has become enhanced in effective participation in locomotion (Rewcastle, 1977, 1980, in press). The origin and significance of such features is reviewed in the next section.

AN EVOLUTIONARY SCENARIO

A major feature of the early evolution of the hindlimb in the diapsid lineage was the consolidation of the tarsal mosaic, involving the loss or co-ossification of tarsal elements. The end result of this process was a proximal tarsal row of two major elements, an astragalus and calcaneum (Peabody, 1951), intimately articulated with one another and with tibia and fibula. Distal tarsalia became gradually reduced in number, and a well-defined mesotarsal joint, between the astragalus and calcaneum, proximally, and the fourth tarsale distally, developed. Occurring contemporaneously with tarsal consolidation was the development of an imbricate metatarsus, in which adjoining proximal ends of the metatarsals come to overlap, and the development of an asymmetrical foot comparable to that seen in modern lizards (Fig. 8). Schaeffer (1941) viewed such developments as reflecting increased dependence on the limbs as the major locomotor organs. Consolidation of the tarsus promoted a structural solidification of the crus, whilst imbrication of the metatarsus permitted the active use of the pes as an additional limb lever (Charig, 1972). However, such changes also reduced the inherent flexibility of the amphibian limb. One consequence was that the foot was carried with the digits directed antero-laterally, reflecting the overall lateral orientation of the limb axis (Schaeffer, 1941).

In modern lizards, the pes is asymmetrical with metatarsal and digit length increasing from the first to the fourth digits. Length increase is such that the tips of the first three metatarsals lie on one straight line, forming an acute angle with the pes long axis (Fig. 8). During locomotion, the pes is placed on the substrate with its long axis between 20° and 50° to a parasagittal plane, and this angle seems to vary between genera (Snyder, 1952), as does the geometry of the foot. Because the foot is asymmetrical, the lines of contact of the metatarsals and the ungual phalanges are near perpendicular to the direction of progression. Thus the foot grips the substrate roughly perpendicular to the presumed vector of force produced by limb retraction.

A second aspect of pedal asymmetry concerns the relationship of the fifth metatarsal to the first four, and the disposition of the pedal

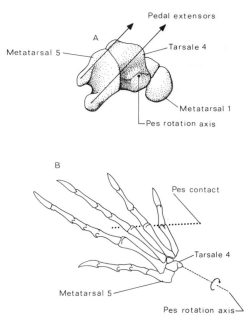

FIG. 8. (A) Proximal end of metatarsus, left limb, *Iguana*. (B) Left foot, dorsal view, *Varanus*. Pes contact (dotted line) is the contact of the first three metatarsals, upon partial pedal extension, with the substrate. Pedal extension is accompanied by rotation at the mesotarsal joint about the pes rotation axis.

extensor musculature. The fifth metatarsal articulates with the fourth tarsal, and lies lateral to the long axis of the pes, which also defines a major rotation axis of the mesotarsal joint (Fig. 8) (Rewcastle, 1980, in press). Onto the fifth metatarsal insert three major pedal extensor muscles, the m. femoral gastrocnemius, the m. peroneus longus and the m. peroneus brevis, and as Robinson (1975) has shown, the fifth metatarsal plays a role analogous to the mammalian calcaneal heel; that is, it provides a lever for the pedal extensors. However, pedal extension is combined with a pronation movement, or mesial rotation about the pes long axis. Since the fifth metatarsal lies lateral to a rotation axis of the mesotarsal joint, the pedal extensors tend to raise the lateral border of the pes, depressing the mesial border and rotating the pes about its long axis (Fig. 8). Thus the pes is lifted onto the mesial digits (Russell & Rewcastle, 1979; Rewcastle, in press). Owing to this motion, the fifth and fourth digits lose contact with the substrate at an early stage of pedal extension, and would appear to be functionally superfluous in locomotion. However, Robinson (1975) has shown that the fifth digit is capable of divergence from the pes long axis and, upon flexion, comes into opposition with the first. The fourth digit also shows marked powers of divergence, especially in anoles. Thus a mechanism for producing a tong-like grasp exists.

An opposable fifth digit is characteristic of the great majority of limbed lizards, and testifies to the importance of grip in the group. Indeed, it seems to indicate widespread scansorial/arboreal adaptation within the Lacertilia, and may represent a key innovation (Rewcastle, in press). One species of lizard, the agamid *Sitana ponticeriana*, lacks a fifth digit, but the fifth metatarsal remains and retains its "heel-bone" role (Russell & Rewcastle, 1979). *Sitana* is a facultative biped, and inhabits sandy areas with little vegetation cover (Subba Rao & Rajabai, 1972), and in such a habitat grip is presumably of minor importance.

Asymmetrical feet of lacertilian type are characteristic of a wide variety of diapsids, including eosuchians (e.g. *Youngia*, Broom, 1914) and thecodonts (e.g. *Euparkeria*, Ewer, 1965; *Proterosuchus*, Cruickshank, 1972; *Lagerpeton* and *Lagosuchus*, Romer, 1971; *Chanaresuchus*, Romer, 1972a; *Gracilisuchus*, Romer, 1972b). There may also be a tendency for the development of a "hooked" fifth metatarsal in these forms, comparable to that of lizards. However, in thecodonts the fifth digit is reduced or lost, and the fifth metatarsal is represented as a splint-like element adjacent to the fourth metatarsal. In crocodiles also the fifth metatarsal remains and receives major insertions of the pedal extensors, despite the presence of a calcaneal heel and an ankle joint comparable to that of mammals. In respect of the latter, it is unfortunate that so little information is available for the crocodilian ankle joint. Preliminary studies indicate that the crocodile ankle is a "double" joint with a lacertilian-type mode of operation for sprawling locomotion, and a mammalian-type operation for "semi-erect" locomotion. Kemp (1978) has reached similar conclusions in a study of the ankle of the therapsid *Regisaurus*.

A picture begins to emerge of diapsids diverging into two different adaptive zones. In lepidosaurs, the opposable fifth digit was retained, and the most striking subsequent elaborations seem to relate to the enhancement of the gripping role of the foot, as in chamaeleons or in the sub-digital "adhesive" pads of geckoes and anoline iguanids. Retention of the fifth digit indicates primarily scansorial/arboreal habits, in which case the stability conferred by the sprawling limb posture is of prime value. In the thecodont lineage, loss of the fifth digit may reflect a primarily ground-dwelling and cursorial existence. Advanced thecodonts, notably *Lagerpeton* and *Lagosuchus* (Romer, 1971) possessed gracile limbs with intramembral limb proportions approaching those of cursorial mammals. *Lagerpeton* and *Lagosuchus* are of especial interest in the presence of a mesially-inturned femoral head in combination with a lacertilian-type asymmetrical pes and possibly a lizard-like ankle.

TABLE I

Limb proportions of thecodonts, with comparative data for lizards. Forelimb length expressed as a percentage of hindlimb length. Note that, in lizards, the hindfoot may be as long as the crus, especially in fast runners. The disproportion between fore- and hindlimbs may therefore be greater than indicated

	$\dfrac{\text{Humerus} + \text{radius}\ (\%)}{\text{Femur} + \text{tibia}}$	Source
Thecodonts		
Euparkeria	67	Ewer (1965)
Proterosuchus	66	Romer (1972a)
Hesperosuchus	67	Colbert (1952)
Chanaresuchus	68	Romer (1972a)
Gracilisuchus	60	Romer (1972b)
Lizards		
Heloderma suspectum	95	
Varanus griseus	85	
Lacerta lepida	57	
Basiliscus sp.	52	

Early tetrapods possessed the sprawling limb posture and a travelling wave lateral undulation system as an inheritance. Locomotion involving lateral undulation may have favoured retention of the sprawling limb posture (Edwards, 1977), but there was a shift to increasing limb dependence. This was reflected by a structural consolidation of the limb, and changes in vertebral construction, notably a change from horizontal to inclined zygapophyses (Olson, 1976). The early reptiles were lizard-sized (Carroll & Baird, 1972; Carroll & Clark, 1973), and small size may have been related to the origin of the amniote egg (Carroll, 1970). Co-evolution of reptiles and insects, with the former exploiting the latter as a food resource (Olson, 1976), may have favoured small body size, and Carroll (1977) suggests that small body size in lizards reflects an insectivorous dietary preference. Clearly, size has a direct effect on locomotor capability, both physically and physiologically, and there seem good grounds for correlating small size, the sprawling limb posture and climbing adaptations. The shift to erect stance occurred in small tetrapods, initially as a cursorial adaptation, but also providing the possibility for dramatic size increase. Aside from questions of support, the erect posture seems adaptive in cursors since it promotes vertical displacements of the centre of mass, and hence generates a suspended phase. Effective utilization of a suspended phase in the thecodont-archosaur lineage may have demanded a shift to bipedal stance since these forms apparently lacked the faculty for sagittal vertebral flexion, unlike mammals. Thecodonts were pre-adapted for bipedality in that they inherited disproportionately long hindlimbs, a functional requirement for fast running sprawling tetrapods.

REFERENCES

Alexander, R. McN. (1976). Estimates of speeds of dinosaurs. *Nature, Lond.* **261**: 129—130.

Alexander, R. McN. & Jayes, A. S. (1978a). Vertical movements in walking and running. *J. Zool., Lond.* **185**: 27—40.

Alexander, R. McN. & Jayes, A. S. (1978b). Optimum walking techniques for idealized animals. *J. Zool., Lond.* **186**: 61—81.

Alexander, R. McN., Maloiy, G. M. O., Hunter, B., Jayes, A. S. & Nturibi, J. (1979). Mechanical stresses in fast locomotion of buffalo (*Syncerus caffer*) and elephant (*Loxodonta africana*). *J. Zool., Lond.* **189**: 135—144.

Alexander, R. McN., Maloiy, G. M. O., Njau, R. & Jayes, A. S. (1979). Mechanics of running of the ostrich (*Struthio camellus*). *J. Zool., Lond.* **187**: 169—178.

Bakker, R. T. (1971). Dinosaur physiology and the origin of mammals. *Evolution* **25**: 636—658.

Bakker, R. T. (1972). Locomotor energetics of lizards and mammals. *Physiologist, N.Y.* **15**: 76.

Barclay, O. R. (1946). The mechanics of amphibian locomotion. *J. exp. Biol.* **23**: 177—203.

Barnett, C. H., Davis, D. V. & MacConaill, M. A. (1961). *Synovial joints: their structure and mechanics*. London: Longmans.

Bennett, A. F. & Dalzell, B. (1973). Dinosaur physiology: A critique. *Evolution* **27**: 170—174.

Bennett, A. F. & Dawson, W. R. (1976). Metabolism. In *Biology of the Reptilia* **5**: 127—223. Gans, C. & Dawson, W. R. (Eds). London & New York: Academic Press.

Brink, A. S. (1954). Speculations on some advanced mammalian characteristics in the higher mammal-like reptiles. *Palaeont. Afr.* **4**: 77—96.

Broom, R. (1914). A new thecodont reptile. *Proc. zool. Soc. Lond.* **1914**: 1072—1077.

Carleton, A. (1941). A comparative study of the inferior tibio-fibular joint. *J. Anat.* **76**: 45—53.

Carroll, R. L. (1970). Quantitative aspects of the amphibian-reptilian transition. *Forma & Functio* **3**: 165—178.

Carroll, R. L. (1977). The origin of lizards. In *Problems in vertebrate evolution* (Linn. Soc. Symp. Ser. No. 4): 359—396. Andrews, S. M., Miles, R. S. & Walker, A. D. (Eds). London & New York: Academic Press.

Carroll, R. L. & Baird, D. (1972). Carboniferous stem-reptiles of the Family Romeriidae. *Bull. Mus. comp. Zool., Harv.* **143**: 321—364.

Carroll, R. L. & Clark, J. (1973). Romeriid reptiles from the lower Permian. *Bull. Mus. comp. Zool. Harv.* **144**: 353—407.

Charig, A. J. (1972). The evolution of the archosaur pelvis and hind-limb: An explanation in functional terms. In *Studies in vertebrate evolution*: 121–155. Joysey, K. A. & Kemp, T. S. (Eds). Edinburgh: Oliver & Boyd.

Colbert, E. H. (1952). A pseudosuchian reptile from Arizona. *Bull. Am. Mus. nat. Hist.* **99**: 565–592.

Coombs, W. P., Jr. (1978). Theoretical aspects of cursorial adaptations in dinosaurs. *Q. Rev. Biol.* **53**: 393–418.

Cott, H. B. (1961). Scientific results of an enquiry into the ecology and economic status of the Nile crocodile (*Crocodylus niloticus*) in Uganda and Northern Rhodesia. *Trans. zool. Soc. Lond.* **29**: 211–356.

Cruickshank, A. R. I. (1972). The proterosuchian thecodonts. In *Studies in vertebrate evolution*: 89–119. Joysey, K. A. & Kemp, T. S. (Eds). Edinburgh: Oliver & Boyd.

Daan, S. & Belterman, Th. (1968). Lateral bending in locomotion of some lower tetrapods. *Proc. K. ned. Akad. Wet.*, (C.) **71**: 245–266.

Desmond, A. J. (1975). *The hot-blooded dinosaurs*. London: Blond & Briggs.

Dodson, P. (1974). Dinosaurs as dinosaurs. *Evolution* **28**: 494.

Edwards, J. L. (1977). The evolution of terrestrial locomotion. In *Major patterns in vertebrate evolution*: 553–577. Hecht, M. K., Goody, P. C. & Hecht, B. M. (Eds). New York: Plenum Publishing Corporation.

Essex, R. (1927). Studies in reptilian degeneration. *Proc. zool. Soc. Lond.* **1927**: 879–945.

Ewer, R. F. (1965). The anatomy of the thecodont reptile *Euparkeria capensis* Broom. *Phil. Trans. R. Soc.* (B.) **248**: 379–435.

Gans, C. (1975). Tetrapod limblessness: Evolution and functional corollaries. *Am. Zool.* **15**: 455–467.

Gray, J. (1944). Studies in the mechanics of the tetrapod skeleton. *J. exp. Biol.* **20**: 88–116.

Gray, J. (1968). *Animal locomotion*. London: Weidenfeld & Nicholson.

Gregory, W. K. (1912). Notes on the principles of quadrupedal locomotion and on the mechanism of the limbs in hoofed animals. *Ann. N.Y. Acad. Sci.* **22**: 267–294.

Gregory, W. K. & Camp, C. L. (1918). Studies in comparative myology and osteology III. *Bull. Am. Mus. nat. Hist.* **38**: 447–563.

Haines, R. W. (1942). The tetrapod knee joint. *J. Anat.* **76**: 270–301.

Haines, R. W. (1952). The shoulder joint of lizards and the primitive reptilian shoulder mechanism. *J. Anat.* **86**: 412–422.

Hildebrand, M. (1959). Motions of the running cheetah and horse. *J. Mamm.* **40**: 481–495.

Hildebrand, M. (1974). *Analysis of vertebrate structure*. New York: John Wiley & Sons.

Hoffstetter, R. & Gasc, J. P. (1969). Vertebrae and ribs of modern reptiles. In *Biology of the Reptilia* **1**: 201–310. Gans, C., Bellairs, A. d'A. & Parsons, T. S. (Eds). London & New York: Academic Press.

Howell, A. B. (1944). *Speed in animals*. Chicago: University of Chicago Press.

Hughes, B. (1963). The earliest archosaurian reptiles. *S. Afr. J. Sci.* **59**: 221–241.

Jenkins, F. A. (1971). Limb posture and locomotion in the Virginia opossum (*Didelphis marsupialis*) and in other non-cursorial mammals. *J. Zool., Lond.* **165**: 303–315.

Kemp, T. S. (1978). Stance and gait in the hindlimb of a therocephalian mammal-like reptile. *J. Zool., Lond.* **186**: 143–161.

Lande, R. (1978). Evolutionary mechanisms of limb loss in tetrapods. *Evolution* 32: 73–92.

Liem, K. F. & Osse, J. W. M. (1975). Biological versatility, evolution and food resource exploitation in African cichlid fishes. *Am. Zool.* 15: 427–454.

MacConaill, M. A. (1966). The geometry and algebra of articular kinematics. *Bio.-Med. Eng.* 1: 205–212.

MacConaill, M. A. & Basmajian, J. V. (1969). *Muscles and movements, a basis for human kinesiology.* Baltimore: Williams & Wilkins Co.

Muybridge, E. (1887). *Animals in motion.* Dover edition (1957), Brown, L. D. (Ed.). New York: Dover Publications.

Nauck, E. T. (1924). Die Beziehungen zwischen Beckenstellung und Gliedmassenstellung bei tetrapoden Vertebraten. *Morph. Jb.* 53: 1–47.

Olson, E. C. (1976). The exploitation of land by early tetrapods. In *Morphology and biology of reptiles* (Linn. Soc. Symp. Ser. No. 3): 1–30. Cox, C. B. & Bellairs, A. d'A. (Eds). London & New York: Academic Press.

Ostrom, J. H. (1969a). Terrestrial vertebrates as indicators of Mesozoic climates. *Proc. N. Am. Paleont. Convent. Chicago*, 1969. Part D: 347–376.

Ostrom, J. H. (1969b). Osteology of *Deinonychus antirrhopus*, an unusual theoropod from the lower Cretaceous of Montana. *Bull. Yale Peabody Mus. nat. Hist.* No. 30: 1–165.

Peabody, F. E. (1951). The origin of the astragalus of reptiles. *Evolution* 5: 339–344.

Peabody, F. E. (1959). Trackways of living and fossil salamanders. *Univ. Calif. Publs Zool.* 63: 1–72.

Regal, P. J. (1978). Behavioral differences between reptiles and mammals: An analysis of activity and mental capabilities. In *Behavior and neurology of lizards*: 183–202. Greenberg, N. & MacLean, P. D. (Eds). U.S. Dept. of Health, Education and Welfare (National Institute of Mental Health).

Renous, S. & Gasc, J. P. (1977). Étude de la locomotion chez un vertébré tétrapode. *Annls Sci. Nat.* (Zool.) (12) 19: 137–186.

Rewcastle, S. C. (1977). *The structure and function of the crus and pes in extant Lacertilia.* Ph.D. thesis: University of London.

Rewcastle, S. C. (1980). Form and function in lacertilian knee and mesotarsal joints; a contribution to the analysis of sprawling locomotion. *J. Zool., Lond.* 191: 147–170.

Rewcastle, S. C. (In press). Fundamental adaptations in the lacertilian hind-limb: A partial analysis of the sprawling limb posture and gait. *Copeia.*

Robinson, P. L. (1975). The functions of the hooked fifth metatarsal in lepidosaurian reptiles. *Colloques int. Cent. natn. Rech. scient.* No. 218: 461–483.

Romer, A. S. (1922). The locomotor apparatus of certain primitive and mammal-like reptiles. *Bull. Am. Mus. nat. Hist.* 46: 517–606.

Romer, A. S. (1942). The development of the tetrapod limb musculature – the thigh of *Lacerta. J. Morph.* 71: 251–298.

Romer, A. S. (1956). *The osteology of the reptiles.* Chicago: University of Chicago Press.

Romer, A. S. (1971). The Chañares (Argentina) Triassic reptile fauna. X. Two new but incompletely known long-limbed pseudosuchians. *Breviora* No. 378: 1–10.

Romer, A. S. (1972a). The Chañares (Argentina) Triassic reptile fauna. XII. The postcranial skeleton of the thecodont *Chanaresuchus. Breviora* No. 385: 1–21.

Romer, A. S. (1972b). The Chañares (Argentina) Triassic reptile fauna. XIII. An early ornithosuchid pseudosuchian, *Gracilisuchus stipanicicorum*, gen. et sp. nov. *Breviora* No. 389: 1—24.

Romer, A. S. (1972c). The Chañares (Argentina) Triassic reptile fauna. XIV. *Lewisuchus admixtus*, gen. et sp. nov., a further thecodont from the Chañares beds. *Breviora* No. 390: 1—13.

Romer, A. S. (1972d). The Chañares (Argentina) Triassic reptile fauna. XV. Further remains of the thecodonts *Lagerpeton* and *Lagosuchus*. *Breviora* No. 394: 1—7

Romer, A. S. & Byrne, F. (1931). The pes of *Diadectes*: Notes on the primitive tetrapod limb. *Paleobiologica* 4: 25—48.

Russell, A. P. (1975). A contribution to the functional analysis of the foot of the tokay, *Gekko gecko* (Reptilia: Gekkonidae). *J. Zool., Lond.* 176: 437—476.

Russell, A. P. (1976). Some comments concerning interrelationships amongst gekkonine geckoes. In *Morphology and biology of reptiles* (Linn. Soc. Symp. Ser. No. 3) 217—244. Bellairs, A. d'A. & Cox, C. B. (Eds). London & New York: Academic Press.

Russell, A. P. (1979). Parallelism and integrated design in the foot structure of gekkonine and diplodactyline geckoes. *Copeia* 1979: 1—21.

Russell, A. P. & Rewcastle, S. C. (1979). Digital reduction in *Sitana* (Reptilia: Agamidae) and the dual roles of the fifth metatarsal in lizards. *Can. J. Zool.* 57: 1129—1135.

Savage, R. J. G. (1977). Evolution in carnivorous mammals. *Palaeontology* 20: 237—271.

Schaeffer, B. (1941). The morphological and functional evolution of the tarsus in amphibians and reptiles. *Bull. Am. Mus. nat. Hist.* 78: 395—472.

Schmidt-Nielsen, K. (1972). Locomotion: Energy cost of swimming, flying and running. *Science, Wash.* 177: 222—228.

Simpson, G. G. (1953). *The major features of evolution*. New York: Columbia University Press.

Slijper, E. J. (1946). Comparative biologic-anatomical investigations on the vertebral column and spinal musculature of mammals. *Verh. K. ned. Akad. Wet.* 42(5): 1—128.

Smith, J. M. & Savage, R. J. G. (1955). Some locomotory adaptations in mammals. *J. Linn. Soc. (Zool.)* 42: 603—622.

Snyder, R. C. (1949). Bipedal locomotion of the lizard *Basiliscus basiliscus*. *Copeia* 1949: 129—137.

Snyder, R. C. (1952). Quadrupedal and bipedal locomotion of lizards. *Copeia* 1952: 64—70.

Snyder, R. C. (1954). The anatomy and function of the pelvic girdle and hindlimb in lizard locomotion. *Am. J. Anat.* 95: 1—36.

Subba Rao, M. V. & Rajabai, B. S. (1972). Ecological aspects of the agamid lizards *Sitana ponticeriana* and *Calotes nemoricola* in India. *Herpetologica* 28: 285—289.

Sukhanov, V. B. (1968). *General system of symmetrical locomotion of terrestrial vertebrates and some features of movement of lower tetrapods*. New Delhi: The Amerind Publishing Co.

Taylor, C. R., Schmidt-Nielsen, K. & Raab, J. L. (1970). Scaling of the energetic cost of running to body size in mammals. *Am. J. Physiol.* 219: 1104—1107.

Tucker, V. A. (1970). Energetic cost of locomotion in animals. *Comp. Biochem. Physiol.* 34: 841—846.

Urban, E. K. (1965). Quantitative study of locomotion in teiid lizards. *Anim. Behav.* **13**: 513–529.

Van Valen, L. (1971). Adaptive zones and the orders of mammals. *Evolution* **25**: 420–428.

Walker, A. D. (1970). A revision of the Jurassic reptile *Hallopus victor* (Marsh), with remarks on the classification of crocodiles. *Phil. Trans. R. Soc.* (B.) **257**: 323–372.

Walker, W. F. (1972). Body form and gait in terrestrial Vertebrates. *Ohio J. Sci.* **72**: 177–183.

Whiting, H. P. (1961). Pelvic girdle in amphibian locomotion. *Symp. zool. Soc. Lond.* No. 5: 43–57.

Zug, G. R. (1974). Crocodilian galloping: An unique gait for reptiles. *Copeia* **1974**: 550–552.

Symp. zool. Soc. Lond. (1981) No. 48, 269–287

The Gaits of Tetrapods: Adaptations for Stability and Economy

R. McNEILL ALEXANDER

Department of Pure and Applied Zoology,
University of Leeds, Leeds, England

SYNOPSIS

The relative merits of different gaits have been investigated by mathematical modelling. The gait used by chelonians seems to be the one which minimizes unwanted displacements (vertical movements, pitching and rolling) for an animal with slow muscles. Gaits only a little different would be unsatisfactory at the very low speeds which chelonians use because the belly would strike the ground. Mammals walk faster and so can tolerate larger departures from equilibrium. The patterns of force they exert with their feet, and the change at a critical speed from walking to running, seem adapted to minimize metabolic power requirements. Quadrupedal mammals use different gaits (usually the trot, canter and gallop) at different speeds. Attempts are made to assess their energy costs but it is uncertain whether the chosen gaits minimize energy costs.

INTRODUCTION

Tetrapods use different gaits at different speeds. Men walk to travel slowly and run to go faster. Horses change successively from walk to trot to canter and finally to gallop as they increase speed. Different tetrapods use different gaits at comparable speeds: for instance, men run but kangaroos hop. This chapter is about the relative merits of different gaits, in different circumstances.

The merits of gaits are not easy to assess by direct observation, since animals cannot easily be persuaded to use one gait in circumstances in which they prefer another. Mathematical models have therefore been used to find out what the consequences would be, of using gaits other than the preferred gaits. The mathematics has been presented elsewhere and only the results are given here. Three specific questions are tackled:

(1) What are the requirements for slow quadrupedal walking?

(2) Why do animals change from walking to running at particular speeds?

(3) Why do most mammals trot at low running speeds but gallop to go faster?

The gaits of mammals have been reviewed by Dagg (1973), by Gambaryan (1974) and most comprehensively by Hildebrand (1976, 1977). The gaits of amphibians and reptiles have been reviewed by Sukhanov (1974) and the gaits of birds by Dagg (1977).

DESCRIPTION OF FOOTFALL PATTERNS

This chapter deals only with regular gaits, in which each foot is set down once in each stride and the same pattern of movements is repeated in successive strides. Such gaits can be described by specifying the footfall pattern and the forces exerted by the feet, in a single stride. The footfall pattern is conveniently specified by stating the duty factor and relative phase of each foot (McGhee, 1968). The duty factor is the fraction of the duration of the stride for which the foot is on the ground. The relative phase indicates the stage of the stride at which the foot is set down. The stride is deemed to start with the setting down of an arbitrarily-chosen reference foot, so the relative phase of the reference foot is zero and those of the other feet are fractions of the duration of a stride. McGhee (1968) assigned zero relative phase to the left fore foot but this convention is apt to give the impression that gaits which are mirror images of each other are grossly different. In this paper, the reference foot of a quadruped is whichever fore foot is set down less than half a stride before the other. If this is the left fore foot, let the relative phases of the feet be 0, γ, δ, ϵ, as indicated in Fig. 1a. If it is the right fore foot let the relative phases be as in a mirror image of Fig. 1a.

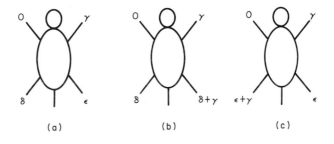

(a)   (b)   (c)

FIG. 1. (a) A diagram showing the symbols used for the relative phases of the feet of a quadruped. Either this diagram or its mirror image applies, whichever makes $\gamma \leqslant 0.5$. (b) and (c) The restricted set of footfall patterns for which calculations are presented are represented by one or other of these diagrams or of their mirror images.

Seven variables (three relative phases and four duty factors) are required to specify all possible quadrupedal footfall patterns. It is not practicable to compute results from the mathematical models presented in this chapter, for all the possibilities. Even if the results had been computed, it would not be practicable to present them in a chapter of this length. This paper considers only the set of footfall patterns for which all four feet have the same duty factor β, and for which the phase difference between the hind feet is the same as the phase difference between the fore feet. Any such footfall pattern is represented by Fig. 1b or Fig. 1c or the mirror image of one of these figures. Most observed footfall patterns resemble a member of this set reasonably closely (Hildebrand, 1976, 1977).

A gait is described as symmetrical if the left and right feet of each pair have equal duty factors and move half a cycle out of phase with each other. In the set of gaits considered in this chapter, all gaits having $\gamma = 0.5$ are symmetrical. They are easily described in Hildebrand's (1976) notation for symmetrical gaits: the percentage of the stride for which each foot is on the ground is $100\,\beta$ and the percentage of the stride by which each fore footfall follows the hind footfall of the same side is $100\,(1 - \delta)$. Hildebrand (1977) used a different and necessarily more complex notation for asymmetrical gaits. For the set of gaits considered in this paper, Hildebrand's fore and hind contact intervals are both $100\,\beta$, his fore and hind leads are both $100\,\gamma/\beta$ and his mid-time lag is $100\,\delta$ (for gaits represented by Fig. 1b) or $100\,\epsilon$ (for gaits represented by Fig. 1c).

Dagg (1979, and earlier papers) describes footfall patterns by tabulating the percentage of the stride during which different combinations of feet are on the ground. Transformation of data from the system used in this paper to her system is possible but tedious. The reverse transformation cannot, in general, be effected unambiguously because her system does not distinguish between a footfall pattern and its reverse in time.

COMPARISON OF SPEEDS

Mammals change gaits at well-defined speeds which depend on their size. For instance, a small boy breaks into a run at a lower speed than an adult man. The physical concept of dynamic similarity is useful in comparisons of animals of different sizes (Alexander, 1976).

Two shapes are said to be geometrically similar if one could be made identical to the other by a uniform change of scale. In the same way, two motions are said to be dynamically similar if one could be made identical to the other by uniform changes of one or both of the scales of length and time. For instance, the motions of pendulums of different lengths swinging through the same angle are dynamically similar. It can be shown that if gravity and inertia are important (as they are for pendulums and for walking animals), two motions can only be dynamically similar if they have equal Froude numbers u^2/gl (Duncan, 1953). Here u is the speed, g is the acceleration of free fall and l is some characteristic length. In studies of gaits, this length is taken to be the height h of the hip joint from the ground in normal standing, so the Froude number chosen becomes u^2/gh.

Stride length λ is the distance travelled by the animal in a stride, and relative stride length is λ/h. Even if two animals are running with equal relative stride lengths and identical duty factors and relative phases, their motions cannot be dynamically similar unless they have equal Froude numbers.

Tetrapods generally take longer strides at higher speeds but mammals of different sizes generally use approximately equal relative stride lengths at equal Froude numbers (Alexander, 1976). Also, different mammals tend to make corresponding changes of gait at equal Froude numbers (Alexander, 1977, where the parameter $\hat{u}$ is the square root of the Froude number). Most mammals change from a walk to a trot or (for man) a run at a Froude number of about 0.6. Most quadrupedal mammals change from a trot to a canter or gallop at some Froude number between 2 and 4.

FORCES EXERTED BY FEET

Account will have to be taken of differences in the patterns of forces exerted by the feet in different gaits. Figure 2 shows forces exerted by a man's foot, in typical walking and running steps. The vertical component F_Y shows two main maxima in walking but only one in running. The maxima are most distinct in fast walking. The horizontal component F_X acts first forward and then back in every case.

Consider a foot which is set down on the ground at time $-T/2$ and lifted at $T/2$. The vertical component of force F_Y which it exerts at time t in this interval can be described quite generally by a Fourier series.

$$F_Y = a_1 \cos(\pi t/T) + b_2 \sin(2\pi t/T) + a_3 \cos(3\pi t/T)$$

$$+ b_4 \sin(4\pi t/T) + \ldots \qquad (11.1)$$

This equation does not imply any assumption about the force pattern. The series has no even-numbered cosine terms or odd-numbered

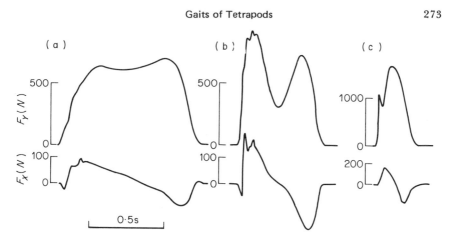

FIG. 2. Records of forces exerted on the ground by one foot of a 70 kg man (a) walking at
0.9 m s⁻¹; (b) walking at 1.9 m s⁻¹ and (c) running at 3.3 m s⁻¹. F_X is the longitudinal
component of the forces and F_Y the vertical component. The time scale applies to all the
records. The records of F_Y can be imitated by equation 11.2 using (a) $r = 0.05, q = 0.22$; (b)
$r = -0.10, q = 0.55$; (c) $r = -0.24, q = 0.12$. From Alexander & Jayes (1978a).

sine terms because F_Y must be zero at times $-T/2$ and $T/2$. Observed
force patterns can be simulated quite closely by using only the first
few terms of the series (Alexander & Jayes, 1980). In this paper only
the first three terms are used and the equation is re-written.

$$F_Y = A[\cos(\pi t/T) + r \sin(2\pi t/T) - q \cos(3\pi t/T)] \quad (11.2)$$

Here r and q are parameters affecting the shape of a graph of F_Y
against t, and A is the factor required to make the leg support the required fraction of body weight. The negative sign is introduced to
conform with the convention of Alexander & Jayes (1978a).

Figure 3a shows that when r and q are both zero, equation 11.2
represents a force which rises to a single maximum and falls again
symmetrically. Figure 3b shows that a positive value of r skews a
graph of F_Y against t towards the end of the step. Negative values of
r skew the graph towards the beginning of the step. Figure 3c shows
that a sufficiently large positive value of q gives F_Y two maxima (as
in Fig. 2a and b). Negative values of q make graphs of F_Y against t
bell-shaped. For animals travelling over non-adhesive ground, F_Y
must always be positive. This restricts the ranges of possible values of
r and q.

It is found that the horizontal components of force shown in Fig.
2, and in other force records of walking and running animals, can be
imitated closely by giving the constant k an appropriate value in the
equation.

$$F_X = utF_Y/k \quad (11.3)$$

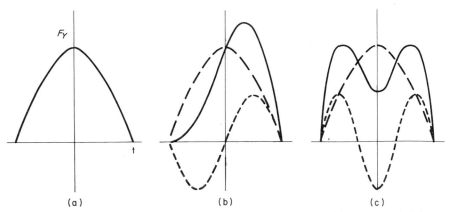

FIG. 3. Graphs of F_Y against time simulated by equation 11.2 using (a) $r = 0$, $q = 0$; (b) $r = 0.5$, $q = 0$; (c) $r = 0$, $q = 0.5$. The broken lines in (b) and (c) show the contributions of individual terms from equation 11.2.

where u is the speed of locomotion. (A backward force on the ground tends to accelerate the body and is treated as positive.) This equation implies that the force exerted by the foot tends to keep in line with some point moving with the speed of the body, at a height k from the ground. Force records of men, dogs and sheep show that k is generally about 1.7 h, where h is hip height (Jayes & Alexander, 1978; Alexander & Jayes, 1978b).

VERY SLOW WALKING

To walk satisfactorily, an animal will presumably have to keep its belly off the ground throughout the stride. This requirement limits the ranges of rising and falling, pitching and rolling which are acceptable. The animal must either use a high stride frequency so that any unwanted movement is quickly corrected by a subsequent footfall, or it must keep itself nearly in equilibrium throughout the stride.

Consider an animal walking with stride frequency f so that the duration of a stride is $1/f$. If it fell freely for this time, starting from rest, it would fall $g/2f^2$. The distance it can fall before hitting the ground is rather less than h. Hence the dimensionless parameter $g/2f^2 h$ can be used as a measure of the need to maintain equilibrium. For a dog galloping it is 1 or less and for a dog walking very slowly it is about 5 (data of Jayes & Alexander, 1978). For the turtle *Geoemyda* walking at its normal speed it is about 200 (data of Jayes & Alexander, 1980), and it would be similarly large for other chelonians. The requirement to keep near equilibrium is particularly stringent for chelonians.

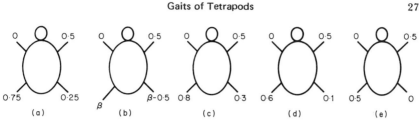

FIG. 4. (a–e) Diagrams showing the relative phases of the feet of a quadruped, in gaits which are discussed in the text. (d) represents the gait generally used by chelonians.

Gray (1944) pointed out that a quadruped can maintain equilibrium throughout the stride if it walks so as always to have at least three feet on the ground. This implies that the mean of the duty factors of the feet must be at least 0.75. Further, the feet must have relative phases such that a vertical line through the centre of mass always intersects the triangle defined by the three feet on the ground. If the duty factor of every foot is 0.75, the only possible relative phases are as shown in Fig. 4a. McGhee & Frank (1968) showed that if all the feet have the same duty factor β (> 0.75) the most stable footfall pattern is the one shown in Fig. 4b.

Chelonians generally walk with duty factors of about 0.8 and so might be expected to walk as indicated in Fig. 4c. They actually walk approximately as shown in Fig. 4d (Jayes & Alexander, 1980). There are only two feet on the ground at times, so equilibrium cannot be maintained throughout the stride.

Jayes & Alexander (1980) sought to explain why chelonians do not use the ideal gait. They made force-platform records of chelonians walking and found that the horizontal components of the forces exerted by the feet were small. Hence it seemed sufficiently realistic to attempt an explanation which assumed that the forces were always vertical. Consider a tortoise walking with a duty factor of 0.75 for every foot. Let the feet move with the relative phases required for equilibrium (Fig. 4a) and let them exert vertical forces so as to keep the animal always in equilibrium. It is easily shown that the forces must be as shown in Fig. 5a. The force exerted by each foot would have to make large, instantaneous changes at appropriate stages of the step. This could not be achieved or even approached in practice. Tortoise leg muscles are very slow (and, probably for that reason, are very economical of energy (Woledge, 1968).)

Jayes & Alexander (1980) devised a model of tortoise locomotion to take account of the slowness of the muscles. They postulated initially that the time course of the force on each foot was as shown in Fig. 3a (i.e. they put $r = 0$. $q = 0$ in equation 11.2). They made calculations to discover how much the tortoise would rise and fall,

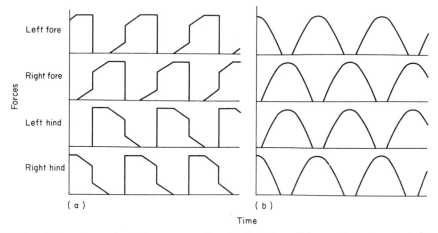

FIG. 5. Schematic graphs of forces exerted by the feet of a walking quadruped, against time. It is supposed that the feet exert only vertical forces. (a) shows the forces required to maintain equilibrium throughout the stride when the duty factor is 0.75. Similar patterns (with some scope for variation when all four feet are on the ground) are required at higher duty factors. (b) shows the forces exerted in the optimum footfall pattern, when the force exerted by each foot is required to be as in equation 11.2 with $r = 0$, $q = 0$. The duty factor is 0.83. From Jayes & Alexander (1980).

pitch and roll in gaits from the set defined by Fig. 1b. The results for a duty factor of 0.83 are as shown in Fig. 6. The only gait for which the amplitudes of vertical movement, pitching and rolling are all small is $\gamma = \delta = 0.5$, $\epsilon = 0$ (Fig. 4e). The same conclusion was reached when the set of gaits defined by Fig. 1c was explored (the gait shown in Fig. 4e is a member of both sets). Thus this is the only feasible gait for very low speeds, for a quadruped which exerts forces as in Fig. 3a. It would cease to be feasible at excessively low speeds because it involves small departures from equilibrium. This gait is only slightly different from the observed gaits of chelonians (Fig. 4d) and seems at first sight very different from the gait previously suggested as ideal for the same duty factor (Fig. 4c). However Fig. 5 shows that it resembles a gait which would give perfect equilibrium, in that diagonally opposite feet exert large forces simultaneously.

Records of the vertical components of the forces exerted by the feet of the turtle *Geoemyda* are not symmetrical, as in Fig. 3a, but rather skew as in Fig. 3b. They can be imitated closely by equation 11.2 using $q = 0$, $r = 0.1$ (for fore feet) or -0.3 (for hind feet). Jayes & Alexander (1980) made their model more realistic by allowing non-zero values of r, but considered only cases in which the fore and hind feet had numerically equal values of r of opposite sign. They found that unwanted displacements were minimized when $r \simeq$

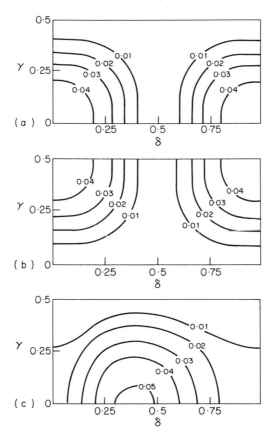

FIG. 6. Ranges of unwanted displacements calculated for a chelonian walking with a duty factor of 0.83 using footfall patterns of the type shown in Fig. 1b. The forces on the feet were calculated from equation 11.2 with $r = 0$, $q = 0$. The graphs show relative phase γ plotted against relative phase δ, with contours showing quantities proportional to (a) the range of vertical displacement of the centre of mass; (b) the anugular range of rolling; and (c) the angular range of pitching. From Jayes & Alexander (1980).

0.16 for fore feet, −0.16 for hind feet, and when $\gamma = 0.5$, $\delta = 0.58$, $\epsilon = 0.08$. This is very similar indeed to the observed gaits of chelonians.

It seems that chelonians cannot maintain equilibrium as they walk because their muscles are slow, but use the gait which in these circumstances minimizes vertical movements, pitching and rolling.

WALKING AND RUNNING

In faster walking, animals do not need to keep so near equilibrium. A wide range of gaits may be feasible but one particular gait may be

preferable to others if it enables the animal to travel with less expenditure of energy. When a man changes from walking to running, the
duty factor falls abruptly and there is an abrupt change in the pattern of forces exerted by the feet (Fig. 2). The time course of the
vertical component of force can be imitated by equation 11.2, using
appropriate positive values of the parameter q for walking (see Fig.
3c) and values around -0.1 for running. Similar changes have been
demonstrated for various birds and quadrupedal mammals (Cavagna,
Heglund & Taylor, 1977; Jayes & Alexander, 1978; Schryver, Bartel,
Langrana & Lowe, 1978), and in the quadrupeds there is in addition
a change in the relative phases of the feet. The changes are less
abrupt for sheep than for the other species.

Alexander & Jayes (1978b) used a mathematical model to estimate the energy cost of walking or running for a biped, using different values of the duty factor β and the parameter q. Alexander
(1980) formulated a more satisfactory model which avoided a source
of error and which was applicable to symmetrical quadrupedal as
well as bipedal gaits. The walking gaits and the slower running gaits
(trot and pace) of quadrupedal mammals are symmetrical.

Tortoises walk with their feet far lateral to the median plane, and
account had to be taken of this in the model of walking which was
applied to them. Mammals and birds walk with their feet under the
body, very close to the median plane. It is assumed in the model to
be introduced now that the feet are set down in the median plane.
Figure 7 illustrates the model. The animal is moving in a system of
Cartesian coordinates. At time t the force on one particular foot has
components F_X, F_Y and the proximal joint of the same leg is at
(x, y). The leg is assumed to have negligible mass so the forces F_X,
F_Y are transmitted to the trunk at (x, y). In a small increment of
time the joint moves to $(x + \delta x, y + \delta y)$ and the leg does work δW
given by

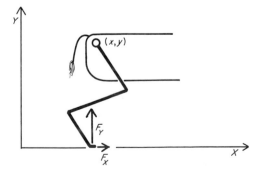

FIG. 7. A diagram illustrating the model of walking and running which is applied to mammals and birds.

$$\delta W \simeq F_X \cdot \delta x + F_Y \cdot \delta y \qquad (11.4)$$

(The equation is approximate but the error is small, in realistic cases). The work δW may be positive or negative. In a complete stride the animal does numerically equal quantities of positive and negative work, which are, for a quadruped

$$\frac{1}{2} \left(\Sigma \mid \delta W_1 \mid + \Sigma \mid \delta W_2 \mid + \Sigma \mid \delta W_3 \mid + \Sigma \mid \delta W_4 \mid \right)$$

The summations are over a complete stride and the subscripts refer to different legs. The modulus sign $\mid \; \mid$ indicates quantities that are to be taken as positive in the formula, irrespective of their sign. Performance of positive and of negative work both consume metabolic energy (Margaria, 1976). Let η be an efficiency such that performance of one unit each of positive and negative work requires $1/\eta$ units of metabolic energy. The stride frequency is f. Thus the metabolic power U required for walking or running is given by

$$U = (f/2\eta) \left(\Sigma \mid \delta W_1 \mid + \Sigma \mid \delta W_2 \mid + \Sigma \mid \delta W_3 \mid + \Sigma \mid \delta W_4 \mid \right)$$
$$(11.5)$$

This power was calculated for symmetrical gaits over a wide range of duty factors. The vertical component of the force on each foot was assumed to be as given by equation 11.2 with $r = 0$ and with q varying from -0.33 to $+1$ (values of q outside this range are unrealistic because they make F_Y negative at some stage of the step). A realistic distribution of weight between the fore and hind quarters was assumed for quadrupeds. The horizontal component of force on each foot was assumed to be as given by equation 11.3.

Figure 8 shows results for a biped, or for a quadruped walking with each fore foot moving in phase with a hind foot. Different graphs have been drawn for different values of the parameter u^2/gk, representing different speeds. They show that for every $u^2/gk < 1$ there is an optimum (β, q) which minimizes power consumption. These optima have $\beta > 0.5$ so the gaits they represent are walks. As u^2/gk approaches 1 the optimum value of q rises towards 1. An abrupt change occurs at $u^2/gk = 1$, and when $u^2/gk > 1$ the optimum gait is a run with β and q as small as possible.

Since the quantity k (equation 11.3) has been shown to be about $1.7\,h$ for various mammals, the parameter u^2/gk can be converted to the equivalent Froude number u^2/gh by multiplying it by 1.7.

The points in Fig. 8a and b are observed values of (β, q) for men walking with the appropriate values of u^2/gk. They lie quite close to the theoretical optima. The theory suggests that men should still be

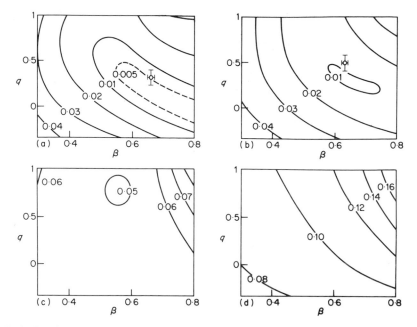

FIG. 8. Graphs showing the metabolic power output U required for walking or running by an animal of mass m, either a biped or a quadruped moving fore and hind legs in phase with each other. Each diagram is a graph of q against β with contours showing calculated values of $U/(mg^2/f\eta)$. They refer to the following values of the speed parameter u^2/gk: (a), 0.05; (b), 0.02; (c), 0.8 and (d), 1.6. The points in (a) and (b) are values for man calculated for the appropriate values of u^2/gk from the empirical regression equations of Alexander & Jayes (1980). They are shown ± two standard errors. From Alexander (1980).

walking at $u^2/gk = 0.8$ (Fig. 8c), corresponding to $u^2/gh = 1.4$ or a speed of about 3.5 m s^{-1} for an adult man. However, men generally run at speeds above 3 m s^{-1}. The discrepancy can be explained by taking account of elastic strain energy.

The model assumes that the work δW is done entirely by the contractile machinery of muscles. However, muscles and especially tendons have elastic properties, so that they do negative work as they are stretched and positive work in a subsequent elastic recoil (Alexander & Bennet-Clark, 1977; Morgan, Proske & Warren, 1978). This elastic work is done without metabolic cost, so metabolic energy can be saved by it. Much larger savings are possible in running than in walking because the forces are larger, and because stretching and recoil of elastic elements occur at more appropriate stages of the stride. Savings are only possible if the elastic elements are being stretched while the leg is doing negative work and recoiling while the leg is doing positive work. This occurs in running with low values of q but the converse occurs for much of the step in walking with high

q (Alexander, 1980). Hence the speed at which running becomes preferable to walking is lower than Fig. 8 suggests.

Alexander (1980) also presented graphs like Fig. 8 for the footfall pattern shown in Fig. 4a. Most quadrupedal mammals walk with footfall patterns very like this (Hildebrand, 1976). The graphs are similar to the graphs in Fig. 8 but the optimum (β, q) for each u^2/gk is approximately the same as for a slightly higher u^2/gk in Fig. 8. The optima they suggest for low u^2/gk are very close to the (β, q) used by sheep walking slowly. The optima they suggest for faster walking involve rather larger values of q than sheep and dogs use, when walking at the corresponding speeds. Dogs and sheep have not been observed to use values of q greater than about 0.4, even at speeds at which the model makes higher q seem advantageous.

QUADRUPEDAL RUNNING

The previous section seems to show why mammals and birds change from walking to running as they increase speed. This one considers the relative merits of different running gaits. Most quadrupedal mammals trot at low running speeds and canter or gallop at higher speeds.

The model described in the previous section has also been used to compare quadrupedal running gaits (Alexander, Jayes & Ker, 1980). All known force records of running show near-zero values of q (Jayes & Alexander, 1978. See also the records in Cavagna, Heglund & Taylor, 1977), so q was taken to be zero. The sets of footfall patterns represented by Fig. 1b and c were explored, including asymmetrical as well as symmetrical gaits.

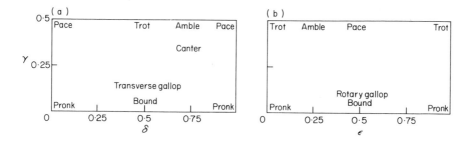

FIG. 9. Diagrams showing the names of running gaits in which the relative phases of the feet are as shown in (a), Fig. 1b; (b), Fig. 1c. From Alexander, Jayes & Ker (1980) with a minor change of labelling made necessary by a difference in the conventions adopted in that paper and this one.

Figure 9 shows the names given to gaits with these footfall patterns. The trot, pace and amble are symmetrical gaits which are used at slow running speeds by most mammals, by camels and by elephants, respectively. (Elephants have no faster gait than the amble.) Many mammals use the canter at intermediate speeds, and most use one or other of the forms of gallop at high speeds. The pronk is a rare gait and the bound seems to be used only by some small mammals (Gambaryan, 1974). In the trot, diagonally opposite feet move together; in the pace, the feet of each side move together; in the bound, the feet of each pair move together and in the pronk all four feet move together. The gallops resemble the bound, but the feet of each pair are slightly out of phase with each other. The amble is a running gait with relative phases as in Fig. 4a.

Alexander, Jayes & Ker (1980) made calculations for two speeds: $u^2/gk = 0.5$, representing a speed close to the minimum for running and $u^2/gk = 10$, representing a fast galloping speed. Realistic values for the duty factor and for the other parameters which were required were assumed for each speed.

Figure 10 shows the instantaneous power outputs of the legs, calculated from equation 11.4 for all stages of the stride for each of

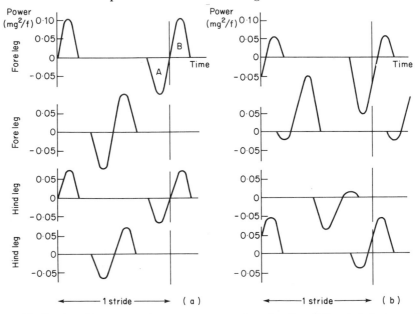

FIG. 10. Graphs of instantaneous power output against time for all four legs in two gaits, calculated from the model described in the text. The masses of the legs are ignored. Power is expressed by the ratio $(dW/dt)/(mg^2/f)$ where m is body mass and f is stride frequency. (a) represents a trot or pace and (b) a canter, both for a low running speed ($u^2/gk = 0.5$). From Alexander, Jayes & Ker (1980).

two gaits at the lower speed. The areas between the graphs and the time axes represent work: for instance, area A represents negative work and area B represents positive work. Each leg does negative followed by positive work. In the trot or pace (Fig. 10a) each leg does numerically equal quantities of negative and positive work so the legs could in principle be replaced by perfectly elastic springs of appropriate stiffness, which would perform the same function without metabolic cost. In the canter (Fig. 10b) each leg does unequal quantities of positive and negative work and only small savings could apparently be made by elastic storage.

If there were no elastic elements in the legs the power required for running would be as given by equation 11.5. If there are elastic elements, a different equation is needed. In an interval of time in which a leg does external work δW and stores elastic strain energy δE, the contractile elements of its muscles do work $(\delta W + \delta E)$. Hence equation (5) must be altered to

$$U = (f/2\eta)\left(\sum | \delta W_1 + \delta E_1 | + \sum | \delta W_2 + \delta E_2 | + \ldots (11.6) \right.$$

(U will be less than in the inelastic case if δW and δE tend to be opposite in sign.) The values of δE depend on the forces on the feet and also on the stiffness of the elastic elements of the legs. Animals might be expected to evolve legs of optimum stiffness, but no one stiffness would be optimal for all speeds. Calculations were made on the assumption that the elastic elements had the stiffness required for the greatest possible savings of energy at the high speed. This implied that they were too stiff to be fully effective at the low speed.

If the elastic elements are assumed to be perfectly elastic, it is found that some gaits can be executed with no metabolic cost whatsoever. The assumption, and the result, are unrealistic. Legs with imperfectly elastic elements were therefore considered. The rebound resilience of an elastic material is the energy recovered in an elastic recoil, as a fraction of the work previously done deforming the material. For a perfectly elastic material it is 1. The results for imperfect elasticity presented in this paper assume a rebound resilience of 0.6, a value based on experiments with tendons in the feet of sheep (Cuming, Alexander & Jayes, 1978). However, more recent experiments by Dr R. F. Ker show that some tendons have much higher rebound resiliences.

The calculations described so far are based on equation 11.4, which ignores the mass of the leg. Since legs have mass, positive work is needed to accelerate them and negative work to decelerate them as they swing backward and forward. This was calculated and found to be large for fast running but small for slow running (and also for walking).

Results for various gaits at the two speeds are shown in Table I. At the lower speed the amble seems to be the most economical gait, whichever assumption is made about elasticity. The trot and canter are a little more expensive and the other gaits much more expensive. Few mammals amble: most use the trot for slow running. The mathematical model may be misleading because it assumes the feet are set down precisely in the median plane. They would actually be set down a little to either side, and transverse movements or rolling would occur in the amble with associated energy costs. These movements could be avoided in the trot, in which a left and a right foot are always on the ground simultaneously. A more realistic model might show that the trot was more economical than the amble.

The differences between the power requirements of different gaits, shown in Table I, are relatively smaller at the higher speed, but the amble still seems the best gait. Consider the imperfectly elastic case, which is probably the most realistic. The assumptions made so far make the power requirement $0.34\ mg^2/f$ for the amble and $0.41\ mg^2/f$ for the gallop, excluding in each case the power needed for swinging the legs. However, the speed represented is one at which quadrupedal mammals gallop.

The results shown without parentheses in the Table depend on equations 11.2 (with $r = 0$, $q = 0$) and 11.3. Together these equations imply that each foot exerts a forward impulse on the ground followed by an equal backward impulse. This makes individual legs do unequal quantities of positive and negative work in asymmetrical gaits such as the canter (Fig. 10b) and gallop, limiting the energy that can be saved by elastic storage. An alternative assumption modifies equation 11.3, so that the forward and backward impulses may be unequal but are adjusted to make each leg do equal quantities of positive and negative work. This requires only small changes in the forces exerted at high speeds but greatly reduces the estimated power requirement for galloping. The results in parentheses in the Table show that it makes the amble, trot, canter and gallop about equally economical. Mammals may prefer the gallop because it enables them to use back muscles as well as leg muscles, and so to spread the power requirement over a greater mass of muscle. However, any back movements would increase the power requirement above the predictions of the model. Mammals generally keep the two feet of each pair slightly out of phase with each other and so use the gallop rather than the bound, which would be more expensive of energy.

The alternative assumption helps to explain the popularity of the gallop. It is unfortunately uncertain whether it is realistic. The

TABLE I

Estimates of metabolic power required for various gaits, at two speeds ($u^2/gk = 0.5$ and 10). Estimates based on three different assumptions about the elastic elements in the legs are given. The power U which would be required if the legs had no mass is given for each gait. The power V required to swing the legs, on account of their mass, is the same for all gaits at given speed. The total power requirement is (U + V). Data from Alexander, Jayes & Ker (1980)

	Power ÷ (mg^2/f), at low speed			Power ÷ (mg^2/f), at high speed		
	Inelastic	Imperfectly elastic	Perfectly elastic	Inelastic	Imperfectly elastic	Perfectly elastic
U for						
Amble	0.16	0.09	0.08	1.03	0.34	0
Trot	0.21	0.14	0.13	1.08	0.36	0
Canter	0.21	0.17	0.16	1.06	0.35 (0.35)[a]	0.16 (0)
Gallop	0.29	0.27	0.26	1.04	0.41 (0.34)	0.28 (0)
Bound	0.40	0.33	0.32	1.27	0.42	0
Pronk	0.51	0.44	0.43	1.38	0.46	0
V for all gaits	0.02	small	small	0.81	0.51	0

[a] Values in parentheses refer to the alternative assumption about forces (see text)

assumption requires one foot of a pair to exert a net forward impulse and the other to exert a net backward impulse, in each stride. To discover whether this occurs, good separate records would be needed of the forces exerted by individual feet in the same galloping stride. No such records are available, for any species.

CONCLUSIONS

The mathematical models described in this paper help to explain why animals use the gaits they do, by demonstrating some of the consequences of other gaits. Chelonians seem to use the gait which minimizes unwanted displacements, so far as their slow muscles allow. Gaits only a little different would be unsatisfactory at their very low speeds, because the carapace would hit the ground at some stage of the stride.

Mammals walk faster, with lower values of the parameter $g/2f^2h$, and so can tolerate much larger departures from equilibrium. They seem to walk so as to minimize metabolic power requirements. The different patterns of force exerted by men walking at different speeds, and the abrupt changes which occur as men and other mammals break into a run, seem explicable in this way.

The attempt to explain the different running gaits which quadrupedal mammals use at different speeds is less satisfactory. The model suggested the amble as the most economical running gait, but most mammals prefer the trot. It was shown that the gallop could be made as economical as other gaits at high speeds, but it was not possible to show whether mammals gallop in the manner required for this.

REFERENCES

Alexander, R. McN. (1976). Estimates of speeds of dinosaurs. *Nature, Lond.* **261**: 129–130.

Alexander, R. McN. (1977). Terrestrial locomotion. In *Mechanics and energetics of animal locomotion*: 168–203. Alexander, R. McN. & Goldspink, G. (Eds). London: Chapman & Hall.

Alexander, R. McN. (1980). Optimum walking techniques for quadrupeds and bipeds. *J. Zool., Lond.* **192**: 97–117.

Alexander, R. McN. & Bennet-Clark, H. C. (1977). Storage of elastic strain energy in muscle and other tissues. *Nature, Lond.* **265**: 114–117.

Alexander, R. McN. & Jayes, A. S. (1978a). Vertical movements in walking and running. *J. Zool., Lond.* **185**: 27–40.

Alexander, R. McN. & Jayes, A. S. (1978b). Optimum walking techniques for idealized animals. *J. Zool., Lond.* **186**: 61–81.

Alexander, R. McN. & Jayes, A. S. (1980). Fourier analysis of forces exerted in walking and running. *J. Biomechan.* **13**: 383–390.

Alexander, R. McN., Jayes, A. S. & Ker, R. F. (1980). Estimates of energy cost for quadrupedal running gaits. *J. Zool., Lond.* **190**: 155–192.

Cavagna, G. A., Heglund, N. C. & Taylor, C. R. (1977). Mechanical work in terrestrial locomotion: two basic mechanisms for minimizing energy expenditure. *Am. J. Physiol.* **233**: R243–R261.

Cuming, W. G., Alexander, R. McN. & Jayes, A. S. (1978). Rebound resilience of tendons in the feet of sheep (*Ovis aries*). *J. exp. Biol.* **74**: 75–81.

Dagg, A. I. (1973). Gaits in mammals. *Mammal Rev.* **3**: 135–154.

Dagg, A. I. (1977). The walk of the Silver gull (*Larus novaehollandiae*) and of other birds. *J. Zool., Lond.* **182**: 529–540.

Dagg, A. I. (1979). The walk of large quadrupedal mammals. *Can. J. Zool.* **57**: 1157–1163.

Duncan, W. J. (1953). *Physical similarity and dimensional analysis.* London: Arnold.

Gambaryan, P. P. (1974). *How mammals run. Anatomical adaptations.* New York: Wiley.

Gray, J. (1944). Studies in the mechanics of the tetrapod skeleton. *J. exp. Biol.* **20**: 88–116.

Hildebrand, M. (1976). Analysis of tetrapod gaits: general considerations and symmetrical gaits. In *Neural control of locomotion*: 203–236. Herman, R. M., Grillner, S., Stein, P. S. G. & Stuart, D. G. (Eds). New York: Plenum.

Hildebrand, M. (1977). Analysis of asymmetrical gaits. *J. Mammal.* **58**: 131–156.

Jayes, A. S. & Alexander, R. McN. (1978). Mechanics of locomotion of dogs (*Canis familiaris*) and sheep (*Ovis aries*). *J. Zool., Lond.* **185**: 289–308.

Jayes, A. S. & Alexander, R. McN. (1980). The gaits of chelonians: walking techniques for very low speeds. *J. Zool., Lond.* **191**: 353–378.

McGhee, R. B. (1968). Some finite state aspects of legged locomotion. *Math. Biosci.* **2**: 67–84.

McGhee, R. B. & Frank, A. A. (1968). On the stability properties of quadrupedal creeping gaits. *Math. Biosci.* **3**: 331–351.

Margaria, R. (1976). *Biomechanics and energetics of muscular exercise.* Oxford: Clarendon.

Morgan, D. L., Proske, U. & Warren, D. (1978). Measurements of muscle stiffness and the mechanism of elastic storage of energy in hopping kangaroos. *J. Physiol., Lond.* **282**: 253–261.

Schryver, H. F., Bartel, D. L., Langrana, N. & Lowe, J. E. (1978). Locomotion in the horse: kinematics and external and internal forces in the normal equine digit in the walk and trot. *Am. J. vet. Res.* **39**: 1728–1733.

Sukhanov, V. B. (1974). *General system of symmetrical locomotion of terrestrial vertebrates and some features of movement of lower tetrapods.* New Delhi: Amerind Publishing Co.

Woledge, R. C. (1968). The energetics of tortoise muscle. *J. Physiol., Lond.* **197**: 685–707.

Symp. zool. Soc. Lond. (1981) No. 48, 289–304

Recruitment of Muscles and Fibres Within Muscles in Running Animals

R. B. ARMSTRONG

Department of Physiology, Oral Roberts University, Tulsa, Oklahoma 74171, USA

SYNOPSIS

The purpose of this article is to describe (1) the characteristics of mammalian skeletal muscle fibre types; (2) the general distribution of the fibre types within and among muscles in the extensor and flexor muscle groups of terrestrial quadrupeds; and (3) patterns of recruitment of fibres within muscles and muscles within groups as a function of terrestrial locomotory speed and gait. Mammalian skeletal muscle fibres may be functionally classified as fast-twitch-oxidative-glycolytic (FOG), fast-twitch-glycolytic (FG), or slow-twitch-oxidative (SO) (Peter *et al.*, 1972). The deepest muscles of extensor groups are predominantly composed of SO fibres. Deep portions of more superficial extensor muscles are primarily of the SO and FOG types, whereas the most superficial portions are predominantly composed of FG fibres. Evidence from studies on animal locomotion employing (1) electromyography (e.g., Smith *et al.*, 1977), (2) determination of forces exerted on tendons by muscles (e.g., Walmsley, Hodgson & Burke, 1978), and (3) loss of glycogen in fibres (e.g., Armstrong, Marum *et al.*, 1977) appears mutually supportive and provides information on how muscles and fibres within muscles are recruited during locomotion. Several general conclusions may be drawn from these various experiments: (1) during quiet standing, muscular force is provided almost entirely by the deep, slow-twitch extensor muscles; (2) when animals walk the deep slow muscles continue to provide most of the force, although SO and FOG fibres are also recruited in the more superficial fast muscles; (3) with increasing running and galloping speeds, this peripheral recruitment of fibres within muscles and muscles within groups continues, until during high-speed galloping (or maximal vertical jumping) nearly all fibres within extensor groups may be recruited to produce force; and (4) extensor muscle SO fibres continue to be active even at the highest galloping speeds, suggesting the different fibre types may play specific functional roles in different contractile phases of the step cycle.

INTRODUCTION

Our understanding of the function of the various types of fibres in mammalian skeletal muscles has broadened considerably over the past several years. It now seems feasible to construct a general scheme of how muscles within groups and fibres within muscles are recruited during locomotion, and to relate these physiological observations to the known anatomical arrangement of the fibre types within and among the muscles of functional groups. The purpose of this chapter will be (1) to briefly review the basic contractile and metabolic characteristics of mammalian skeletal muscle fibre types, (2) to describe the distribution of the fibre types within and among muscles in extensor and flexor locomotory muscle groups, and (3) to discuss patterns of recruitment of fibres within muscles and muscles within groups during terrestrial locomotion.

MAMMALIAN SKELETAL MUSCLE FIBRE TYPES

Three reasonably distinct types of fibres may be recognized in the skeletal muscles of most mammals (Ariano, Armstrong & Edgerton, 1973; Armstrong, 1980; Close, 1972; Peter *et al.*, 1972). A number of classification systems have been suggested in the literature for identification of the fibre types (see review by Close, 1972); one system that directly refers to the functional characteristics of the fibres is that proposed by Peter and co-workers (1972), in which fibres are classified as fast-twitch-oxidative-glycolytic (FOG), fast-twitch-glycolytic (FG), or slow-twitch-oxidative (SO).

These three fibre types may be identified from enzyme histochemistry as illustrated in Fig. 1. Fibres with a high (dark staining) myofibrillar ATPase (Padykula & Herman, 1955) activity are classified as fast-twitching, and those with a low (light staining) activity as slow-twitching (SO). From the stain for NADH-oxidoreductase (Novikoff, Shin & Drucker, 1961), fast-twitching fibres (as identified from the ATPase stain) can be categorized as either high-oxidative (dark staining) (FOG) or low-oxidative (light staining) (FG). Muscle enzyme biochemistry has demonstrated that both types of fast-twitching fibres (FOG and FG) have relatively high potentials for glycolysis (Baldwin, Winder, Terjung & Holloszy, 1973; Ianuzzo & Armstrong, 1976; Peter *et al.*, 1972; Saubert, Armstrong, Shepherd & Gollnick, 1973).

Several important physiological and metabolic characteristics of the three fibre types are presented in Fig. 2. The data included in this figure were obtained from experiments on cat skeletal muscle motor units (Burke & Edgerton, 1975); similar relative differences among the three fibre types have been reported for several other laboratory animals (Barnard, Edgerton, Furakawa & Peter, 1971; Close, 1967;

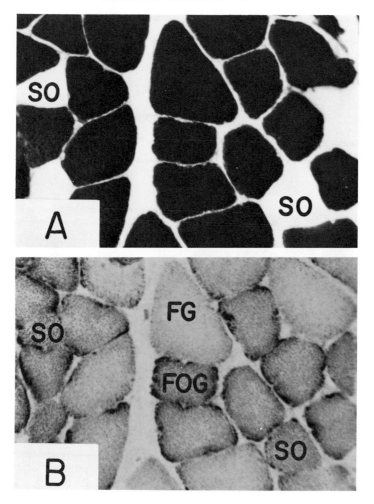

FIG. 1. Serial sections of gastrocnemius muscle from a rat kangaroo (*Bettongia lesueuri*) stained for myofibrillar ATPase (A) and NADH-oxidoreductase (B) activities. One FG, one FOG, and two SO fibres are identified on section B. The two SO fibres are also indicated on section A. Magnification × 325.

Peter *et al.*, 1972). As demonstrated in Fig. 2, SO motor units have relatively slow contraction—relaxation times, but are very resistant to fatigue. On the other hand, fast-twitching units develop high forces rapidly, but fatigue quickly during continuous activity. Because of their high oxidative potential, FOG fibres are more fatigue-resistant than FG fibres.

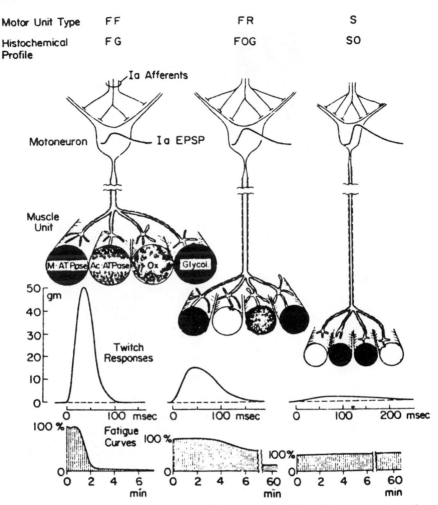

FIG. 2. Histochemical, twitch, and fatigueability properties of three fibre type motor units in cat skeletal muscles. Shading in the fibres represents relative staining intensities for the histochemical stains. Nomenclature: FF, fast-twitch, fatigueable (synonymous with FG type); FR, fast-twitch, fatigue resistant (synonymous with FOG type), S, slow-twitch (synonymous with SO type). Reproduced from Burke & Edgerton (1975).

DISTRIBUTION OF THE FIBRE TYPES AMONG AND WITHIN MUSCLES

Distribution among Muscles

General patterns of fibre type distribution among locomotory muscles may be described for terrestrial bipedal and quadrupedal mammals (Armstong, 1980). As illustrated in Fig. 3, which shows SO fibre density in the muscles of arm and thigh in dog and wallaby, the

Slow-twitch fiber content

Arm (Brachium)

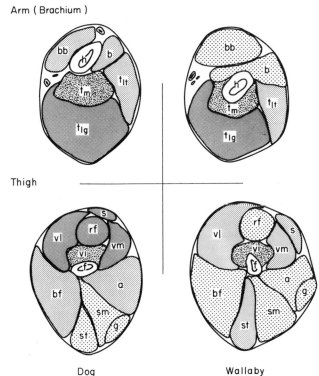

Thigh

Dog   Wallaby

FIG. 3. Cross-sections of arm and thigh muscles of a 10-kg dog (*Canis familiaris*) and a 10-kg wallaby (*Wallabia dama dama*). Density of shading represents the proportion of muscle composed of SO fibres. SO populations ranged from 100% in medial head of triceps and vastus intermedius muscles of wallaby down to 4% in rectus femoris muscle of wallaby. Muscles of the arm: biceps brachii (bb); brachialis (b); triceps brachii, medial (t_m), lateral (t_{lt}), and long (t_{lg}) heads. Muscles of the thigh: sartorius (s); vastus lateralis (vl); rectus femoris (rf); vastus medialis (vm); vastus intermedius (vi); biceps femoris (bf); semitendinosus (st); semimembranosus (sm); adductor (a): gracilis (g). Bones: humerus (h) in the arm and femur (f) in the thigh.

deepest muscle in each extensor group has a high proportion of SO fibres (80-100%). More superficial extensor muscles have fewer SO fibres (20-50%), but generally possess more than the antagonistic flexor muscles (5-25%). In some species the deep extensor muscle may be composed of 100% SO fibres, e.g., medial head of triceps brachii and vastus intermedius muscles in wallaby, or soleus muscle in guinea pigs and cats (Ariano *et al.*, 1973). Generally even the most superficial extensor and flexor muscles contain at least a small percentage of SO fibres, although some are pure fast-twitch, e.g., tensor fascia latae muscle in guinea pigs and rats (Ariano *et al.*, 1973).

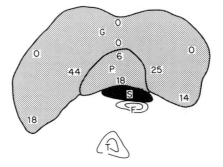

FIG. 4. SO fibre composition of the ankle extensor muscles of the laboratory rat, and the distribution of SO fibres within the muscles. Soleus (S) muscle has about 85% SO fibres, whereas gastrocnemius (G) and plantaris (P) muscles have less than 10% SO fibres. The muscles are shown in relation to tibia (T) and fibular (F). Reproduced from Armstrong (1980).

Distribution within Muscles

The deep slow extensor muscles normally have homogeneous distributions of the fibre types throughout their cross-sections. However, the other muscles in extensor groups show distinct stratification of fibres: the deep portions contain relatively high proportions of SO fibres, whereas the more superficial portions contain fewer SO fibres. Figure 4 demonstrates this "layering" of the fibre types in the calf muscles of the laboratory rat. FOG fibres are distributed within muscles similarly to SO fibres, i.e. there are relatively large populations in the deep portions with smaller numbers in the more superficial parts. Stratification of the fibre types is generally not as pronounced in flexor muscles.

RECRUITMENT OF MUSCLES AND FIBRES WITHIN MUSCLES DURING TERRESTRIAL LOCOMOTION

Three different types of experiments will be cited to construct a scheme for muscle and fibre recruitment during locomotion: (1) electromyography (EMG); (2) force recording from tendons of locomotory muscles; and (3) patterns of glycogen loss in locomotory muscle fibres. The purpose of the discussion will not be to provide a comprehensive review of the literature in each of these experimental areas, but to integrate information from pertinent representative studies of each into a reasonable picture of how muscle fibre activity changes as a function of speed and gait.

In EMG experiments electrodes are positioned in the muscles to record changes in electrical potential resulting from depolarization of muscle cell membranes (sarcolemma and tubular and reticular membranes) and, to a lesser extent, nerve cell membranes that are located in the muscles. Although the amplitude of the EMG may be used to estimate magnitude of the muscular contractions, the most useful information from these recordings is whether the muscle is "on" or "off" at different times during the step cycle. Even this interpretation must be made carefully because of the widely different twitch and relaxation times of the different fibre types (Fig. 2) and the variability in electrode placement in relation to the active fibres. Also, in locomotion experiments it is not possible to resolve the types of fibres that are active from EMG recordings unless the muscles are type-pure.

Force recordings from strain gauges attached to the tendons of muscles during locomotion (Walmsley, Hodgson & Burke, 1978) provide important information about (1) the magnitude of forces generated by the muscles and (2) the sequence, or timing, of the produced forces by different muscles during the step cycle. With this procedure it is not possible to distinguish between active tension produced by isometric or shortening active muscle fibres and the passive tension of eccentric muscle contractions. Also, like EMG recording, the technique does not provide any information about the types of fibres within muscles that are contributing to force production unless the muscle is composed of only one fibre type.

The third experimental procedure that will be referred to is that in which glycogen loss is used to estimate muscle fibre activity. Caution must be observed in the interpretation of glycogen loss data (as discussed in Armstrong, Saubert et al., 1974; Burke & Edgerton, 1975; Gollnick et al., 1973), but under the proper experimental conditions it may be used to estimate whether or not fibres have been active during a bout of exercise. Because of the differences in metabolic capacity and substrate preference of the different fibre types (Baldwin, Klinkerfuss et al., 1972; Baldwin, Winder et al., 1973; Ianuzzo & Armstrong, 1976; Peter et al., 1972) it is not possible to quantify relative amounts of activity among fibres with any degree of confidence. Nor can one determine precisely when during the step cycle the fibres that lost glycogen were recruited. However, the technique does provide important information about the types and numbers of fibres that are active in muscles during locomotion.

Standing

Postural maintenance during quiet standing results primarily from force produced by deep, slow-twitch extensor muscles, as demonstrated by EMG (Campbell, Biggs, Blanton & Lehr, 1973; Smith,

Edgerton, Betts & Collatos, 1977; Walmsley *et al.*, 1978) and muscle force (Walmsley *et al.*, 1978) recordings. Interestingly, soleus muscle in cats (pure SO fibre composition, Ariano *et al.*, 1973) is electrically as active during quadrupedal standing as during running or jumping (Smith *et al.*, 1977), and generates nearly as much force (Walmsley *et al.*, 1978). Thus, all soleus muscle motor units may be recruited simultaneously for postural support. More superficial extensor muscles, on the other hand, contribute minimally to force production during postural maintenance. For example, gastrocnemius muscle shows only occasional low-amplitude bursts of EMG activity during standing, if at all (Campbell *et al.*, 1973; Smith *et al.*, 1977), indicating that relatively few motor units in this muscle are recruited to produce tension. The highest force generated by the medial head of cat gastrocnemius muscle in quiet standing is only about 5% of that produced by the muscle during vertical jumping (Walmsley *et al.*, 1978).

Because soleus muscle in cat is composed entirely of SO fibres (Ariano *et al.*, 1973), and contributes most of the force at the ankle during quadrupedal standing (Walmsley *et al.*, 1978), it is clear that SO fibres are primarily responsible for posture at this joint. In fact, it is probable that the small forces produced in gastrocnemius muscle result from recruitment of SO motor units in that muscle. It seems reasonable to predict that the postural support function within all extensor groups in an animal would follow this same scheme because of the generality of the pattern of fibre distribution in the antigravity muscle groups (i.e. deep slow muscle, superficial fast muscles).

Locomotion

At all locomotory speeds EMG activity (Engberg & Lundberg, 1969; Smith *et al.*, 1977; Walmsley *et al.*, 1978) in extensor muscles commences just prior to foot contact with the ground (E_1). It then continues at varying levels through the active lengthening phase (E_2) to the shortening phase (E_3), which immediately precedes the point in time when the foot leaves the ground. Peak force on the tendons of the extensor muscles within each step cycle occurs during the active lengthening phase (E_2) when the animals's centre of mass is undergoing negative acceleration (Walmsley *et al.*, 1978). Active lengthening is known to markedly enhance muscular force from in vitro muscle physiology experiments (Cavagna & Citterio, 1974).

Walking

EMG and force recordings for soleus and gastrocnemius muscles in a walking cat are presented in Fig. 5. EMG amplitude (Smith *et al.*,

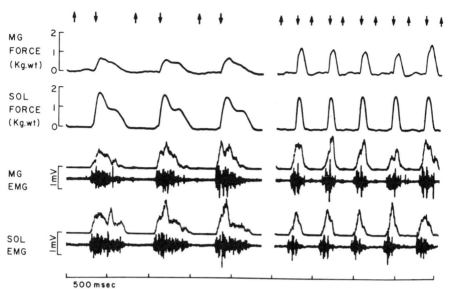

500 msec

FIG. 5. Medial head of gastrocnemius (MG) and soleus (SOL) muscle force and EMG recordings from a cat during slow (left traces) and fast (right traces) walking. EMG recordings show raw (lower traces) and integrated (upper trace) data. Arrows indicate foot down (↓) and foot lift (↑). Reproduced from Walmsley *et al.* (1978), by permission from the American Physiological Society.

1977) and peak force produced (Walmsley *et al.*, 1978) in cat soleus muscle during walking are similar to values both for quiet standing and for high speed running. This indicates that force modulation during locomotion is not accomplished by increasing or decreasing the active cross-sectional area of the deep, slow extensor muscles. On the other hand, force produced by cat gastrocnemius muscle increases as a function of walking speed (Fig. 5). Thus, force modulation in an extensor muscle group during locomotion primarily results from recruitment or derecruitment of motor units in superficial "mixed" muscles. Glycogen loss data from lions (Armstrong, Marum *et al.*, 1976) indicate the major contribution to force production in superficial extensor muscles (in this case, long head of triceps brachii) during fast walking is from SO fibres, with FOG fibres making a smaller contribution (Fig. 6).

Running
Both EMG amplitude (Smith *et al.*, 1977; Walmsley *et al.*, 1978) and muscle force (Walmsley *et al.*, 1978) increase in cat gastrocnemius muscle as a function of running speed (Fig. 7). As indicated above, cat soleus muscle does not change the peak force it develops within the step cycle or the amplitude of its EMG activity as the animal goes

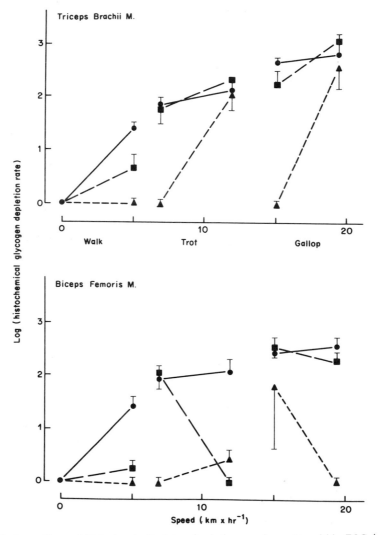

FIG. 6. Logarithms of histochemically-determined glycogen loss rates within FOG (■), FG (▲), and SO (●) fibre populations of muscles as a function of speed. Values represent means ± SEM. Reproduced from Armstrong, Marum *et al.* (1977), by permission from the American Physiological Society.

from standing to walking to running. However, the rate of force development increases in soleus muscle with increasing running speed (Walmsley *et al.*, 1978). This is probably due to the elevated rate of muscle lengthening that occurs with increasing running speed (Goslow, Reinking & Stuart, 1973) against a similar stiffness. Also, as in walking, peak forces in both soleus and gastrocnemius muscles

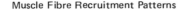

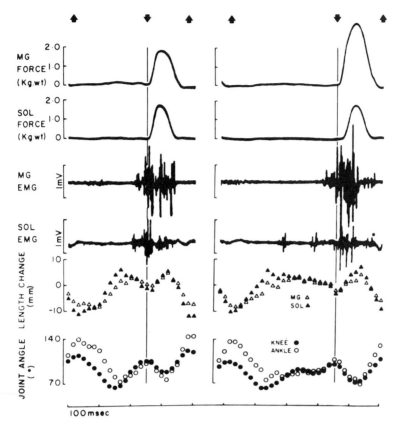

FIG. 7. Records from two step cycles during fast running from the same cat as in Fig. 5. Step cycle on the left was typical of this running speed; that on the right was atypical. Reproduced from Walmsley *et al.* (1978), by permission from the American Physiological Society.

occur during the active lengthening phase (E_2) (Walmsley *et al.*, 1978). Unlike walking, when force was measurable prior to foot-down (E_1) (Fig. 5), force in the extensor muscle tendons during running did not rise until after foot placement (Fig. 7). However, as noted before, EMG activity is apparent in both soleus and gastrocnemius muscles well before the foot touches the ground (Smith *et al.*, 1977; Walmsley *et al.*, 1978). In dogs, the length of time of EMG activity in extensor muscles before foot contact increases with increasing locomotory speed (C. R. Taylor, personal communication). Although no consistent differences occur in the time of onset of EMG activity between soleus and gastrocnemius muscles in cats prior to foot contact, the amplitude in gastrocnemius muscle consistently peaks 20-40 ms later than in soleus muscle (Smith *et al.*, 1977). This

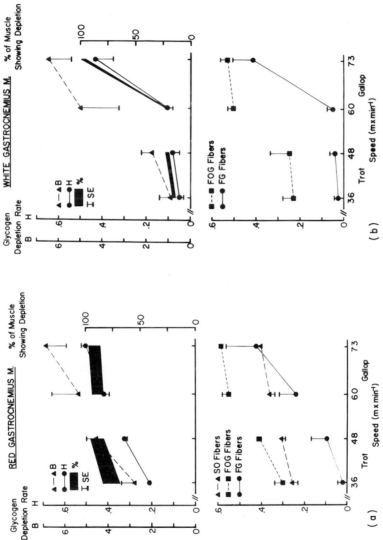

FIG. 8. Biochemical and histochemical estimations of glycogen loss in (a) red (deep) and (b) white (superficial) portions of rat gastrocnemius muscle as a function of running speed. *Upper graph:* biochemically (B) and histochemically (H) determined glycogen loss rates (*left ordinate*), and percentages of muscle cross-sectional areas (shaded areas) showing glycogen loss (*right ordinate*). Ordinate units are mmol glucose units • kg^{-1} (wet muscle weight) • min^{-1} (B) and histochemical units • min^{-1} (H). *Lower graph:* histochemically determined glycogen loss rates within fibre types. Ordinate units are histochemical units • min^{-1}. Reproduced from Sullivan & Armstrong (1978), by permission from the American Physiological Society.

difference is particularly evident at the higher locomotory speeds.

During slow trotting muscular force appears to be produced almost entirely by SO and FOG motor units with minimal contribution by FG fibres (Armstrong, Marum *et al.*, 1977; Sullivan & Armstrong, 1978). This is illustrated in Fig. 6 for long head of triceps muscle in lion and in Fig. 8 for rat gastrocnemius muscle. However, with increasing trotting speed, FG fibres are recruited, and make a marked contribution to force production at maximal trotting speeds (Figs 6 and 8). These relationships are idealized in Fig. 9. No change in the area of soleus muscle showing glycogen loss occurs with increasing trotting speeds, which agrees with the EMG (Smith *et al.*, 1977) and force (Walmsley *et al.*, 1978) data. Another factor that might be involved in increases in muscular force with running speed is increasing spike frequency to the same active motor units within the muscle (temporal summation). Although this possibility exists, evidence indicates that individual units in cats continue to operate at similar activation frequencies with increasing running speeds (Zajac & Young, 1976).

A consistent observation in studies using the glycogen loss technique (e.g., Armstrong, Marum *et al.*, 1977; Armstrong, Saubert *et al.*, 1974; Sullivan & Armstrong, 1978) has been that with increasing

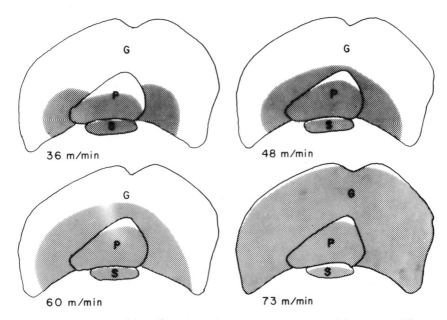

FIG. 9. Idealized active (shaded) cross-sectional areas of rat soleus (S), plantaris (P), and gastrocnemius (G) muscles during trotting (36 and 48 m min⁻¹) and galloping (60 and 73 m min⁻¹).

running speed, when fast-twitch fibres are used in significant numbers, SO fibres continue to be recruited. One explanation for this apparent lack of efficiency in muscular control is that SO fibres continue to be recruited to participate in the lengthening phase (E_2) of the step cycle when the muscles are storing elastic energy (Cavagna, Heglund & Taylor, 1977), whereas the fast-twitch fibres are primarily active during the active shortening phase (E_3) immediately prior to the time the foot leaves the ground. This interpretation receives support from the EMG data referred to above demonstrating the "fast-twitch" gastrocnemius muscle electrical activity peaks later in the step cycle than that for the "slow-twitch" soleus muscle (Smith et al., 1977).

Galloping

Increases in force with increasing galloping speed primarily result from increases in FG fibre recruitment (Armstrong, Marum et al., 1977; Sullivan & Armstrong, 1978). When rats gallop at near maximal speeds, glycogen loss patterns indicate that almost all fibres in the triceps surae extensor group are recruited (Figs 8 and 9). On the other hand, during fast galloping in lions only about 45% of the cross-section of long head of triceps muscle loses glycogen (Fig. 10). Also, interestingly, a marked discontinuity in FG fibre recruitment occurs in this muscle at the trot—gallop transition. Marked FG fibre activity is observed during fast trotting, but at the slow gallop these fibres are not recruited. These data suggest that when the lion changes from trotting to galloping it increases force by recruiting different muscles (e.g., trunk muscles), but must "re-recruit" FG motor units in triceps muscle with increasing galloping speed.

EMG (Smith et al., 1977) and force (Walmsley et al., 1978) data for cats suggest that only during maximal vertical jumps are all motor units in the triceps surae group recruited. Walmsley and co-workers (1978) in fact concluded from their force recordings that even during fast running no FG units would be required to produce the recorded tension in gastrocnemius muscle (medial head) tendon, which is less than 25% of the maximal isometric force the muscle is capable of producing in situ. However, if fibres are recruited for different tasks during the step cycle as discussed above (e.g., SO fibres for eccentric contractions, fast fibres for concentric), extrapolating from tendon force recordings at any given time during the cycle to proportions of active fibres becomes very difficult, if not impossible. This problem is illustrated by the observation that muscles produce greater forces during jumping than when stimulated to perform maximal isometric tetanic contractions (Walmsley et al., 1978).

% OF MUSCLE
CROSS-SECTION SHOWING
GLYCOGEN DEPLETION

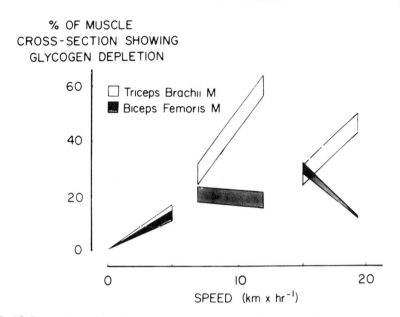

FIG. 10. Proportions of muscle cross-sectional areas showing glycogen loss after running as a function of speed. These values were calculated from percentages of fibres of each type showing any glycogen loss extrapolated to whole muscle cross-sectional area. Reproduced from Armstrong, Marum *et al.* (1977), by permission from the American Physiological Society.

REFERENCES

Ariano, M. A., Armstrong, R. B. & Edgerton, V. R. (1973). Hindlimb muscle fiber populations of five mammals. *J. Histochem. Cytochem.* 21: 51–55.

Armstrong, R. B. (1980). Properties and distribution of the fiber types in the locomotory muscles of mammals. In *Comparative physiology: primitive mammals*: 243–254. Schmidt-Nielsen, K. (Ed.).

Armstrong, R. B., Marum, P., Saubert, C. W. IV, Seeherman, J & Taylor, C. R. (1977). Muscle fiber activity as a function of speed and gait. *J. Appl. Physiol.* 43: 672–677.

Armstrong, R. B., Saubert, C. W. IV, Sembrowich, W. L., Shepherd, R. E. & Gollnick, P. D. (1974). Glycogen depletion in rat skeletal muscle fibers at different intensities and durations of exercise. *Pflügers Arch.* 352: 243–256.

Baldwin, K., Klinkerfuss, G. H., Terjung, R. L., Molé, P. A. & Holloszy, J. O. (1972). Respiratory capacity of white, red, and intermediate muscle: adaptive response to exercise. *Am. J. Physiol.* 222: 373–378.

Baldwin, K., Winder, W. W., Terjung, R. L. & Holloszy, J. O. (1973). Glycolytic enzymes in different types of skeletal muscle: adaptation to exercise. *Am. J. Physiol.* 225: 962–966.

Barnard, R. J., Edgerton, V. R., Furakawa, T. & Peter, J. B. (1971). Histochemical, biochemical and contractile properties of red, white, and intermediate fibers. *Am. J. Physiol.* 220: 410–414.

Burke, R. E. & Edgerton, V. R. (1975). Motor unit properties and selective involvement in movement. In *Exercise and sport sciences reviews* 3: 31—81. Wilmore, J. H. & Keogh, J. F. (Eds). New York & London: Academic Press.

Campbell, K. M., Biggs, N. L., Blanton, P. L. & Lehr, R. P. (1973). Electromyographic investigation of relative activity among four components of the triceps surae. *Am. J. Phys. Med.* 52: 30—41.

Cavagna, G. A. & Citterio, G. (1974). Effect of stretching on the elastic characteristics and the contractile component of frog striated muscle. *J. Physiol., Lond.* 239: 1—14.

Cavagna, G. A., Heglund, N. C. & Taylor, C. R. (1977). Mechanical work in terrestrial locomotion: two basic mechanisms for minimizing energy expenditure. *Am. J. Physiol.* 233: R243—261.

Close, R. I. (1967). Properties of motor units in fast and slow skeletal muscles of the rat. *J. Physiol., Lond.* 193: 45—55.

Close, R. I. (1972). Dynamic properties of mammalian skeletal muscles. *Physiol. Rev.* 52: 129—197.

Engberg, I. & Lundberg, A. (1969). An electromyographic analysis of muscular activity in the hindlimb of the cat during unrestrained locomotion. *Acta physiol. scand.* 75: 614—630.

Gollnick, P., Armstrong, R. B., Saubert, C. W. IV, Sembrowich, W. L. & Shepherd, R. E. (1973). Glycogen depletion patterns in human skeletal muscle fibers during prolonged work. *Pflügers Arch.* 244: 1—12.

Goslow, G. E., Reinking, R. M. & Stuart, D. G. (1973). The cat step cycle: hind limb joint angles and muscle lengths during unrestrained locomotion. *J. Morph.* 141: 1—42.

Ianuzzo, C. D. & Armstrong, R. B. (1976). Phosphofructokinase and succinate dehydrogenase activities of normal and diabetic rat skeletal muscle. *Horm. Metab. Res.* 8: 244—245.

Novikoff, A. B., Shin, W. & Drucker, J. (1961). Mitochondrial localization of oxidative enzymes: staining results with two tetrazolium salts. *J. Biophys. Biochem. Cytol.* 9: 47—61.

Padykula, H. A. & Herman, E. (1955). The specificity of the histochemical method of adenosine triphosphatase. *J. Histochem. Cytochem.* 3: 170—195.

Peter, J. B., Barnard, R. J., Edgerton, V. R., Gillespie, C. A. & Stempel, K. E. (1972). Metabolic profiles of three fiber types of skeletal muscle in guinea pigs and rabbits. *Biochemistry* 11: 2627—2633.

Saubert, C. W. IV, Armstrong, R. B., Shepherd, R. E. & Gollnick, P. D. (1973). Anaerobic enzyme adaptations to sprint training in rats. *Pflügers Arch.* 341: 305—312.

Smith, J. L., Edgerton, V. R., Betts, B. & Collatos, T. C. (1977). EMG of slow and fast ankle extensors of cat during posture, locomotion, and jumping. *J. Neurophysiol.* 40: 503—513.

Sullivan, T. E. & Armstrong, R. B. (1978). Rat locomotory muscle fiber activity during trotting and galloping. *J. appl. Physiol.* 44: 358—363.

Walmsley, B., Hodgson, J. A. & Burke, R. E. (1978). Forces produced by medial gastrocnemius and soleus muscles during locomotion in freely moving cats. *J. Neurophysiol.* 41: 1103—1216.

Zajac, F. E. & Young, J. L. (1976). Discharge patterns of motor units during cat locomotion and their relation to muscle performance. In *Neural control of locomotion*: 789—793. Herman, R. M., Grillner, S., Stein, P. S. G., & Stuart, D. G. (Eds). New York: Plenum Press.

Symp. zool. Soc. Lond. (1981) No. 48, 305–329

Locomotor Loading and Functional Adaptation in Limb Bones

L. E. LANYON

Department of Veterinary Anatomy, University of Bristol, Park Row, Bristol, England and School of Veterinary Medicine, Tufts University, Boston, USA

SYNOPSIS

Functional activity influences bone remodelling to produce and maintain the "normal" shape and mass of each bone. It has been supposed that the controlling influence arises from the functional strain pattern within the bone, and that the response which this engenders is directed towards producing an optimum mechanical structure. This structure must provide not only adequate fatigue life with respect to the strains produced within its tissue by co-ordinated activity, but also sufficient strength to withstand a reasonable proportion of the loading "accidents" which it is likely to encounter.

If the predominant structural "goal" of bones were maximum tissue economy in relation to the loads of co-ordinated activity, then their shape would be directed towards restricting the manner of this loading to axial compression, and the thickness of their cortices would be adjusted proportionally to its amount. These two effects would result in a uniform (longitudinal compressive) functional strain level throughout the shaft of the bone. This level would be sufficiently high to be economical in the use of tissue, but sufficiently low to ensure adequate safety factors to monotonic failure, and an adequate fatigue life in relation to the animal's lifespan, activity level, and rate of secondary bone remodelling.

Strain gauges attached to the radius and tibia in horses and dogs showed that the manner of locomotor loading of these bones remains almost constant not only during each stride, but also over the animal's range of speed from a walk to a gallop. Although gait transition from a walk to a trot involves increased limb bone loading, that from a trot to a canter results in a reduction. Constancy in the manner of loading means that one structure is equally appropriate for each gait. Reduction in the amount of loading from a trot to a gallop means that increase in speed can be achieved without progressive erosion of safety factors to failure.

Strain gauge studies on other bones have shown that their individual manner of loading is also constant throughout their locomotor range. However, whereas in some bones all the cortices are under compression, in others there is enough bending for the whole thickness of opposite cortices to be either in tension or compression. Furthermore the development of curvatures in some long bones contributes to rather than reduces this functional bending. Cortical thickness is also not proportional to load so that functional strain levels differ throughout individual bones, and between different bones. Nevertheless although the functional strain levels on opposing cortices of a bone may differ by almost a factor of two, experimentally induced strain reorganization resulting in increases in peak walking strain of only 10-20% results in adaptive remodelling and additional bone formation. Each bone must therefore be considered as an organ of which the form, architecture and mass are the result of mechanically responsive remodelling directed towards attaining and maintaining an appropriate relationship between structure and loading. The results of this relationship are the dynamic functional strains within the tissue. The resulting variation in level and distribution of strain throughout one bone and between different bones suggest that any fundamental optimum relationship between the strain environment and the properties and arrangement of bone tissue, may be subordinated to wider considerations relevant to the bone's overall organization as a structure.

MECHANICAL FUNCTION AND BONE FORM

During normal development mechanical function provides a stimulus for the formation of the characteristic detailed architecture of each bone. Bones in paralysed limbs show that in the absence of this mechanical stimulus a basic structure develops of which the form is broadly recognizable as a tibia or femur but in which the girth and cortical thickness will be less than normal, and certain lines, crests, tuberosities and curvatures will be absent or poorly developed (Howell, 1917; Washburn, 1947; Gillespie, 1954; Moore, 1965; Murray, 1936; Ralis et al., 1976; Lanyon, 1980). Thus the normality of each bone in the adult is the product of a normal functional environment having acted upon a basic genetically determined bone structure receptive to mechanical influences. The presence of each anatomical feature is likely to provide some mechanical advantage to the musculo-skeletal unit concerned, and the development of each feature is likely to be related to the aspect of the bone's mechanical situation involved. The reversibility of this adaptive process, and the extent to which features that required mechanical function for their development will regress when normal function is removed, is difficult to assess since the bone remodelling process is so slow, particularly in the mature animal. Compared with the benefits conferred by its presence in the functional situation there seems little benefit in actively removing a tuberosity just because it is no longer required. Nevertheless, while gross features such as this may persist in the absence of continued function, a striking but reversible loss of bone mass does accompany bed rest, immmobilization and space flight, all of which involve reduced levels of functional activity (Donaldson et al., 1970; Uhthoff & Jaworski, 1979; Whedon et al., 1977).

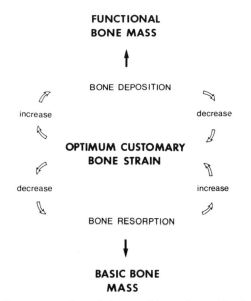

FIG. 1. A schematic representation of the possible strain-sensitive feed-back loops which influence bone remodelling, control bone mass and shape, and maintain presumably optimum strain levels. The mass of bone tissue in normal functional bones is achieved and maintained by the effect of functional strains acting originally on a basic genetically determined structure. It is probably the removal of this functional stimulus which allows regression back towards the genetic structure, but it is possible that resorption following disuse is a positive reaction to reassert the optimum functional strain level.

This loss in bone tissue is achieved by reduction in the amount of cancellous bone and a decline in the girth and cortical thickness of the bone's shaft (Uhthoff, Jaworski & Liskova-Kiar, 1979).

This remodelling response may be a simple regression towards the genetically determined base-line value of bone mass which had been elevated by the effect of normal activity, or it may be an active attempt to restore customary strain levels in the light of declining functional loads (Fig. 1).

If there are any mechanical advantages to be gained from reducing bone mass in the face of decreased functional activity they are likely to be marginal. The need to limit the level of strain engendered during function is, on the contrary, essential. The level of normal functional strain which exists within bone tissue has presumably been attained, and is maintained, by adjusting bone mass to a compromise level where adequate safety factors to failure or damage are preserved with reasonable tissue economy. Thus, the professional tennis players who have developed 35% more bone in the humerus on their playing than on their non-playing side (Jones et al., 1977), can be considered to be preserving the same safety factors to failure or damage within

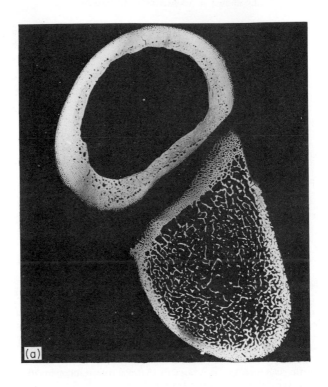

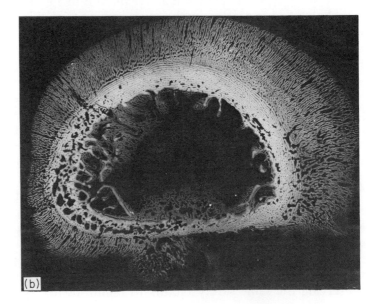

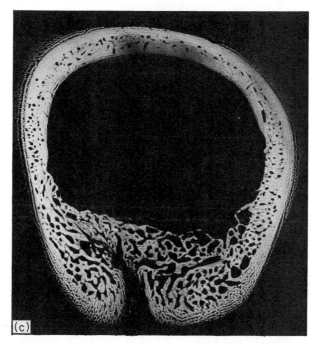

FIG. 2. (a) Microradiograph of 100 micron section taken from the mid-shaft of a radius and ulna in a 4-month old growing pig. The area of bone in the radius is approximately equal to that in the ulna. (b) The radius undergoing rapid "explosive" adaptive hypertrophy 3 weeks after removal of the ulna. (c) A mid-shaft section of the radius 12 weeks after removal of the ulna. Bone is now being laid down in a normal circumferential arrangement. The area of bone in the radius equals that in the radius and ulna on the opposite (unoperated) side. The peak walking strain values on the two radii have also equalized. From Goodship, Lanyon & McFie (1979).

their two humeri by increasing the mass of the one that receives the more rigorous mechanical battering.

Qualitative evidence that it is intermittent and not static loading which influences bone mass has been provided by artificial loading of rabbit tibiae *in vivo* (Hert, Liskova & Landa, 1971). Quantitative verification that this response is linked to the level of functional strain was impossible until the means for determining bone strain during functional activity were provided by the development of the technique for attaching strain gauges to bone surfaces (Lanyon & Smith, 1970; Lanyon, 1976). This technique has been used to quantify the functional bone strain/bone mass relationship in growing pigs where the radius was overloaded by removal of the ulnar shaft (Goodship, Lanyon & McFie, 1979) (Fig. 2). The compressive overstrain which this produced in the radius during walking was 2—2.5 times normal, an increase which immediately engendered new bone formation in

the radius. This new bone formation was sufficient to completely compensate for the loss of the ulna by restoring the total area of bone in
the forearm cross section. Since the radius was loaded in compression
this restoration of total bone area resulted in equilibration of functional strain levels at the gauge site between radii on operated and
unoperated sides. This supports the hypothesis that bone mass is not
only influenced by, but is perhaps also regulated by, the level of customary intermittent strain produced within it by functional loading.

LOCOMOTOR LOADING OF LIMB BONES

The strain within each limb bone during locomotion results from the
interaction of loads transmitted through the limb segment and those
engendered within it. Most of the animals whose limb bones have been
instrumented with strain gauges have been digitigrade. In these
animals the manner of limb bone loading is indicated by the angle of
the principal strains on the surface of each cortex, and the strain
magnitudes on opposite cortices. These remain practically constant
during the stance period of the stride (Lanyon & Baggott, 1976;
Lanyon & Bourn, 1979) (Fig. 3). It has also been shown in dogs and
horses (Rubin & Lanyon, 1979; Lanyon & Rubin, 1980) that this
manner of loading remains essentially constant throughout the
animal's range of speed (the ratio of strain due to bending and compression changes by only 10%) (Fig. 4). Thus the activity of the
intrinsic limb musculature (Tokuriki, 1973a, b, 1974) conspires with
the pattern of external loading to present to the limb bones a
dynamic strain regime which is restricted in character. Since normal
adult bone form results at least in part from functionally adaptive
bone remodelling, and since this process is sensitive to dynamic
rather than static loads (Hert *et al.*, 1971), it follows that bone
morphology is likely to be predominantly appropriate to the functional loading situation which influenced its development.

   While the manner of bone loading is kept fairly constant through
an animal's locomotor range, its magnitude, and thus the peak strains
induced at each stride, increases with speed but not with a simply proportional relationship. In the horse and dog change in gait from a
walk to a trot results in a sharp increase in peak strain, but the
change from a trot to a canter results in a substantial decrease (Fig.
4). In the tibia the values of peak strain achieved during a fast trot
are re-attained at a fast gallop, but in our treadmill experiments at
least they were not re-attained in the radius.

   Thus while co-ordination of muscle activity and external forces

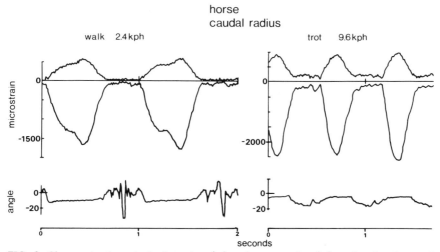

FIG. 3. Changes in the principal tension (+) and compression (−) strains, in microstrain (×10⁻⁶) and the angle of principal compressive strain to the long axis of the horse radius (degrees) recorded during a walk and a trot from gauges attached to the bone's midshaft in a 137 Kg animal. The angle of the strain to the bone's long axis remains practically constant during the stance phase at both speeds and is similar at both speeds. From Rubin & Lanyon (in preparation).

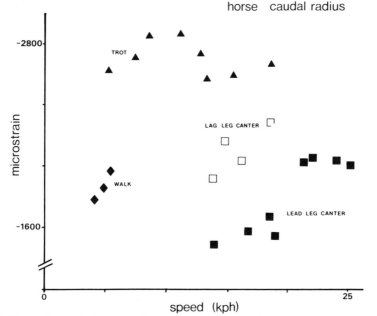

FIG. 4. The peak principal compressive strain during each stride on the caudal surface of the horse radius plotted against the animal's speed on a treadmill. The bone strain increases steeply with speed during the walk, there is a sharp increase at the walk—trot transition, it remains almost constant during the trot, and drops steeply at the trot—canter transition. From Rubin & Lanyon (in preparation).

restricts the manner of skeletal loading, the sequencing of limbs and redistribution of loads which result from gait change regulate its magnitude. These two features narrow the bone's range of commonly encountered loading situations in both size and manner, and so provide a restricted design problem which should permit a unique, and economic, structural solution.

One of the primary design requirements for limb bones is to be able to withstand normal functional loading. The strain induced by this loading is also the stimulus under the influence of which the bone develops. Thus, while the general form of the bone is genetically determined, a precise individual match between each structure and its mechanical environment should result from the influence which the manner and magnitude of loading has on both the mass and orientation of bone tissue. Although peak strain magnitude at each stride does not increase in simple proportion to locomotor speed, the increase in strain rate is almost linear (Rubin & Lanyon, in prep.). The strain related remodelling response of bone is profoundly influenced by the rate at which the strain is applied. Overstrain at high strain rates induces far more new bone formation than similar overstrain at low rates (Lanyon & O'Connor, 1980). Thus while the manner of bone loading, and therefore the induced strain pattern, is similar at high and low speeds, locomotion at high speed would be far more effective at increasing bone mass than that at low speeds even if the peak strains were less.

Normal bone form reflects the influence of dynamic functional loading on a genetically determined structure. Since the manner of locomotor bone loading is restricted, one structure is appropriate in shape for each gait, but high speed locomotion which involves high bone strain rates will have a greater effect in determining bone mass.

BONES AS INTERNAL FORCE TRANSDUCERS

The information on locomotor bone loading provided by the use of bone-bonded strain gauges is relevant not only to the architecture of the bone, but also to locomotor mechanics as a whole. A bone with two or more strain gauges attached round its circumference acts as a force transducer within the animal which is sensitive to all the bending, compressive, and torsional loads engendered within, and transmitted through, that segment of the limb. In the horse maximum loading in the radius occurs at a trot whereas in the tibia trotting levels are also achieved at a high speed gallop. The maximum stresses within these bones due to both bending and axial compression were determined from the recorded strains and compared with the estimates calculated by Alexander *et al.* (1979) for fast moving animals. Alexander's estimates for the buffalo tibia were 4.7×10^7 N m^{-2} for stress due to bending and 1.1×10^7 N m^{-2} for that due to axial compression. Our data for the horse tibia provide remarkably similar values of 4.6×10^7 for bending and 0.7×10^7 for compression. The reduction in bone strain which occurs in both the radius and tibia of horses and dogs at the trot/canter transition reflects a decrease in the amount of limb bone loading, not an alteration in its manner. This reduced loading is associated with the reduction in peak vertical force on the ground shown between these two gaits (Cavagna, Heglund & Taylor, 1977) which may be made possible by recruitment of the musculature of the trunk and a more efficient forward use of movement of the animal's centre of mass. While this is naturally beneficial for the animal's energetics as a whole, allowing employment of additional muscles and redistribution of elastic storage from the limbs to the trunk (Cavagna *et al.*, 1977; Taylor, 1978), for the limb bones it means that high speeds can be achieved without progressive erosion of safety margins either to yield strain or fatigue failure.

DESIRABLE OBJECTIVES IN BONE DESIGN

To fulfil their structural role bones must not only provide an adequate fatigue life in relation to the repetitive predictable loads of normal function, but also be able to withstand a reasonable proportion of those uncoordinated loading "accidents" which could cause monotonic failure. To achieve high safety factors either to monotonic fracture or fatigue failure it is necessary to keep the stress within the bone tissue as low as possible. The presence of crests and tuberosities appears to be an obvious local reaction to the need to provide substantial anchorage for muscle groups and yet avoid stress concentrations. The functional dependence of these structures suggests a causal relationship between loading and bone mass which is possibly mediated by the repetitive strain induced within the tissue as a result of normal functional activity. For instance, tension at undeveloped muscular insertions will cause unacceptably high strain levels, which by inducing new bone formation locally within the genetically determined basic bone structure, engender crests and tuberosities whose size is sufficient to reduce tissue strains to an acceptable (and presumably optimum) level.

It is conventional modern engineering practice, especially when manufacturing objects in large numbers, to rank economy in the materials used almost as high in importance as the mechanical competence of the structure produced. It has generally been assumed that the design of the skeleton carries similar priorities. The strategies available to reduce the levels of functional stress within a bone's tissue are to adjust its material properties, its mass, its shape, and its orientation with respect to its manner of loading. For a long bone the most economical loading situation occurs when it is axially compressed. This ensures that the bone tissue is loaded in its strongest mode, and that the least strain is induced for the greatest load. Certainly in cancellous bone the precise, and apparently mechanically advantageous, orientation of trabeculae in the direction of the principal tensile and compressive strains has been regarded as an adaptation whereby these elements can achieve the advantages of axial loading and avoid the high strains associated with bending (Koch, 1917; Lanyon, 1974; Pauwels, 1973). However, whereas it is possible for the individual elements of a bone's internal cancellous architecture to be arranged to avoid bending loads, the position and loading situation of whole bones makes it inevitable that some bending moments are applied to them. Many of the bones within the limbs also develop characteristic longitudinal curvatures, the significance of which has been considered by Frost (1973) to be an adaptation to neutralize the effect of this external bending.

Frost (1973) followed traditional thought by proposing that since bone is strongest in compression and most likely to fail in bending, and since a strut can carry the greatest load with the least strain if it is loaded axially, it would be logical for bones to adjust their shape so as to be normally loaded in this economically advantageous manner. He therefore proposed that bone curvatures are actually developed to establish axial loading, and minimize bending, along a shaft the position and manner of loading of which make the application of some bending moments inevitable. Frost considered that axial loading would be achieved when the bending in the bone arising from contraction of its adjacent musculature is precisely matched by that induced by end-on loading of a structure which is curved. This arrangement could be achieved if dynamic flexural strains within the bone influenced its cortices to drift towards the concavity so induced. Such cortical relocation is certainly feasible and once it had established compressive strain across the bone's section regulation of strain magnitude could be accomplished by adjusting cortical thickness according to the level of local functional loading. In this way the situation could be produced where all the tissue in a long bone shaft is customarily strained at a uniform, and presumably optimum, strain level (the "critical stress" level proposed by Pauwels (1973) and the "optimum customary strain" proposed by Lanyon, Goodship & Baggott (1976)). In areas of local stress concentrations this level would be achieved by local strain-engendered new bone formation which would increase cortical thickness. The strain or stress level thus established throughout the bone would naturally be such that there would be an optimum balance between tissue economy on the one hand, and an acceptable level of fatigue damage and incidence of gross fracture on the other.

In the absence of any *in vivo* strain data it seemed possible that adult bone form was the result of loads within the bone tissue acting locally to produce a minimum structure with a uniform functional strain level. The simplicity, suitability and fundamental elegance of such a mechanism made it attractive; and since bone can adjust its shape in relation to its mechanical situation the means whereby it could be achieved seemed to be available.

THE SIGNIFICANCE OF BONE CURVATURE

Despite the simplicity and apparent advantages of the relationship just outlined, the experimental evidence does not support its existence. Although some bones (i.e. the horse and sheep metacarpus) are loaded during locomotion so that all their cortices are under compression, others (the radius, tibia and humerus) are subjected to sufficient bending that some of their cortices are in tension while others are in compression (Lanyon & Baggott, 1976; Lanyon & Bourn, 1979; Lanyon *et al.*, unpublished data) (Fig. 5). Furthermore, although the position of these bones within the limbs makes it inevitable that some external bending moments are applied to them, the longitudinal curvatures that they develop are not appropriate to neutralize this bending. In dogs, sheep and horses the radius, for instance, which is subjected during locomotion to cranio-caudal bending has a slight, cranially convex, longitudinal curvature. The principal muscles adjacent to the radius which are active (Tokuriki, 1973a, b, 1974) to support the animal during the stance phase of the limb are the digital and carpal flexors which lie on the bone's caudal surface. Loading of the radius through its joint surfaces would bend the bone's shaft to produce longitudinal tension on the cranial (convex) surface and compression on the caudal (concave) one. Tension in the digital and carpal flexors during the stance phase of locomotion would accentuate rather than counteract this bending (Fig. 5).

There seems no obvious mechanical advantage relevant to co-

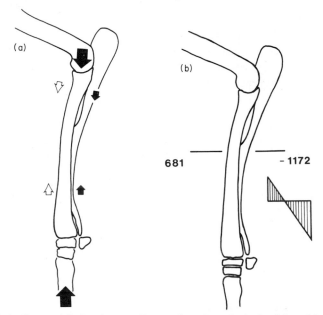

FIG. 5. (a) A diagram of the sheep radius to show its principal origins of load. Loading through its joint surfaces (large black arrows) will cause bending with the cranial (convex) surface in longitudinal tension and the caudal (concave) surface in longitudinal compression. Tension in the flexor muscles on the concave surface (small black arrows) supports the animal during its stance phase and contributes to this bending. Only tension in the extensor musculature on the convex surface (open arrows) or relocation of the bone's cortices towards the compressive surface could neutralise this bending. (b) The strain situation on the sheep radius at peak strain during walking. The cranial (convex) surface of the radius is under longitudinal tension (681×10^{-6}), the caudal (concave) surface is under longitudinal compression (-1172×10^{-6}). The bone's natural curvature accentuates the bending within the bone which arises owing to compression through the joint surfaces and tension in the flexor musculature. From Lanyon, Magee & Baggott (1979).

ordinated functional loading to be gained by an arrangement which, by ensuring bending loads, narrows the safety margins to failure and decreases fatigue life, or increases the amount of bone tissue necessary to prevent strains becoming unacceptably high. Yet despite its apparent disadvantages, the curvature of the radius was probably not developed in spite of, but under the influence of, its normal mechanical environment.

In the rat tibia development of longitudinal bone curvature has been shown to depend on functional activity of the limb. Neurectomy of the muscle groups adjacent to the tibia in growing rats results in this normally S-shaped bone growing to be almost straight (Fig. 6) (Lanyon, 1980). Unfortunately it is not possible to determine whether the functional stimulus for the development of this curvature stems from the effect on the periosteum of an active musculature adjacent to it, or alternatively from the effects of the

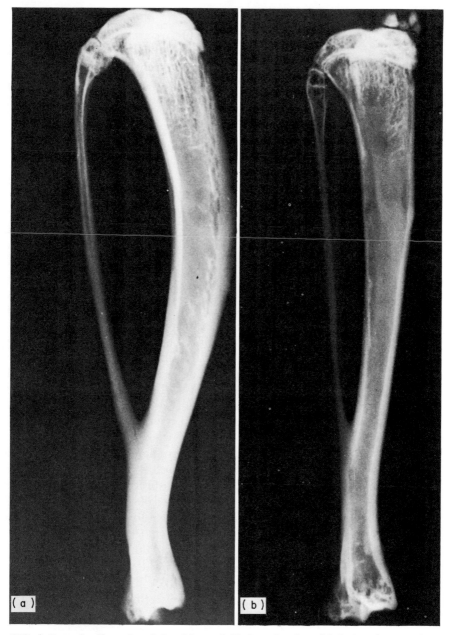

FIG. 6. Lateral radiographs of the tibiae and fibulae taken from (a) the leg of a normal rat, and (b) the leg of a rat which had had a unilateral sciatic neurectomy on that side during its growing period. The pictures were taken at the same radiographic exposure so their mineral content is reflected by the blackening of the film. The length of the two bones is similar but the normal bones are wider, their cortices are thicker, they contain more cancellous bone and their curvatures are developed. The bone on the neurectomized side, deprived of normal functional influences, is nearer the basic genetically determined bone structure. From Lanyon (1980).

strain within the bone tissue which result from normal load bearing. Pressure on the periosteum from tumours, aneurysms and blood vessels can certainly cause local bone resorption, and muscle pressure could induce drift of bone cortices resulting in a curvature. One possible advantage of this curvature would be to provide space and a straight uninterrupted direction of pull for the muscle bellies. The penalty to the bone from this arrangement would, however, be the high strains induced by bending and thus the greater amount of tissue necessary to preserve adequate safety factors to failure. This would seem a high penalty to the bone for a small benefit to the muscle, and so it is reasonable to seek alternative advantages relevant to the bones themselves. If the formation of a bone curvature is not related to its adjacent musculature but rather to the functional strains within its tissue this would imply that the high bending strains are perhaps not a penalty to the bone but a remodelling objective. Thus curvatures are perhaps developed not regardless of the high strains they induce, but specifically to engender them. This concept is directly contrary to conventional ideas on bone morphology which have concentrated on the advantages of tissue economy. However, intermittent strain in bone may not be the entirely deleterious phenomenon that it would be in an inanimate object since it may provide some benefit to the physiology of the tissue. The rapid removal of bone tissue which follows disuse may be a simple regression towards the basic bone structure in the absence of continued function, a reaction which would have some advantages since transport of superfluous tissue is wasteful of energy to the whole organism. Alternatively the absence of a sufficiently high strain level may involve a positive disadvantage to bone tissue. This disadvantage could arise from failure of movement of tissue fluid, or failure in the generation of strain-related electrical potentials, either or both of which may be beneficial to bone cell physiology (Lanyon & Hartman, 1977; Starkebaum, Pollack & Korostoff, 1979). If a continuing, intermittent strain environment is not simply the stimulus for maintaining bone mass, but is also beneficial to bone tissue, then the bone loss which follows reduced loading could be a positive attempt to maintain pre-existing strains, or it could reflect the absence of a benefit derived from higher strains.

A FUNCTIONAL STRAIN RANGE FOR BONE TISSUE

If both low and high functional strains are, for different reasons, detrimental to bone tissue this implies that there is an optimum or tolerable functional strain range, the lower limits of which are determined by physiological criteria and the upper limits by mechanical criteria. The lower limit could be when the cellular population is inadequately supplied with nutrients or suitable stimulation. The upper limit (which if exceeded induces new bone formation) would be the level which allows reasonable tissue economy and a realistic fatigue life in relation to the animal's life-span and level of activity. If such a tolerable strain range exists it would be an intrinsic property of bone tissue. The upper limit of the tissue's tolerable strain range should be encountered in locations where there has been strong selection pressure for a minimum structure as would be expected in the wings of birds. Measurements in the goose humerus during flight (Lanyon, unpublished data) showed that the peak strains engendered with each wing beat were -2600×10^{-6}. Although high, this is no higher than the peak strains induced within the radius with each stride during trotting in the horse (Lanyon & Rubin, 1980). In the horse metacarpus, however, which is compressed across its whole surface, the peak strain during trotting is only -1200×10^{-6}. The radius is loaded in compression and bending, the metacarpus almost in axial compression. It is significant that the radius has not utilized its remodelling capability to adjust its curvature and be loaded in axial compression and that the metacarpus has not capitalized on its advantageous manner of loading to reduce its mass and establish strains as high as in the radius. The tolerance to mechanical strain of the tissue in the horse radius is unlikely to be any different from that in the metacarpus, so the reason for the difference in functional strain between these two bones occurs not as a result of considerations relevant to their tissues but rather considerations relevant to their structure, location, or mechanical responsibility.

The possible effect on skeletal morphology if bones did maintain a high uniform strain level but aligned themselves to be loaded in axial compression is illustrated by the radius and ulna of the sheep (transverse section Fig. 7b). The peak principal strain on the cranial surface of the radius during walking is some 700×10^{-6} (tension), that on the caudal surface -1200×10^{-6} (compression); the strain due to bending is therefore 950×10^{-6}, that due to compression -250×10^{-6}. The ulna is subject to compression across its whole surface and the strain is -1400×10^{-6}. Since the area of the radius is some five times that of the ulna (120 compared to 20 mm^2) the compressive load carried by the two bones is almost equal. Thus if bending moments could be eliminated the compressive loads of locomotion could be supported with acceptable functional strain levels by two solid bones of the dimensions of the ulna.

The lack of stability and the small surface available for muscle attachment that this arrangement would provide is evident. Thus resistance to buckling and lateral bending, and provision of adequate surfaces for both joints and muscle attachments are requirements which may dictate a certain minimum size, shape or mass of bone tissue. Under these constraints the longitudinal functional strain in the mid-shaft region of some bones could be sub-optimal if they were loaded axially, but would be brought within the optimal strain range if a curvature were present to engender strain due to bending.

STRAIN DISTRIBUTION IN BONE STRUCTURES

While considerations of the physiology and safety factors to failure of the tissue determine the possible range of functional bone strain throughout the skeleton, it appears that "whole bone" considerations distort this simple "minimum tissue" or "maximum safety factor" design. The variation in these "whole bone" considerations in different locations must account for some of the diversity in the relationship between bone form and customary loading. Thus the need for the radius to accommodate adjacent musculature may impose the curvature which makes high strains due to bending inevitable. Alternatively the need for large areas for muscle attachment imposes a certain size in which optimum strains may only be achieved if substantial bending occurs. The metacarpus with neither of these needs but a high functional compressive loading can be straight and loaded axially. However, the metacarpus being in a more accident prone situation than the radius cannot afford the economically low tissue mass that its advantageous manner of customary loading should make possible if strains as high as those in the radius were acceptable. Thus the metacarpus, which as a structure has to be sufficiently massive to withstand accidents, consequently has a normal (functional) level of strain within its tissue which is practically half that in the radius. This level, although it may be at the lower end of the tolerable strain range, must clearly still be above any minimum strain level necessary for the bone's physiology.

Such explanations must remain speculative until we have enough information better to appreciate the structural goal of bones *vis-à-vis* their customary mechanical environment. At present it is reasonable to assume that the final conformation of limb bones, including their curvature, girth and cortical thickness, is achieved when an acceptable structural compromise is reached between various possibly conflicting anatomical, mechanical, and biological requirements. The extent to which each bone's unique structural goal is reached as a result of the influence of its customary mechanical environment remains to be determined.

THE MECHANICALLY ADAPTIVE REMODELLING RESPONSE OF BONE

It has already been suggested that bone remodelling adjusts bone mass to achieve and maintain a certain optimum level of customary functional strain but that structural considerations may distort any universal relationship between the tissue and its mechanical circumstances. In the experiments previously mentioned the ulna was removed in growing pigs and this caused a substantial overstrain (2–2.5 times) in the radius. Adaptive remodelling resulted in a compensatory increase in bone mass which equalized functional strains (and bone cross-sectional areas) on operated and unoperated limbs. In this situation the pig radius clearly restored a pre-existing relationship between load and strain by adjusting its girth and cortical thickness. The sensitivity of mechanically related bone remodelling in relation to a much smaller reorganization in customary strain in a bone with a non-uniform strain distribution was illustrated in an essentially similar experiment in sheep. In this animal the radius is loaded during locomotion in compression and bending so that its cranial, longitudinally convex, surface is subjected to longitudinal tension, and its caudal concave surface to compression. The caudal cortex is slightly thinner than the cranial and the functional strain on its surface is 1.8 times as much (Lanyon, Magee & Baggott, 1979). In the proximal third of the forearm, radius and ulna are separate although the ulna, which is much the smaller of the two bones, is fused to the radius both proximally and distally. In this experiment part of the ulnar shaft was removed at the level of the interosseus space (Fig. 7). The mechanical effect of this was to increase walking strains in the radius by 15–20% on the cranial (tension) cortex and 5–10% on the caudal (compression) cortex. The animals were confined in pens except for one hour every day, during which they were walked round a circular track. The new bone formation which resulted from the mechanical reorganization was marked with fluorescent labels. Despite the small increase in the bone's customary (walking) strain induced by ulnar osteotomy (an increase which could easily have been achieved by walking slightly faster) the radius responded by forming new subperiosteal bone predominantly on the compressive cortex (Fig. 7) (Lanyon, Goodship, O'Connor & Pye, 1980). After 12 months the amount of new bone formation produced by this adaptive reaction resulted in the radius restoring the area of bone in the forearm lost by removal of the ulna. The effect of this was to equalize the strain due to compression in osteotomized and unosteotomized radii. The strain due to bending was reduced, however, so that the total strain on the cranial (tension) cortex of the radius during walking was 20%

L. E. Lanyon

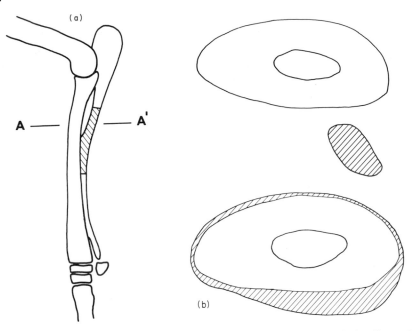

FIG. 7. (a) A lateral outline traced from a radiograph of a sheep radius and ulna illustrating the cranially convex curvature of the radius and the area (shaded) of the ulna removed in the overload experiment described in the text. (b) Transverse outlines traced from sections taken at the level A-A' on Fig. 7a. The upper section shows the normal radius with the much smaller ulna adjacent to the lateral edge of this caudal surface. The new bone formation which resulted over a one-year period following removal of the ulna is shaded on the lower section. Although the cranial (upper) surface experienced the greater (20%) functional overstrain the new bone formation was predominantly deposited on the caudal surface which had only a 10% overstrain but a higher absolute normal strain (see Fig. 5b). From Lanyon, Goodship, O'Connor & Pye (1980).

less than in normal unosteotomized limbs, while the strains on the caudal cortex were 10% less. It is, of course, impossible to discount the possibility that this new bone formation was induced by factors other than the change in mechanical circumstances of the radius. However, assuming that mechanical factors were involved, this experiment and the strain data from normal bones in sheep and other animals demonstrate a number of features on which possible hypotheses can be based. These features are:

(1) The normal functional loading patterns of most but not all bones involve sufficient bending that the whole thickness of the cortices on opposite sides of the bone may be subjected to longitudinal tension or compression. A uniform strain distribution across the section of these bones is therefore impossible.

(2) Local cortical thickness is not proportional to local functional

load so in addition to the sign of the functional strain (tensile or compressive) being different in different locations, the magnitude is different also. (In the sheep radius the peak compressive longitudinal locomotor strain on the caudal surface is 1.8 times the tensile longitudinal strain on the cranial surface.)

(3) Despite the large discrepancy between normal functional strain values between different locations on the same bone and between different bones, strain reorganization involving only slight increases in customary strains results in new bone formation. In the sheep radius experiment this reaction cancelled (indeed overcompensated for) the initial overstrain. However, although the strain due to compression was restored, and that due to bending overcompensated for, the new bone was not deposited evenly around the section nor at the site of greatest proportional overstrain but at the site of greatest absolute strain. Adaptive remodelling in this situation appeared to be more probably a co-ordinated response to the strain imbalance than to the slight increase in strain level. The site of the new bone was strategically placed in the region normally subjected to the highest functional strain.

(4) Although strain levels which are optimum in some locations would lead to adaptive new bone formation or bone loss in others, the customary functional strain at each location appears to be maintained within very close limits indeed. It seems possible that bone can distinguish between strain due to bending and that due to compression. In order for it to be able to do this it must be able to integrate for strain across its whole section. The adaptive reaction responsible for establishing this situation cannot be regulated by a simple, locally controlled process directed to achieving a uniform optimum strain level for bone tissue. Instead it must be directed towards establishing an individually appropriate structural situation at each location throughout the skeleton.

HYPOTHESIS ON THE RELATIONSHIP BETWEEN FUNCTIONAL STRAIN, BONE MASS AND FORM

A possible hypothesis based on these features could be that genetic direction and functional adaptation interact to determine the shape, external proportions and some features of the internal architecture of each bone. In long bones these effects do not result in a uniform loading mode, nor do they result in the thickness of each cortex being individually adjusted according to the loads placed upon it.

Functional loading of each bone therefore results in an uneven strain distribution throughout its structure. During vigorous activity the strains in the highly strained areas of some bones approach the "maximum tolerable" strain level of bone tissue with respect to acceptable degrees of fatigue damage or minimum pre-determined safety margins between functional strain levels and yield or ultimate strains. This "maximum tolerable" strain is the property of the tissue and will always act as a threshold level which, if exceeded, will induce new bone formation to increase the mass of the structure of which it is a part. It follows that this "maximum tolerable" tissue threshold will also be the highest strain value engendered anywhere throughout the skeleton during customary functional activity (data from the goose wing and horse radius suggest $2500-3000 \times 10^{-6}$).

Influences other than regulation of normal functional strain result in some bones being more massive than the minimum structures in which normal functional strain is also the "maximum tolerable" tissue strain. In these bones the normal functional strain values are far below this threshold. If adaptive hypertrophy were to occur only when the "maximum tolerable" tissue strain levels were reached then these "overdesigned" bones would withstand substantial functional overstrain without inducing an adaptive response. This seems less likely than if different bones were each to have an individual "critical" strain situation relevant to their particular position and structure. The components of this "critical strain situation", which would be a property of the structure, are the absolute strain levels and the strain levels relative to those in other locations throughout the structure. If either the absolute "critical" strain levels were exceeded, or the relative strain ratios disturbed, adaptive bone re-modelling would be engendered. In this way the mass and, to some extent, the shape of each individual bone would be related to the customary functional strains within it. However, since each individual structure (bone) has a different set of mechanical priorities their critical strains will differ also. Only in those bones in which tissue economy is the predominant influence will functional strain levels as determined by the "critical" strains within the structure coincide with the "maximum tolerable" strain of the tissue.

Thus, in the sheep ulnar osteotomy experiment, if it is assumed that the remodelling in the radius was the result of the strain reorganization caused by ulnar osteotomy, then the small (10—20%) increase in functional (walking) strain on the cranial and caudal surfaces either exceeded the "critical" strain threshold or disturbed the "critical strain ratios" of this bone so inducing new bone formation. Since the highest functional (walking) strain on the sheep radius (-1200×10^{-6}) which stimulated new bone formation was well below the highest functional strains recorded in other limbs (goose humerus, horse, -2500×10^{-6}) the "critical" strain level of this bone is below the "maximum tolerable" tissue level (assuming that the material properties of goose, horse and sheep bones are similar and are designed to withstand the same number of strain reversals).

It is noteworthy that in these experiments the sheep radius responded to such a small overstrain during walking at one particular speed but yet was never repetitively subjected to strains which could not easily have been attained at a faster walk or during a trot. The mechanism necessary to sense such a small increase or reorganization in strain, and control appropriate remodelling, must be a formidable one. Since mechanical strain levels which are normal for some bones induce adaptive change in others there must be a genetic blueprint of what is acceptable for each individual bone. If this is the case adaptive remodelling will occur whenever the prevailing mechanical situation is outside the genetically determined limits.

Much circumstantial evidence suggests that strain-generated electrical potentials are involved in presenting the bone's cellular population with its "appreciation" of the mechanical circumstances within the bone structure as a whole (Lanyon & Hartman, 1977). Since the size of the potentials is related to both strain and strain rate this is consistent with the observation that strains at high strain rates are far more potent at inducing new bone formation than similar strains at low rates (Lanyon & O'Connor, 1980). Thus a bone's mass will be affected disproportionately according to the animal's routine activity which involves the highest strain rates.

Although hypotheses on the mechanism of adaptive remodelling must remain speculative there is only support for the idea that "normal" bone mass, and to some extent shape, can be affected positively or negatively by altering that bone's mechanical circumstances. This mechanical influence on bone remodelling can only be derived from, or related to, those strains which are actually encountered. The predominant functional influence is most likely to arise from commonly encountered functional strains rather than from infrequent loading "accidents". Nevertheless the manner of loading differs in different bones and strains that are normal in some locations induce adaptive remodelling in others. Since the tissue properties in limb bones vary little between one another each bone must therefore be considered a unique organ in which any underlying relationship between the tissue and its mechanical circumstances is subordinated to wider considerations relevant to the structure of which it is part. These considerations include the relationship of the bone with respect to structures attached or adjacent to it, its loading mode, position in the limb, and the degree of robustness necessary to withstand a reasonable proportion of the loading accidents to which it is subjected as a result of its location. The degree of importance of these factors will vary in different animals and different locations. Attempts to interpret bone structure solely in relation to tissue properties and customary loading must be considered inadequate.

CONCLUSIONS

The normal shape and mass of each bone reflect the influence of functional activity on a basic, genetically determined, bone structure which although recognizable as an individual bone lacks the normal characteristic lines, crests, tuberosities, curvature, cortical thickness and cancellous architecture. It has generally been assumed that the influence on bone remodelling is engendered by the functional strains induced within the bone tissue, and that function-dependent features are engendered as a result of a local response to regulate strain levels. The higher the functional strains the narrower the safety margins with respect to yield or ultimate strength, and the greater the level of fatigue damage which requires reparative secondary re-modelling. There is likely therefore to be an effective upper physio-logical limit for functional strain values in bone tissue (the "maxi-mum tolerable" tissue strain) which experiments to date suggest to be $2500-3000 \times 10^{-6}$. There may also be a strain level below which the physiology of the bone tissue is sub-optimal.

Since bone remodelling is sensitive only to intermittent strain, and can only respond to those strain situations which it encounters, it is likely to be predominantly influenced by the loading regime imposed during normal co-ordinated activity. This activity in the limb bones of quadrupeds is locomotion. The forces which act on these bones are engendered either within the limb from its intrinsic musculature, or transmitted through it between the trunk and the ground. These two sets of forces interact to maintain a practically constant loading axis during the stand phase of each stride. This constancy in the manner of loading is preserved throughout the animal's entire speed range. Its magnitude, and thus the peak strains induced, increase with speed although not with a simply proportional relationship. The change in gait from a trot to a canter results in a reduction in limb loading which, although probably not the purpose of the gait change, permits increased speed without progressive erosion of safety factors to failure or higher requirements of mass within the limb bones. Thus, both the manner and magnitude of limb loading are restricted so maximizing the opportunity for tissue economy.

If the predominant remodelling "goal" of bone were economy of tissue in relation to the loads of co-ordinated activity then long bone curvatures (which develop in response to function) would be directed towards neutralizing external bending moments so producing axial compression. Cortical thickness would also be proportional to load, so producing a uniform (longitudinal compressive) functional strain level just low enough to ensure an adequate fatigue life in relation to the animal's life span, activity and rate of secondary bone remodelling. This functional strain level would be the "maximum tolerable" tissue strain level. The manner of loading of different bones is such that in some (the horse and sheep metacarpus), all the cortices are in compression, whereas in others (the radius and tibia), bending is sufficient that the whole thickness of their cranial cortices is in longitudinal tension. In some locations the bone curvatures which develop as a result of function contribute to, rather than reduce, functional bending. For whole bones, tissue economy in relation to the loads of repetitive functional activity cannot therefore have a high priority. Despite a non-uniform load distribution cortical thickness in bones is not proportional to load. Furthermore, although functional strain levels differ substantially between different locations small re-organizations in functional strain lead to adaptive bone remodelling.

There are no known or suspected discrepancies in the properties of bone tissue which could account for substantially different functional strain levels in different bones. It is reasonable to assume therefore that bone does not remodel to obliterate these functional strain discrepancies either because there is no benefit from uniform functional strain, or because remodelling is directed towards considerations different from, or additional to, those concerned with co-ordinated activity. If this is the case then each bone must be considered to some extent as a unique structure in which adjustment of form and regulation of mass are directed to maintaining individual strain levels which are "critical" throughout the structure but which, except in some locations, do not approach the "maximum tolerable" tissue strain level. Attempts to interpret bone structure solely in relation to tissue properties and customary loading without some knowledge of these considerations must be considered inadequate.

ACKNOWLEDGEMENTS

The work performed by the author and his colleagues referred to in this chapter has been supported over the years by the Medical

328 L. E. Lanyon

Research Council, the Arthritis and Rheumatism Council, the Royal Society, the National Institute of Health, the Horserace Betting Levy Board and the Wellcome Trust. We are grateful to all these bodies. The work has been performed in the Department of Anatomy, University of Bristol, UK, and the Concord Field Station, Harvard University, USA. I am grateful to all my colleagues and in particular Clinton Rubin, much of whose work is referred to but whose contribution to the form of this manuscript and the ideas it contains is also substantial.

REFERENCES

Alexander, R. McN., Malioy, G. M. O., Hunter, B., Jayes, A. S. & Nturibi, J. (1979). Mechanical stresses in fast locomotion of buffalo (*Syncerus caffer*) and elephant (*Loxodonta africana*). *J. Zool., Lond.* **189**: 135–144.

Cavagna, G. A., Heglund, N. C. & Taylor, C. R. (1977). Mechanical work in terrestrial locomotion, two basic mechanisms for minimising energy expenditure. *Am. J. Physiol.* **233**: R243–R261.

Donaldson, C. L., Hulley, S. B., Vogel, J. M., Hattner, R. S., Bayers, J. H. & McMillan, D. E. (1970). Effect of prolonged bed rest on bone mineral. *Metabolism* **19**: 1071–1084.

Frost, H. M. (1973). *Bone modelling and skeletal modelling errors* (Orthopaedic Lectures, Vol. IV). Springfield, Illinois: Charles C. Thomas.

Gillespie, J. A. (1954). The nature of the bone changes associated with nerve injuries and disuse. *J. Bone Jt Surg.* **36B**: 464–473.

Goodship, A. E., Lanyon, L. E. & McFie, H. (1979). Functional adaptation of bone to increased stress. *J. Bone Jt Surg.* **61A**: 539–546.

Hert, J., Liskova, M. & Landa, J. (1971). Reaction of bone to mechanical stimuli. Part 1. Continuous and intermittent loading of tibia in rabbit. *Folia morph.* **19**: 290–300.

Howell, J. A. (1917). An experimental study of the effect of stress and strain on bone development. *Anat. Rec.* **13**: 233–252.

Jones, H. H., Priest, J. D., Hayes, W. C., Tickenor, C. C. & Nagel, D. A. (1977). Humeral hypertrophy in response to exercise. *J. Bone Jt Surg.* **59A**: 204–208.

Koch, J. C. (1917). Laws of bone architecture. *Am. J. Anat.* **21**: 177–298.

Lanyon, L. E. (1974). Experimental support for the trajectorial theory of bone structure. *J. Bone Jt Surg.* **56B**: 160–166.

Lanyon, L. E. (1976). The measurement of bone strain *in vivo*. *Acta orthop. belg.* **42** (suppl. 1): 98–108.

Lanyon, L. E. (1980). The influence of function on the development of bone curvature. An experimental study on the rat tibia. *J. Zool., Lond.* **192**: 457–466.

Lanyon, L. E. & Baggott, D. G. (1976). Mechanical function as an influence on the structure and form of bone. *J. Bone Jt Surg.* **58B**: 436–443.

Lanyon, L. E. & Bourn, S. (1979). The influence of mechanical function on the development and remodelling of the tibia. An experimental study in sheep. *J. Bone Jt Surg.* **61A**: 263–273.

Lanyon, L. E., Goodship, A. E. & Baggott, D. G. (1976). The significance of bone strain *in vivo*. *Acta orthop. belg.* **42** (suppl. 1): 109—123.

Lanyon, L. E., Goodship, A. E., O'Connor, J. A. & Pye, C. J. (1980). A quantitative study on the functional adaptation of bone. *Trans. orthop. Res. Soc.* 1980: 296.

Lanyon, L. E. & Hartman, W. (1977). Strain-related electrical potentials recorded from bone *in vitro* and *in vivo*. *Calcif. Tissue Res.* **22**: 315—327.

Lanyon, L. E., Magee, P. T. & Baggott, D. G. (1979). The relationship of functional stress and strain to the processes of bone remodelling. An experimental study on the sheep radius. *J. Biomech.* **12**: 593—600.

Lanyon, L. E. & O'Connor, J. A. (1980). Adaptation of bone artificially loaded at high and low physiological strain rates. *J. Physiol.* **303**: 36P.

Lanyon, L. E. & Rubin, C. T. (1980). Loading of mammalian long bones during locomotion. *J. Physiol.* **303**: 72P.

Lanyon, L. E. & Smith, R. N. (1970). Bone strain in the tibia during normal quadrupedal locomotion. *Acta orthop. scand.* **41**: 238—248.

Moore, W. J. (1965). Masticatory function and skull growth. *J. Zool., Lond.* **146**: 123—137.

Murray, P. D. F. (1936). *Bones. A study of the development and structure of the vertebrate skeleton*. Cambridge: University Press.

Pauwels, F. (1973). Kurzer Überblick uber die mechanische Beanspruchung des Knochens und ihre Bedeutung für die funktionelle Anpassung. *Z. Orthop. Grenzgeb.* **3**: 681—705.

Ralis, Z. A., Ralis, H. M., Randall, M., Watkins, G. & Blake, P. D. (1976). Changes in shape, ossification and quality of bones in children with spina bifida. *Devl. Med. Child Neurol.* **18** (Suppl. 37): 29—41.

Rubin, C. T. & Lanyon, L. E. (1979). Bone strain as a function of speed. *(Proc. Am. Physiol. Soc.) Physiologist, Wash.* **22**: 109.

Starkebaum, W., Pollack, S. R. & Korostoff, E. (1979). Micro electrode studies of stress generated potentials in four-point bending of bone. *J. Biomed. Mater. Res.* **13**: 729—751.

Taylor, C. R. (1978). Why change gaits? Recruitment of muscles and muscle fibre as a function of speed and gait. *Am. Zool.* **18**: 153—161.

Tokuriki, M. (1973a). Electromyographic and joint mechanical studies in quadrupedal locomotion. I. Walk. *Jap. J. vet. Sci.* **35**: 433—446.

Tokuriki, M. (1973b). Electromyographic and joint mechanical studies in quadrupedal locomotion. II. Trot. *Jap. J. vet. Sci.* **35**: 525—533.

Tokuriki, M. (1974). Electromyographic and joint mechanical studies in quadrupedal locomotion. III. Gallop. *Jap. J. vet. Sci.* **36**: 121—132.

Uhthoff, H. K. & Jaworski, Z. F. G. (1979). Bone loss in response to long term immobilisation. *J. Bone Jt Surg.* **60B**: 420—429.

Uhthoff, H. K., Jaworski, Z. F. & Liskova-Kiar, M. (1979). Age specific activity of bone envelopes in experimental disuse osteoporosis and in its reversal. *Trans. orthop. Res. Soc.* 1979: 125.

Washburn, S. L. (1947). The relation of the temporal muscle to the form of the skull. *Anat. Rec.* **99**: 239—243.

Whedon, G. D., Lutwak, L., Rambant, P. C., Whittle, M. W., Smith, M. C., Reed, J., Leach, C., Stadler, C. R. & Sanford, R. D. (1977). Mineral and nitrogen metabolic studies. Experiment MO71. *Results from Skylab.* Chapter 18. Johnston, R. S. & Dictleer, L. F. (Eds). NASA.

Symp. zool. Soc. Lond. (1981) No. 48, 331—358

The Allometry of Primate Body Proportions

LESLIE C. AIELLO

Department of Anthropology, University College London, Gower Street, London WC1, England

SYNOPSIS

Limb length and long bone length in the primates are determined by two factors, body size and locomotor specialization. Allometry (the scaling of characteristics to body size) is more efficient than indices in separating the effects of these two variables. However, conclusions drawn from allometric analyses are dependent on the nature of the sample used as the basis of the analysis, the significance attributed to the variation observed in the relationships and the assumptions surrounding the dependent and independent variables. With attention to these factors, the following general conclusions are drawn from the allometry of adult anthropoid primate body proportions. First, trunk length is not a constant measure of body weight across the anthropoid primates and, therefore, quite different conclusions can be drawn from analyses using body weight rather than trunk length as the independent variable. Secondly, although allometric analyses of the relationship between trunk length, body weight, limb length and long bone length are potentially useful in mechanical and locomotor analyses, they are unsuitable, by themselves, to answer questions of the commonality in growth processes producing adult body proportions. This results from the fact that neither trunk length nor body weight has a proven constant relationship with the growth factors that result in adult proportions (speed of growth, time of onset of growth and duration of growth) across the anthropoid primates. Thirdly, allometric analyses of body weight and bone cross-section size are difficult to interpret in mechanical terms because of the largely unknown relationship between body weight and the stress experienced by the bone under different loading conditions on the one hand, and the unknown relationship between the external measurements of the cross-section and cross-section strength on the other.

Allometry should be viewed as a framework for interpretation. Enthusiasm for the often clear linear relationships produced by these analyses should not obscure the need for attention to be given to the assumptions used in the construction of the plots or the assumptions made in the interpretation of given relationships.

INTRODUCTION

The use of allometry (scaling of characteristics to body size) has shown a recent revival in the analysis of morphological specialization,

as well as in other areas of analysis where size is an important correlate of the variable under study (Gould, 1966, 1975; Pilbeam &
Gould, 1974; Jerison, 1973; Sacher, 1970). Change in shape of a
morphological element may occur as a correlate of change in size of
an animal or as the result of a particular specialization independent
of body size. It is desirable in both taxonomic and functional
analyses to separate the effects of these two variables.

Allometry is normally approached through a logarithmic plot of
any one of the variables under analysis against body weight. The resulting bivariate display separates those specimens which show a constant linear relationship with body weight from those specimens
which deviate from the linear relationship. Taxonomic or functional
interpretations based on these allometric relationships are dependent
on the assumptions made in the construction of the bivariate plot
and on the statistical interpretation of the observed relationships.
There are a number of specific factors which influence the interpretation of allometric relationships. These include the nature of the
sample, the validity of the statistical techniques, the significance of
the variation around the linear trend, and the validity of the relationship between the measurements used and the dependent or independent variables they are assumed to represent. Of these factors, most
attention has been directed toward the validity of alternative statistical techniques available for the characterization of the observed
linear relationships. Least squares regression analysis is the most frequently used technique for this purpose. However, this technique
only considers variation in one of the two variables employed in the
analysis. As a result, it can produce misleading results when the correlation coefficient between these variables is low. Statistical techniques such as reduced major axis and principal axis, which take into
consideration variation in both parameters, are becoming increasingly
popular as techniques that provide an accurate picture of the linear
relationship. However, for interspecific analyses which normally produce correlation coefficients in excess of 0.95 the differing results of
these various statistical techniques are seldom significant.

Little attention has been drawn to the remaining factors involved
in the interpretation of allometric relationships; the nature of the
sample, the significance of the observed variation and the assumptions surrounding both the dependent and the independent variables.
It is the purpose of this chapter to examine these factors in relation
to the analysis of the allometry of the adult primate postcranial
skeleton.

MATERIALS AND METHODS

The bivariate plots are constructed on the basis of the means for each sex of each species. A total of 275 individual skeletons representing 32 species and 17 genera of anthropoid primates were measured (Table I). The skeletal measurements are presented and described in detail elsewhere (Aiello, in preparation). Body weights were primarily taken from the literature and represent wild weights wherever poss- ible, but in the majority of the cases they do not represent the weights of the individual animals from which the skeletal measure- ments were taken. Both this factor and the small size of some of the samples could be expected to increase the variation expressed in the bivariate relationships. However, the consistency found within genera and between primates of similar locomotor groupings suggests that these are not significant factors.

The locomotor classification of the primates included in the study is derived primarily from Rose (1973) and is presented in Table I. Discussion of the locomotion of extant and fossil anthropoid pri- mates is also presented elsewhere (Aiello, in press).

In this analysis the Principal Axis is used to characterize the ob- served linear allometric relationships. The principal axis is derived according to the following formula:

$$\text{principal axis (b)} = s_{xy}/(\lambda_1 - s_y^2) \tag{14.1}$$

$$\text{where: } s_{xy} = \Sigma xy/N - 1 = \text{ the covariance of X and Y}$$

$$s^2 y = \Sigma y^2/N - 1 = \text{ the variance of Y}$$

$$\lambda_1 = \tfrac{1}{2}\{s^2 x + s^2 y + \sqrt{[(s^2 x + s^2 y)^2 - 4(s^2 x \, s^2 y - s^2 xy)]}\}$$

The equation for the principal axis is:

$$Y = \bar{Y} + b_1(X - \bar{X}) \text{(Sokal \& Rohlf, 1969).} \tag{14.2}$$

The least squares equations are also provided for the allometric plots. In no case are the slopes derived by the principal axis method differ- ent from those derived by the least squares method at the 95% level of significance. Differences between samples, or between individuals and specific samples, were tested by least squares analysis and are presented at the 95% level of significance.

THE ALLOMETRY OF LIMB AND BONE LENGTH IN THE HIGHER PRIMATES

Since the middle of the nineteenth century the necessity of correcting

TABLE I

The locomotor classification of the primates included in the analyses[a]

Locomotor category	Included genera	Included species	Sample size		
			Males	Females	Total
Arboreal quadrupedalism — Medium size	Cebus	C. apella	1	1	2
		C. albifrons		1	1
Arboreal quadrupedalism — Large size	Cercocebus	C. albigena	9	8	17
		C. torquatus	4	2	6
Branch sitting and walking	Cercopithecus	C. aethiops	4	6	10
		C. mitis	11	12	23
		C. mona	7		7
		C. neglectus	3	3	6
		C. talapoin	3	5	8
Arboreal quadrupedalism — Large size	Colobus	C. badius	7	5	12
		C. guereza	6	5	11
		C. polykomos	4	2	6
Old World semibrachiation	Presbytis	P. obscura	9	13	22
Arboreal quadrupedalism — Large size	Alouatta	A. belzebul		1	1
		A. seniculus	1	1	2*
New World semibrachiation	Ateles	A. paniscus			1*
	Brachyteles	B. arachnoides			1*
	Lagothrix	L. lagothricha	1		1
Part terrestrial quadrupedalism and part arboreal quadrupedalism (Branch sitting and walking)	Cynopithecus	C. niger		1	1
	Macaca	M. fascicularis	1	2	3
		M. fuscata		1	1
		M. mulatta	1		1
		M. nemestrina		1	1
		M. sylvanus	1	1	2

Locomotor category	Genus	Species			
Terrestrial quadrupedalism – Ground standing and walking	Papio	P. anubis	3	3	6
		P. cynocephalus	2		2
		P. ursinus	1		1
Brachiation	Hylobates	H. mulleri			6**
Bipedalism	Homo	H. sapiens – caucasian	10	5	15
		H. sapiens – negro	20	20	40
Knuckle walking (Pan and Gorilla)	Gorilla	G. gorilla	9	7	16
	Pan	P. paniscus	12	17	29
	Pongo	P. pygmaeus	10	4	14

[a] The classification of the quadrupedal Ceboidea and Cercopithecoidea is based on the listed primate genera and is after Rose (1973). * = sex unknown. ** = sexes included together in the analyses.

limb and bone length for the effects of body size in comparative analyses has been recognized (T. H. Huxley, 1864; Lucae, 1865; Mivart, 1867). However, until recently this has been achieved in the analysis of the primate postcrania through the use of indices. These early analyses as well as subsequent work (Mollison, 1910; Schultz, 1930, 1933, 1937; Erikson, 1963) have provided a body of data which outlines the general proportional relationships in the anthropoid primates. Conclusions resulting from this work, such as the relatively long fore-limb for trunk length shared by the hominoid primates, have been widely accepted in the interpretation of primate evolutionary history.

Indices, however, are not an ideal means of either correcting for the effects of body size or analysing the relationships between body size and limb or bone length. If the two variables comprising the index do not change in length at the same rate (isometrically) in primates of differing sizes, radically different indices can result. This can occur even though there may be a constant and highly correlated allometric relationship between the variables comprising the index. This problem is accentuated if one of the variables comprising the index is body weight while the other is limb or bone length, because body weight is proportional to body volume which would increase roughly in proportion to the cube of a linear measurement. Therefore, even if there were geometric similarity (identity in proportions) in primates of different sizes, indices would obscure this homogeneity.

Allometry as a technique avoids the pitfalls of indices since it employs logarithmic transformation to correct for differential increases between variables, and bivariate plots to illustrate constant proportional relationships. Although allometry was early recognized as a superior analytical technique in morphological analysis (Snell, 1891; J. Huxley, 1932) it has not been applied systematically to the analysis of adult primate body and limb proportions.

Biegert & Maurer (1972, a widely cited current reference on anthropoid primate limb proportions) employ a modified form of allometry which does not entirely avoid the problems of indices. They plot a raw measurement of trunk length against an index composed of limb length and trunk length. Their analyses suggest that there is a constant relationship between trunk length and hind limb length in the anthropoid primates with the exception of *Homo sapiens* and *Hylobates* which show a long hind limb for their trunk lengths. From this they conclude that the length of the hind limb in the great apes is unspecialized in relation to the other anthropoid primates while the hind limb of man and the gibbons is clearly elongated in relation to this sample. The relationships between trunk length and the index composed of fore limb length and trunk length appear similarly homogeneous. However, the fore limb increases in length in relation to trunk length at a more rapid rate than does the hind limb. In addition, they state that *Homo sapiens* is consistent in this relationship with the majority of the other anthropoid primates while *Hylobates* and *Pongo* deviate from the trend in having long forelimbs for their trunk lengths. Biegert & Maurer (1972) conclude, therefore, that the forelimb is positively allometric in relation to both skeletal trunk length and to the length of the hindlimb. The high intermembral index (forelimb length/hindlimb length × 100) characteristic of the African apes is thus interpreted as the natural result of increase in body size in these primates by comparison with smaller bodied anthropoid primates. Limb proportions in the African apes, therefore, would be determined by body size rather than by locomotor specialization. By inference, the high intermembral index characteristic of *Papio* and *Alouatta* and *Lagothrix* also would be accounted for by a similar explanation.

The pattern of relationships suggested by Biegert & Maurer's (1972) modified allometric analysis is supported by the direct allometric analysis of trunk lenth and limb length using the larger number of species employed here (Fig. 1, Table II). The principal

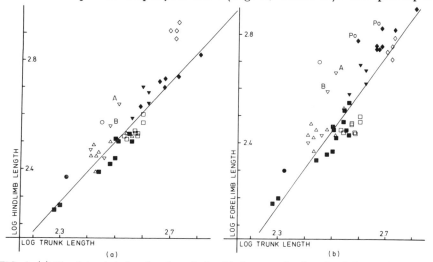

FIG. 1. (a) Bivariate plot showing the relationship between log length of the trunk and log length of the hindlimb. The solid line represents the principal axis. ● = Medium Sized Arboreal Quadrupedal primates (*Cebus*). ■ = Branch Sitting and Walking primates (*Cercocebus* & *Cercopithecus*). □ = Old World Semibrachiating primates (*Colobus* & *Presbytis*). △ = Part Branch Sitting and Walking and Part Ground Standing and Walking primates (*Cynopithecus* & *Macaca*). ▲ = Ground Standing and Walking primates (*Papio*).   = New World Semibrachiating primates (*Alouatta, Ateles, Brachyteles* & *Lagothrix*). ○ = Brachiating primates (*Hylobates*). ◇ = Bipedal primates (*Homo*). ♦ = Pongids. A = *Ateles*. B = *Brachyteles*. (b) Bivariate plot showing the relationship between log length of the trunk and log length of the forelimb. Po = *Pongo*. The remaining symbols are as in Fig. 1a.

TABLE II

Statistics for the bivariate comparisons (Figs 1–13)[a]

Comparison	Sample	N	P.a.	95% C.L.	L.S. equation	r	$s^2 yx$
x = Trunk length y = Hindlimb length	all genera except *Homo, Hylobates, Brachyteles* and *Ateles*	41	1.12	1.00–1.26	y = −0.10 + 1.04x	0.937	0.0018
x = Trunk length y = Forelimb length	All genera except *Pongo, Hylobates, Brachyteles* and *Ateles*	42	1.39	1.21–1.61	y = −0.60 + 1.22x	0.906	0.0046
x = Hindlimb length y = Forelimb length	*Colobus, Presbytis, Cercopithecus, Cercocebus*	17	0.96	0.91–1.01	y = 0.04 + 0.95x	0.995	0.00011
x = Humerus length y = Radius length	All Old World Monkeys	30	1.09	1.04–1.14	y = −0.17 + 1.08x	0.992	0.00022
x = Femur length y = Tibia length	All genera	51	0.88	0.85–0.91	y = 0.23 + 0.88x	0.994	0.00026
x = Body weight y = Trunk length	*Colobus, Presbytis, Cercopithecus, Cercocebus*	21	0.32	0.29–0.36	y = 1.31 + 0.32x	0.969	0.00047
x = Body weight y = Hindlimb length	*Colobus, Presbytis, Cercopithecus, Cercocebus*	16	0.35	0.30–0.40	y = 1.20 + 0.35x	0.962	0.00084
x = Body weight y = Forelimb length	All genera except *Hylobates, Ateles, & Brachyteles*	45	0.34	0.32–0.36	y = 1.16 + 0.34x	0.973	0.0016
x = Body weight v = Humerus length	All genera except *Hylobates & Ateles*	47	0.36	0.34–0.38	y = 0.78 + 0.35x	0.978	0.0013

		N				r	$s^2 yx$
x = Body weight y = Radius length	All genera except *Hylobates* & *Ateles*	46	0.32	0.29–0.35	y = 0.94 + 0.31x	0.956	0.0023
x = Femur transverse diameter y = Femur length	*Colobus, Presbytis,* *Cercopithecus,* *Cercocebus*	21	1.05	0.92–1.20	y = 1.20 + 1.00x	0.955	0.00098
x = Body weight y = Humerus circumference	All genera	48	0.35	0.33–0.37	y = 0.17 + 0.35x	0.979	0.0013
x = Body weight y = Femur circumference	All genera	48	0.36	0.34–0.38	y = 0.17 + 0.34x	0.984	0.0010
x = Femur circumference y = Humerus circumference	All genera except *Ateles* & the hominoids	41	1.02	0.95–1.08	y = −0.03 + 0.99x	0.979	0.00046
x = Humerus transverse diameter y = Humerus sagittal diameter	All genera	52	1.14	1.07–1.21	y = −0.14 + 1.10x	0.972	0.0020
x = Femur transverse diameter y = Femur sagittal diameter	All genera except *Lagothrix, Alouatta,* *Brachyteles, Ateles* & the hominoids	36	0.98	0.93–1.04	y = 0.03 + 0.97x	0.984	0.00040
x = Humerus circumference y = Humerus length	All genera except *Hylobates, Ateles* & *Brachyteles*	50	1.01	0.96–1.06	y = 0.66 + 1.00x	0.984	0.00098
x = Femur circumference y = Femur length	*Colobus, Presbytis,* *Cercopithecus,* *Cercocebus*	20	1.11	1.03–1.20	y = 0.56 + 1.09x	0.986	0.00030

[a] In all cases the statistics are computed on the logarithmic transformations of the indicated variables. N = the sample size. Males and females of each species are included separately in the analyses. P.a. = principal axis. 95% C.L. = the 95% confidence limits for the principal axis. L.S. equation = the least squares equation for the indicated comparison. r = the correlation coefficient. $s^2 yx$ = the variance of "y" left unexplained by "x" (Sokal & Rohlf, 1969).

axis for the relationship between trunk length and hindlimb length is not significantly different from isometry, but *Homo sapiens* as well as *Hylobates* are significantly different from the trend. In addition, the New World primates *Brachyteles* and *Ateles* deviate from the general trend in the direction of a long hindlimb for their trunk lengths.

The principal axis for the relationship between trunk length and forelimb length is positively allometric and both *Pongo* and *Hylobates* are significantly different from this trend. In addition, both *Brachyteles* and *Ateles* deviate from the general trend in the direction of a long forelimb for their trunk lengths.

Although these comparisons support the pattern of relationships suggested by Biegert & Maurer (1972), they do not necessarily support the conclusion that there is a homogeneous relationship in limb proportions characterizing the smaller bodied primates as well as the African apes. Of particular importance is the variation around the principal axis in both comparisons in which the smaller bodied primates cluster in locomotor and taxonomic groupings. This suggests that the principal axes in both these comparisons may be an artefact of the species included in the analysis and not represent a true consistency in proportional relationships across the entire sample.

When the length of the forelimb is compared directly to the length of the hindlimb, clear grade distinctions appear between primates of differing locomotor patterns and clear linear relationships are evident among primates of the same or similar locomotor patterns (Fig. 2). Those primates that are characterized by branch sitting and walking or Old World semibrachiation show a highly correlated isometric relationship between the length of the forelimb and the length of the hindlimb. This indicates that these primates are characterized by constant intermembral proportions throughout their size range. There is no indication that the length of the forelimb increases more rapidly in relation to the hindlimb as Biegert & Maurer (1972) predict. In addition, with the exception of *Homo sapiens*, primates of other locomotor groups are displaced from this trend in the direction of a longer forelimb for their hindlimb lengths.

Total length is a composite measurement made up of the lengths of the bones comprising the limbs; the humerus and the radius for the forelimb and the femur and the tibia for the hindlimb. These bones need not have a constant length relationship with each other across the sample. If they do not, the use of total limb length in allometric analyses may obscure significant proportional differences or similarities in the sample.

The length of the tibia and the length of the femur are highly correlated across the sample, but the length of the femur is positively allometric in relation to the length of the tibia (Fig. 3). This indicates

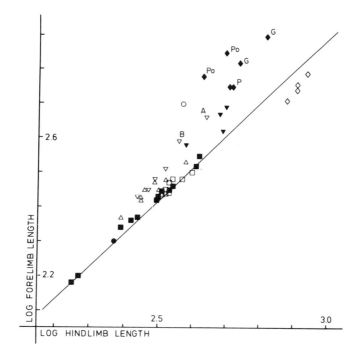

FIG. 2. Bivariate plot showing the relationship between log length of the hindlimb and log length of the forelimb. G = *Gorilla*. P = *Pan*. Other symbols as in Fig. 1.

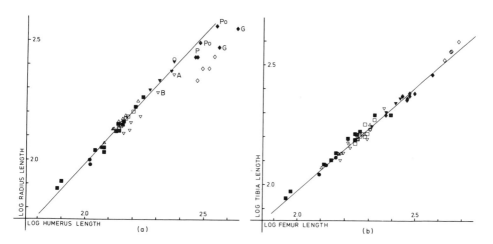

FIG. 3. (a) Bivariate plot showing the relationship between log length of the humerus and log length of the radius. (b) Bivariate plot showing the relationship between log length of the femur and log length of the tibia. Symbols as in Figs 1 and 2.

a constant, although positively allometric, length relationship between these bones in spite of the variation in total hindlimb length
and trunk length in these primates. This relationship is true even for
Homo sapiens, Hylobates, Ateles and *Brachyteles* which are characterized by an unusually elongated hindlimb for their trunk lengths.
This consistency is not true for the relationship between the length
of the humerus and the length of the radius. The Old World monkeys
are characterized by a positive allometric relationship where the
length of the humerus increases at a faster rate than does the length
of the radius (Fig. 3). The New World genera *Alouatta, Lagothrix,
Ateles* and *Brachyteles*, as well as the African apes and man, deviate
from this trend in the direction of a shorter than expected radius for
their humerus lengths. Therefore, total forelimb length is not a homogeneous measurement across the entire sample. The Old World
monkeys are characterized by a constant positive allometric relationship between the length of the humerus and the length of the radius,
in spite of the differences in the intermembral proportions which distinguish the branch sitting and walking primates and the Old World
semibrachiating primates from the remaining species. In addition, the
Old World monkeys and the New World species which share an intermembral index approaching unity are distinguished by differences in
their brachial proportions. The New World primates consistently
have a shorter radius for their humerus lengths than do the Old
World species.

Thus the allometric comparison of trunk length and limb length
across the higher primate sample obscures non-allometric differences
in intermembral and brachial proportions within the sample. The
comparison of the lengths of the limbs with each other and the comparison of the lengths of the bones comprised in each limb suggests
that there is no clear and simple allometric relationship which is
characteristic of the majority of the higher primates as a whole.

There is another factor which must be taken into consideration in
the interpretation of the relationship between trunk length and limb
length or trunk length and bone length. Biegert & Maurer (1972), as
well as others who have employed their results, have assumed that
trunk length is an accurate measurement of body weight across the
higher primates. The direct comparison of body weight with trunk
length shows that this is not the case. The principal axis for the
branch sitting and walking and the Old World semibrachiating primates is isometric, indicating that trunk length in these primates increases one third as fast as does body weight (Fig. 4). However, the
remaining primates, with the exception of *Macaca*, are significantly
different from this trend in the direction of a relatively short skeletal
trunk length for their body weights. (*Macaca* forms a variable group,

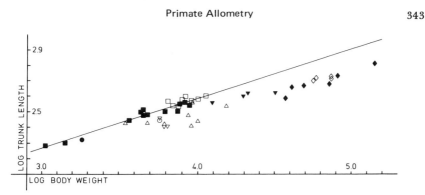

FIG. 4. Bivariate plot showing the relationship between log body weight and log length of the trunk. Symbols as in Fig. 1.

some members of which are significantly different from the trend while others are not.) Therefore, if trunk length is specifically em-polyed as a measurement of body weight in an allometric analysis across the higher primates, results will reflect not only changes in the variable under consideration in relation to trunk length but also changes in the relationship between trunk length and body weight.

The variation in the relationship between the trunk length and body weight is reflected in the different conclusions that can be drawn when body weight is used as the independent variable in the allometric analysis of limb length and bone length (Figs 5 and 6). These differences are particularly apparent in the comparison be-tween body weight and hindlimb length (Fig. 5). In this relationship *Homo sapiens* is consistent with the isometric trend which charac-terizes the smaller bodied primates while the three large apes deviate from this trend in the direction of a short hindlimb for their body weights. This would suggest that the large apes, and not *Homo sapiens*, are specialized in this feature.

In addition, the relationship between total forelimb length and body weight is isometric across the entire sample with the exception of only *Hylobates, Ateles* and *Brachyteles* (Fig. 5). The consistency of all the large apes with the majority of the smaller bodied primates is also apparent in the relationship between body weight and length of the humerus and body weight and length of the radius (Fig. 6). This suggests that the higher intermembral index in all the great apes is primarily a function of the shortening of the hindlimb in relation both to body weight and to total forelimb length.

Therefore, the general conclusions which can be drawn from the use of the body weight as the independent variable are quite differ-ent from those which can be drawn from the use of trunk length. However, both these measures, when used as the independent vari-able, obscure the non-allometric differences in intermembral and

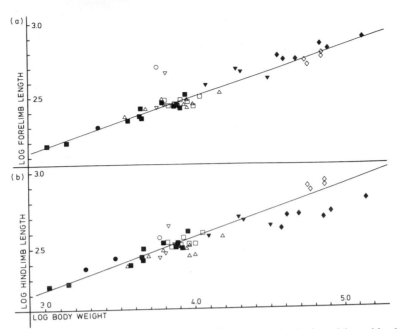

FIG. 5. (a) Bivariate plot showing the relationship between log body weight and log length of the forelimb. (b) Bivariate plot showing the relationship between log body weight and log length of the hindlimb. Symbols as in Fig. 1.

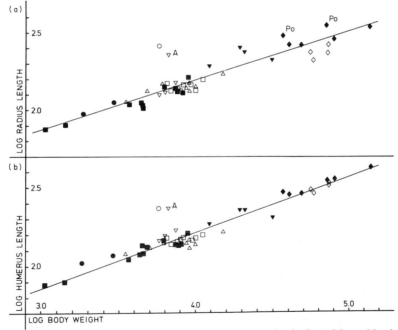

FIG. 6. (a) Bivariate plot showing the relationship between log body weight and log length of the radius. (b) Bivariate plot showing the relationship between log body weight and log length of the humerus. Symbols as in Fig. 1.

brachial proportions across the entire sample. The combined analysis of body weight, trunk length, limb length and bone length is potentially useful and significant in ordering primate species for mechanical and locomotor analyses (Aiello, in press). However, it is not immediately clear how this information can be used to establish primitive or derived proportional patterns and, therefore, become useful in cladistic or phylogenetic analyses.

Differences in body proportions or limb proportions between adult primates result from the differential growth of the body segments during the ontogeny of the animal. There are three factors which affect proportions in the adult animal; the speed of growth of the segments in relation to each other, the time of onset of growth in the different segments and the total duration of growth (Gould, 1977; Lumer, 1939; Lumer & Schultz, 1947).

Ontogenetic studies of the limb growth in primates suggest that differences in limb proportions within species or between some closely related species result from differences in the duration of growth while maintaining the same growth curve, i.e. a constant relationship of speed and onset of growth between segments. Lumer (1939) and Lumer & Schultz (1947) have shown that this is true for different species of *Hylobates* as well as being generally true for *Macaca*. In addition, sexual differences in proportions between males and females of *Gorilla, Pan, Ateles* and *Macaca* have been shown to result from this factor. Differences in proportions between the genera *Gorilla, Pan* and *Pongo* result from differences in the speed and onset of growth of the different segments as well as from differences in the total duration of the growth period (Lumer, 1939; Lumer & Schultz, 1941).

These results indicate that it is not the differences or similarities in body proportions, but the ontogenetic causes of these differences or similarities which are important in establishing morphological affinity. As a result hypotheses which are orientated toward interspecific similarity or difference in growth factors are best tested by ontogenetic analyses of proportional growth.

Static allometric analyses of adult animals are on the whole unsuitable for this type of interpretation. In order for them to be useful in this context, the independent variable in the analysis must have a known and constant relationship to growth across the entire sample. For example, age at skeletal maturity would be a factor that would consistently reflect the duration of the growth process resulting in the adult skeletal proportions. Linear relationships resulting from the application of this factor as the independent variable in allometric analysis would then reflect a common ontogenetic growth curve resulting from similar speeds of growth and time of onset of growth. Deviations from the linear relationship would result from essential differences in the ontogenetic curve.

Age at skeletal maturity is an unknown factor for many primate species and, therefore, there are no comprehensive data for such an analysis. The traditional variables of trunk length and body weight cannot, at present, be accepted as reflections of constant growth relationships across the sample. In relation to trunk length, both the proportions of the individual vertebrae and the proportions of the vertebral column as a whole reflect locomotor specializations across the higher primates (Rose, 1975). These differences in vertebral proportions can be expected to have resulted from the same ontogenetic factors which produce variations in limb proportions across the sample. Total trunk length cannot, therefore, be accepted as a reflection of constant growth across the sample.

Body weight is a measure of total growth in an organism. As such it measures the result of not only skeletal growth but also growth in soft tissues. Consistent linear relationships of skeletal proportions with body weight, therefore, reflect not only consistency in the particular skeletal parameter under consideration, but also consistency in the total skeletal and soft tissue composition of the body. It is not clear how variations in total skeletal and soft tissue composition affect the allometric relationship between body weight and a particular skeletal variable such as hindlimb length or forelimb length. The lack of clear statistical separation in the relationship between body weight and either forelimb or hindlimb length, between those smaller bodied primates distinguished by differences in intermembral proportions, most likely results from this factor.

The relationship between variation in total skeletal and soft tissue composition, body weight and a particular skeletal variable is also significant in the interpretation of the homogeneous relationship between *Homo sapiens* and the smaller bodied primates in the relationship between hindlimb length and body weight. In addition, it is important in the interpretation of the homogeneous relationship between all of the large hominoid primates and the smaller bodied primates in the relationship between total forelimb length, humerus length and radius length and body weight. There are gross differences in the distribution of body weight in man and the great apes. Until the relationship between body weight and growth of the skeleton is established, the question of the apparent similarity in body proportions between man and the small bodied primates cannot be firmly accepted as an indication of shared primitive growth relationships. This similarity could equally well result from convergence of proportions due to different relationships between duration of growth and speed and onset of growth in different segments.

However, the combination of static allometric analyses and comparative morphology of extant and fossil primates can provide indirect evidence relevant to phylogenetic affinity among primates in postcranial morphology (Aiello, in preparation). In relation to the general question of the allometric position of *Homo sapiens* in relation to the smaller bodied primates, analysis of Miocene and Pliocene fossil material suggests that the ancestral hominids need not have passed through a morphological phase similar to that characterizing the modern great apes. This suggests that the modern great apes are indeed specialized, particularly in their hindlimb morphology.

Allometric analysis of body weight in relation to various postcranial skeletal parameters has shown that one of the most consistent variables that is highly correlated with body weight in the higher primates is the transverse diameter of the midshaft of the femur (Aiello, in press). This measurement is positively allometric with body weight and the relationship is constant in spite of the variations in hindlimb characterizing the larger bodied primates. The transverse diameter of the midshaft of the femur, therefore, can be used as a postcranial estimate of body weight in specimens of unknown weight. When femur length is compared directly to the transverse diameter of the midshaft of the femur, the reduced femur length in the great apes is apparent. Selected fossil apes and hominids have been included in Fig. 7. None of these fossils shows the reduced femur length characterizing the extant great apes. In addition, the relationship between the length of the femur and the length of the humerus in the A1 288-1 skeleton from Hadar (Johanson & Taieb, 1976), the only hominid skeleton for which there are sufficient data to include in these allometric comparisons, shows similarity not only with the fossil apes but also with the climbing New World primates, *Alouatta* and *Lagothrix* (Aiello, in press).

Therefore, on the general level of the question of the primitive and derived nature of the hindlimb proportions in the hominoid primates, it is likely that *Homo sapiens* more closely approaches the general ontogenetic patterns of the small bodied anthropoid primates than do the African apes. However, questions of specific similarity or differences in growth processes between any of the extant primates can only be answered from ontogenetic analyses. Static allometric analyses can best be employed as a framework within which to organize these analyses as well as to approach general problems of a morphological and locomotor nature.

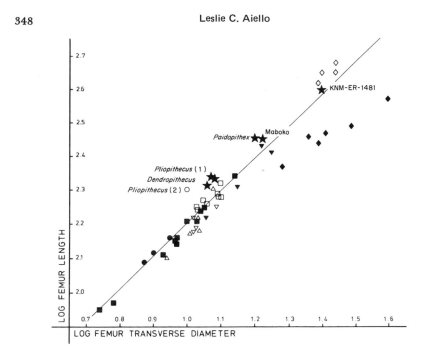

FIG. 7. Bivariate plot showing the relationship between log transverse diameter of the mid-shaft of the femur and log length of the femur. ★ = fossil primates. Other symbols as in Fig. 1. Bone measurements for the fossil specimens are taken from the literature and discussed elsewhere (Aiello, in press, in preparation).

THE ALLOMETRY OF THE CROSS-SECTION OF THE LIMB BONES IN THE HIGHER PRIMATES

The cross-section of a bone must adapt directly to the load to which it is exposed during the locomotion of an animal. It is well established that the observed size and strength of the bone cross-section results not only from genetic factors but also from the stress to which it is exposed during the life of the animal (Evans, 1957). As the result of this the assumptions which must be made in the interpretation of the allometric relationship between body weight and cross-sectional size are quite different from those which are made in the allometric analysis of limb and body proportions. In this case it is not specifically growth, but the manner in which body weight affects the load borne by the bone which is the critical independent variable in the analysis. The actual load borne by the bone is determined by body weight, locomotor pattern, limb length and the various muscular and/or ligamentous arrangements of the postcranial skeleton (Pauwels, 1965; Kummer, 1959). In addition, interpretation is complicated by uncertainty as to the manner in which the cross-section of the bone adapts to the load it must bear. The cross-section can adapt by alterations in the external size of the section, alterations in the cortical thickness and alterations in the shape of the cross-section (Pauwels, 1965; Lovejoy, Burstein & Heiple, 1976; Currey, 1967). There are, therefore, uncertainties in both the manner in which body weight relates to the load borne by the cross-section and the manner in which any single variable taken to represent cross-sectional strength actually relates to the strength of the cross-section.

With these points in mind, it is perhaps surprising that both the midshaft circumference of the humerus and the midshaft circumference of the femur show a highly correlated linear relationship with body weight across the entire sample (Fig. 8). However, when the midshaft circumference of the femur is compared directly to the midshaft circumference of the humerus there is considerable variation in the upper ranges of the distribution (Fig. 9). *Pongo* is significantly displaced in the direction of a large humerus circumference for its femur circumference, while *Homo sapiens* is displaced in the opposite direction. This suggests that body weight is affecting the bones of the forelimb and the bones of the hindlimb in these primates in different fashions.

This is not a surprising conclusion in view of the bipedal locomotion in *Homo sapiens* and the forelimb dominated quadrumanual climbing of *Pongo*. What is surprising, however, is the high correlation of these variables with each other in the smaller-bodied primates. These smaller-bodied primates are characterized by locomotor patterns as diverse as the highly acrobatic brachiation of the gibbons and the fully terrestrial quadrupedalism of *Papio*. In addition, there is considerable variation in the shape of the humerus cross-section among these smaller-bodied primates (Fig. 10) and considerable variation in the shape of the cross-section of the femur in the larger-bodied forms and the New World primates (Fig. 11).

Until it is known how the external measurements of a bone cross-section relate to the strength of a bone and how body weight affects the stresses to which the bone must adapt, it is premature to suggest general models for the mechanical determinants of cross-sectional size.

McMahon (1973) recently suggested within the theoretical model of elastic similarity, that no matter what the conditions of loading, whether they are by buckling, bending, torsion or a combination of these conditions, the length of a bone should be proportional to the two-thirds power of its diameter. In addition, the length of the limb should be proportional to the one-quarter power of body weight and

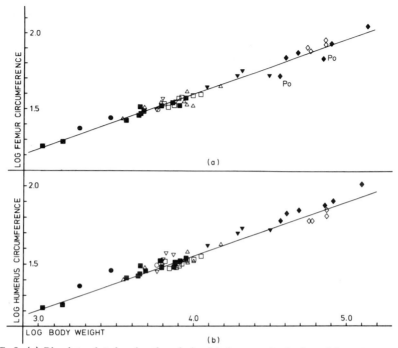

FIG. 8. (a) Bivariate plot showing the relationship between log body weight and log circumference of the midshaft of the femur. (b) Bivariate plot showing the relationship between log body weight and log circumference of the midshaft of the humerus. Symbols as in Fig. 1.

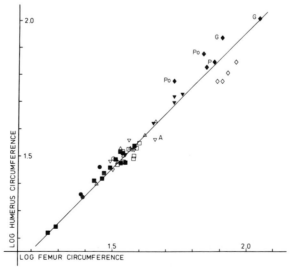

FIG. 9. Bivariate plot showing the relationship between log circumference of the midshaft of the femur and log circumference of the midshaft of the humerus. Symbols as in Figs 1 and 2.

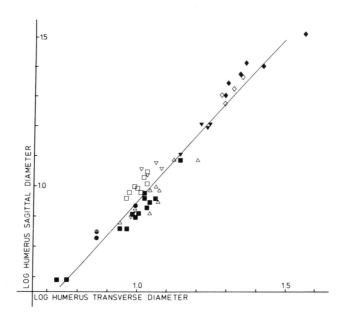

FIG. 10. Bivariate plot showing the relationship between log transverse diameter of the mid-shaft of the humerus and log sagittal diameter of the midshaft of the humerus. Symbols as in Fig. 1.

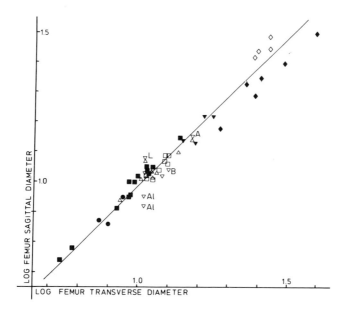

FIG. 11. Bivariate plot showing the relationship between log transverse diameter of the mid-shaft of the femur and log sagittal diameter of the midshaft of the femur. L = *Lagothrix*; Al = *Alouatta*. Other symbols as in Fig. 1.

the diameter of a limb should be proportional to the three-eighths power of body weight. McMahon has supported his argument with empirical data from bovids and other artiodactyls (1975).

The assumptions underlying this model are that the limb is loaded by gravitational self loading and that the length of the limb times its diameter squared is proportional to body weight. Neither of these assumptions can be accepted to be true across the primates in particular or the mammals in general. Alexander, Jayes, Maloiy & Wathuta (1979) have recently shown that McMahon's model is supported by empirical data neither for primates nor for non-primate mammals as a whole. Alexander *et al.* (1979) employ a small sample of primates of different locomotor patterns in their analysis. However, their conclusions are confirmed when the sample is increased in number and restricted to primates of similar locomotor type and body proportions. Table III presents the results for the arboreal Old World monkeys (branch sitting and walking primates and Old World semi-brachiating primates). The length of the forelimb and the length of the hindlimb are isometric with body weight, as are the diameters as well as the circumferences of the midshafts of the bones. The direct comparison of the circumference of the midshaft of the femur and the length of the femur is the only relationship which deviates from isometry and this deviation is in the direction opposite to that predicted by McMahon (1973).

These data also contradict the work of Preuschoft & Weinmann (1973). These authors suggest that as body size increases the cross-section of the bone will become relatively smaller. Their analysis is orientated towards the hypothesis that gracile limb bones, and particularly gracile forelimb bones, in primates do not necessarily indicate a change in function from a quadrupedal form of locomotion to an arm-swinging or brachiating locomotor pattern. The assumption underlying this model is that limb bones are loaded in bending during locomotion and that the diameter of the cross section is proportional to the second moment of inertia of the section. Therefore, the strength of the cross-section would be proportional to body weight to the fourth power. The strength of the cross-section in bending would then increase more rapidly than does body weight. In order to maintain physiological similarity (identity of strength) a larger animal could have a more gracile cross-section in relation to body weight than a smaller animal.

Both these models are simplistic in their assumptions in view of the complex variables involved both in the loading of the cross-section and in the response of the cross-section to the stress it must bear. Simple allometric analysis of body weight, external size of the

TABLE III

Statistics for the indicated bivariate comparison based on the proportionally homogeneous sample composed of the branch sitting and walking and Old World semibrachiating primates (Colobus, Presbytis, Cercopithecus and Cercocebus). Abbreviations are as in Table II

Comparison	Forelimb			Comparison	Hindlimb		
	N	P.a. & 95% C.L.	r		N	P.a. & 95% C.L.	r
x = Body weight y = Forelimb length	16	0.33 (0.29–0.38)	0.965	x = Body weight y = Hindlimb length	16	0.35 (0.30–0.40)	0.962
x = Body weight y = Humerus length	20	0.33 (0.30–0.37)	0.973	x = Body weight y = Femur length	20	0.37 (0.33–0.41)	0.968
x = Body weight y = Humerus circumference	20	0.32 (0.29–0.36)	0.969	x = Body weight y = Femur circumference	20	0.34 (0.31–0.37)	0.968
x = Body weight y = Humerus transverse diameter	20	0.29 (0.23–0.35)	0.907	x = Body weight y = Femur transverse diameter	20	0.35 (0.32–0.39)	0.972
x = Body weight y = Humerus sagittal diameter	20	0.35 (0.31–0.39)	0.972	x = Body weight y = Femur sagittal diameter	20	0.35 (0.31–0.38)	0.974
x = Humerus circumference y = Humerus length	20	1.02 (0.94–1.11)	0.982	x = Femur circumference y = Femur length	20	1.11 (1.03–1.20)	0.986

section and limb length can establish that they do not correspond to empirical data. However, these analyses, in conjunction with allometric analysis of cortical thickness (Aiello, in preparation) can do no more than offer a framework for the more detailed study of bone robusticity and the reaction of the cross-section to stress. In the direct comparison of the circumference of the midshaft of the humerus with the length of the humerus there is an isometric relationship in all primates with the exception of *Hylobates, Ateles* and *Brachyteles* (Fig. 12). In this case, it is the essential homogeneity of humerus robusticity common to primates of varying locomotor patterns which requires explanation. In the comparison between the length of the femur and the circumference of the midshaft of the femur, there is a clear linear trend in the proportions of this bone in the majority of the higher primates (Fig. 13). The apparent separation of the more terrestrially adapted primates is the feature which merits detailed examination. The variation in these relationships should not be ignored in order to construct simplistically based theoretical explanations for the general trends.

CONCLUSIONS

Interspecific allometry of the skeletal proportions of adult primates, therefore, should be viewed as a tool, and not as an end in itself. It is superior in correction for the effects of body size over the use of indices and more clearly lays out the problems for analysis. However, enthusiasm for the often clear linear relationships produced by these analyses should not obscure the need for attention to be given to the assumptions made in the construction of the plot and the assumptions made in the interpretation of given relationships.

The assumptions involved in the interpretation of the allometry of primate limb and bone length are different from those involved in the interpretation of the allometry of the size and shape of the cross-sections of the bones. Adult primate limb and bone lengths are the end result of ontogeny and, as such, are governed by the time onset of growth in a particular segment, the speed of growth in a particular segment and the length of the growth period. Similarities and differences in these ontogenetic factors, and not similarities and differences in the adult proportions themselves, are the important criteria in establishing affinity in proportional patterns in the primates. Neither trunk length nor body weight has a known constant relationship with any of these ontogenetic factors across the entire primate sample and, therefore, allometric analyses employing these factors as inde-

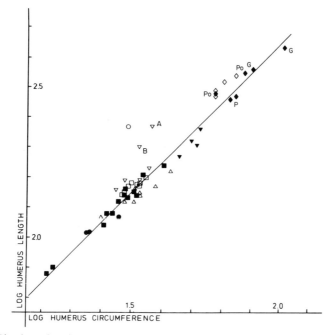

FIG. 12. Bivariate plot showing the relationship between log circumference of the midshaft of the humerus and log length of the humerus. Symbols as in Figs 1 and 2.

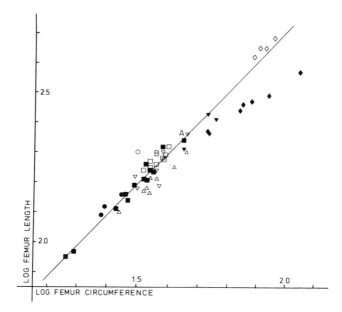

FIG. 13. Bivariate plot showing the relationship between log circumference of the midshaft of the femur and log femur length. Symbols as in Fig. 1.

pendent variables cannot be used to establish consistency in growth relationships. Although the relationships produced by these analyses are useful in ordering species for locomotor and mechanical analyses, they must be supplemented by other forms of analysis if interspecific allometry of adult limb and bone lengths is to be used to answer questions of similarity in the ontogenetic processes which produce the adult proportions. The combination of interspecific allometric analyses and comparative morphology of extant and fossil primates can provide indirect evidence relevant to ontogenetic affinity among primates in postcranial morphology. However, questions of specific similarity or difference in growth processes between any of the extant primates can only be answered from ontogenetic analyses.

In the allometric analysis of the size and shape of the cross-section of the limb bones, assumptions must be made in relation to the manner in which the independent variable employed in the analysis relates to the actual load borne by the bone. Assumptions must also be made in relation to the manner in which the cross-section of the bone adapts to the stresses to which it is exposed. Because of uncertainties in both these factors, such analyses are best used as frameworks for more detailed stress analysis. As a result of the homogeneity of the size of the cross-section of both the humerus and the femur in relation to body weight and bone length across the entire sample, these analyses are only of limited use in distinguishing primates of differing locomotor categories. However, this homogeneity poses interesting questions in relation both to the manner in which limb bones are stressed and to the manner in which they adapt to this stress in primates of highly diverse locomotor patterns.

ACKNOWLEDGEMENTS

I would like to thank Dr Peter Andrews (British Museum (Natural History)) and Professor Michael Day (St. Thomas's Hospital Medical School) for their support and encouragement during the course of the research reported here. In addition, I would like to thank Dr Ben Rudder for the majority of the data on primate body weights and Dr Lawrence Angel (Smithsonian Institution), Prue Napier (British Museum (Natural History)) and Mr Barton (Powell-Cotton Collection) for access to skeletal material in their care.

REFERENCES

Aiello, L. C. (In preparation). *On analysis of shape and strength in the long bones of higher primates*. Ph.D. dissertation: University of London.

Aiello, L. C. (In press). Locomotion in the Miocene Hominoidea. In *Aspects of human evolution*. Stringer, C. (Ed.). London: Taylor and Francis.

Alexander, R. McN., Jayes, A. S., Maloiy, G. M. O. & Wathuta, E. M. (1979). Allometry of the limb bones of mammals from shrews (*Sorex*) to elephant (*Loxodonta*). *J. Zool., Lond.* 189: 305–314.

Biegert, J. & Maurer, R. (1972). Rumpfskelettlänge, Allometrien und Körperproportionen bei catarrhinen Primaten. *Folia primatol.* 17: 142–156.

Currey, J. D. (1967). The failure of exoskeletons and endoskeletons. *J. Morph.* 123: 1–16.

Erikson, G. E. (1963). Brachiation in New World monkeys and in anthropoid apes. *Symp. zool. Soc. Lond.* No. 10: 135–163.

Evans, F. G. (1957). *Stress and strain in bones*. Springfield, Illinois: Charles C. Thomas.

Gould, S. J. (1966). Allometry and size in ontogeny and phylogeny. *Biol. Rev.* 41: 587–640.

Gould, S. J. (1975). Allometry in primates, with emphasis on scaling the evolution of the brain. (In *Approaches to primate paleobiology*. Szalay, F. (Ed.)) *Contr. Primat.* 5: 244–292.

Gould, S. J. (1977). *Ontogeny and phylogeny*. Massachusetts: The Belknap Press of Harvard University Press.

Huxley, J. (1932). *Problems of relative growth*. London: Methuen.

Huxley, T. H. (1864). *Evidence as to man's place in nature*. London: Williams and Norgate.

Jerison, H. J. (1973). *Evolution of the brain and intelligence*. New York & London: Academic Press.

Johanson, D. C. & Taieb, M. (1976). Plio-Pleistocene hominid discoveries in Hadar, Ethiopia. *Nature, Lond.* 260: 293–297.

Kummer, B. (1959). *Bauprinzipien des Säugerskeletes*. Stuttgart: Georg Thieme Verlag.

Lovejoy, C. O., Burstein, A. H. & Heiple, K. G. (1976). The biomechanical analysis of bone strength: A method and its application to Platycnemia. *Am. J. phys. Anthrop.* 44: 489–506.

Lucae, J. C. G. (1865). Die Hand und der Fuss. Ein Beitrag zur vergleichenden Osteologie der Menschen, Affen und Beutelthiere. *Abh. senckenb. naturforsch. Ges.* 5: 275–332.

Lumer, H. (1939). Relative growth of the limb bones in the anthropoid apes. *Hum. Biol.* 11: 379–392.

Lumer, H. & Schultz, A. H. (1941). Relative growth of the limb segments and tail in the macaques. *Hum. Biol.* 13: 283–305.

Lumer, H. & Schultz, A. H. (1947). Relative growth of the limb segments and tail in *Ateles geoffroyi* and *Cebus capucinus*. *Hum. Biol.* 19: 53–67.

McMahon, T. A. (1973). Size and shape in biology. *Science, N.Y.* 179: 1201–1204.

McMahon, T. A. (1975). Allometry and biometrics: limb bones in adult ungulates. *Am. Nat.* 109: 547–563.

Mivart, St. G. (1867). On the appendicular skeleton of the primates. *Phil. Trans. R. Soc.* 157: 299–430.

Mollison, T. (1910). Die Körperproportionen der Primaten. *Morph. Jb*. 42: 79–304.

Pauwels, F. (1965). *Gesammelte Abhandlungen zur funktionellen Anatomie des Bewegungsapparates*. Berlin: Springer Verlag.

Pilbeam, D. & Gould, S. J. (1974). Size and scaling in human evolution. *Science, N.Y*. 186: 892–901.

Preuschoft, H. & Weinman, W. (1973). Biochemical investigations of *Limnopithecus* with special reference to the influence exerted by body weight on bone thickness. *Am. J. phys. Anthrop*. 38: 241–250.

Rose, M. D. (1973). Quadrupedalism in primates. *Primates* 14: 337–358.

Rose, M. D. (1975). Functional proportions of primate lumbar vertebral bodies. *J. Hum. Evol*. 4: 21–38.

Sacher, G. A. (1970). Allometric and factorial analysis of brain structure in insectivores and primates. In *The primate brain*: 245–287. Noback, C. R. & Montagna, W. (Eds). New York: Appleton-Century-Crofts.

Schultz, A. H. (1930). The skeleton of the trunk and limbs of higher primates. *Hum. Biol*. 2: 303–438.

Schultz, A. H. (1933). Die Körperproportionen der erwachsenen catarrhinen Primaten, mit spezieller Berücksichtigung der Menschenaffen. *Anthrop. Anz*. 10: 154–185.

Schultz, A. H. (1937). Proportions, variability and asymmetries of the long bones of the limbs and the clavicles in man and apes. *Hum. Biol*. 9: 281–328.

Snell, O. (1891). Das Gewicht des Gehirns und des Himmantels der Säugetiere in Beziehung zu deren geistigne Fähigkeiten. *Sber. Ges. Morph. Physiol. Münch* 7: 90–94.

Sokal, R. R. & Rohlf, F. J. (1969). *Biometry*. San Francisco and London: W. H. Freeman.

Symp. zool. Soc. Lond. (1981) No. 48, 359–375

Climbing: A Biomechanical Link with Brachiation and with Bipedalism

JOHN G. FLEAGLE, JACK T. STERN, JR, WILLIAM L. JUNGERS, RANDALL L. SUSMAN

Department of Anatomical Sciences, Health Sciences Center, State University of New York at Stony Brook, Long Island, New York 11794, USA

ANDREA K. VANGOR

Department of Anatomy, University of Pennsylvania, School of Veterinary Medicine H1, Philadelphia, Pa. 19104, USA

and

JAMES P. WELLS

West Virginia School of Osteopathic Medicine, Lewisburg, West Virginia 24901, USA

SYNOPSIS

Classical anatomical studies of the past one hundred years have clearly demonstrated a close morphological similarity between humans and living apes. These anatomical similarities, particularly in forelimb and trunk morphology, have led to the generally accepted conclusion that bipedal hominids have evolved from some type of brachiating ancestor. It is much less clear why or how brachiation should give rise to bipedalism. Recent experimental studies of non-human primate locomotion using electromyography, force plate analysis, high speed kinematics and strain gauges suggest that climbing is a biomechanical link between brachiation and bipedalism.

Electromyographic studies of forelimb muscles in large ceboids and in apes indicate that many of the muscles that are especially large and uniquely constructed in "brachiators" are more active in climbing and hoisting behaviours than in brachiation. Likewise, preliminary studies indicate that the direction and magnitude of ulnar bone strain is very similar in brachiation and in climbing by comparison with the patterns seen in quadrupedal walking. Thus many of the biomechanical requirements of the forelimb for climbing are comparable to, or exceed, those for brachiation.

Kinematic studies of hindlimb movements in chimpanzees have demonstrated that vertical climbing involves considerable extension and medial rotation of the thigh comparable to that found in human bipedalism. Electromyographic studies of hip and thigh muscles in large ceboids demonstrate that compared with quadrupedal walking, bipedal walking and vertical climbing invoke similar recruitment patterns and are more similar to the patterns of muscle use seen in human bipedalism. Force plate studies show that the facultative bipedalism of orangutans, chimpanzees and spider monkeys, all adept climbers, is more comparable to that of humans than is the bipedalism of other non-human primates studied. Climbing thus would appear to be pre-adaptive for human bipedalism.

In summary, recent experimental studies of non-human primate locomotion suggest that a human ancestor primarily adapted for climbing would show a forelimb morphology comparable to that normally associated with brachiation and a hindlimb morphology pre-adaptive for human bipedalism.

INTRODUCTION

Although most students of human evolution would generally agree that our most recent non-bipedal ancestor was probably some sort of brachiating hominoid, in the broadest sense, the "brachiationist" theory of human origins has been confronted with several difficulties in recent years. First, as the naturalistic locomotor behaviour of apes and other so-called brachiators (such as the larger atelines) has become studied, it is apparent that brachiation *sensu stricto* is not as frequent or probably as important a part of their daily activities as was previously suspected. Indeed, brachiation appears to be relatively uncommon among great apes or atelines, and even among hylobatids it is less prominent than quadrumanous climbing during most activities (Fleagle, 1976; Mittermeier, 1978; Tuttle, 1977; Susman, Badrian & Badrian, 1980). If brachiation is not a particularly common locomotor activity among these animals, it is difficult to argue that their morphological similarities are due to that activity without invoking extensive retention of previously, but not currently, adaptive morphological features. This difficulty has been further compounded by the documentation of numerous "brachiating" adaptations among other groups of animals, such as lorisoids, in which brachiation is totally unknown (Cartmill & Milton, 1977). These data have led many current theorists to suggest that the morphological similarities shared by humans, apes and other brachiators are perhaps the result of selection for other types of behaviour, in particular quadrumanous climbing (Washburn, 1973; Fleagle, 1976; Tuttle, 1975; Cartmill & Milton, 1977; Mendel, 1976).

Perhaps the greatest conceptual difficulty in accounting for the origin of human bipedalism through brachiation is that it is primarily an argument based on phenetic similarities between humans and "brachiators". Brachiation gave rise to bipedalism because humans share many similarities in limb and trunk anatomy with "brachiators". Functional arguments for how or why brachiation, rather than any other activity, should or did give rise to bipedalism are very hard to find. The most convincing functional reason for morphological similarities between brachiators and bipeds is that brachiation pre-adapted our ancestors for upright posture (Keith, 1923; Morton, 1926). Keith has convincingly shown how many common aspects of trunk and visceral anatomy among hominoids are adaptations for orthograde locomotion. In their attempts to provide alternative explanations for the brachiation adaptations of hominoids, Cartmill & Milton (1977) agreed that orthograde posture remained the best explanation for the visceral peculiarities of hominoids.

The second frequently offered explanation as to why brachiation should predispose an early hominid to bipedalism is that brachiators have long arms designed for suspension rather than "support", and such animals make ungainly quadrupeds. This is almost certainly why gibbons walk bipedally on both large arboreal supports and on the ground, and it has been argued that our ancestors first adopted habitual bipedalism because their arms were too long and too far adapted for suspension for use in quadrupedal walking (Stern, 1976).

The brachiation theory of human evolution was largely formulated in a time when comparative anatomy was the primary tool available to students of human evolution, and it is no accident that the founders and major proponents of this theory were all great anatomists. The diversity of information available to students of human evolution is now much greater. Naturalistic studies of primate behaviour have helped clarify the importance of particular locomotor activities in the daily life of living species, and experimental laboratory studies may deepen our understanding of the biomechanics of locomotion (Fleagle, 1979). The recent application of such highly technological approaches as electromyography, high speed kinematics, cineradiography, force plate studies, and *in vivo* strain gauges have produced a wealth of data on primate locomotion that help clarify many of the difficulties surrounding brachiation theory and provide a remarkably consistent picture of the type of locomotor behavior that almost certainly paved the way for the evolution of hominid bipedalism. These studies suggest that many aspects of forelimb anatomy that have previously been identified as brachiating adaptations can be explained as well or better as adaptations to vertical climbing. Furthermore, they indicate that vertical climbing is the one activity found among non-human primates that would functionally pre-adapt the hindlimb musculature for human-like bipedal walking.

THE FUNCTION OF PRIMATE FORELIMBS

The numerous similarities between the shoulder and forelimb musculature of humans and that of extant hominoids and other brachiators
have been thoroughly documented by many workers during the past
century (e.g. Keith, 1923; Miller, 1932; Inman *et al.*, 1944; Ashton &
Oxnard, 1963), and the derived morphological features shared by
these animals have been attributed to the requirements of brachiation
and forelimb suspension. Only recently, through the development of
telemetered electromyography, has it become possible to examine
precisely the way in which non-human primates actually use their
muscles during locomotor activities, including brachiation. Because
of their phylogenetic importance, forelimb and shoulder muscles
have received a high priority in these investigations.

Stern, Wells, Vangor & Fleagle (1977) examined the firing patterns
of five shoulder muscles in spider monkeys (*Ateles* sp.) and woolly
monkeys (*Lagothrix lagotricha*) during arm-swinging, pronograde
quadrupedalism and vertical climbing. They found that, generally,
these muscles were more active in arm-swinging than in quadrupedal
walking and running, supporting the hypothesis that their large size
and specialized morphology could be interpreted as brachiating adaptations. However, they also found that in most instances these
muscles were as active, or even more active, in vertical climbing as in
arm-swinging (Fig. 1). This observation, that the shoulder muscles of
brachiators are more active in vertical climbing than in arm-swinging
activities, has been confirmed by more detailed studies on pectoralis
major (Stern, Wells, Jungers, Vangor & Fleagle, 1980) and serratus
anterior (Stern, Wells, Jungers & Vangor, 1980) in a wide range of primates. Perhaps the most dramatic demonstration of the importance
of climbing in forelimb muscle function comes from a study by
Jungers & Stern (1980) on brachiation and climbing in gibbons
(*Hylobates* sp.). When they examined the activity of six shoulder and
forelimb muscles during vertical climbing, slow brachiation, and fast
(ricocheting) brachiation, they found that climbing and hoisting required much higher levels of muscle recruitment than either type of
brachiation and that faster brachiation actually involved very little
activity in the muscles tested (Fig. 2).

Unfortunately, the studies of Tuttle & Basmajian (reviewed in
Tuttle, Basmajian & Ishida, 1979) on the shoulder and forelimb
muscles of great apes, are not directly comparable since these
workers did not investigate either arm-swinging or vertical climbing,
but concentrated on quadrupedal locomotion and suspensory posture. Nevertheless, they repeatedly found the shoulder and forelimb

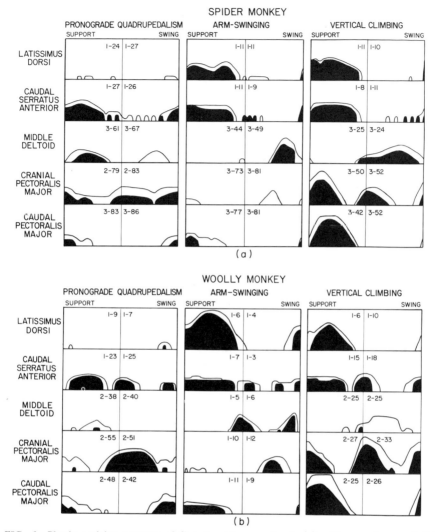

FIG. 1. Phasic activity patterns of five shoulder muscles in (a) spider monkeys and (b) woolly monkeys during pronograde quadrupedalism, arm-swinging, and vertical climbing. Black areas indicate very consistent presence of EMG activity during that part of the loco-motor cycle; the white areas indicate frequent, but less consistent activity. The heights of these areas indicate relative amplitude of EMG activity. Two numbers appear in each box. The first is the number of animals from which data were obtained for this phase; the second is the number of activity cycles. For details of data analysis see Stern, Wells, Jungers & Vangor (1980). Note the high levels of activity in vertical climbing.

muscles to be largely quiescent during passive suspension and most active during hoisting.

The experimental findings have major implications for interpreting the evolution of large, often uniquely constructed, shoulder and fore-

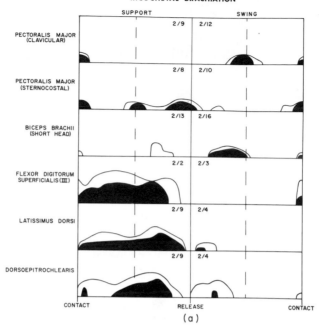

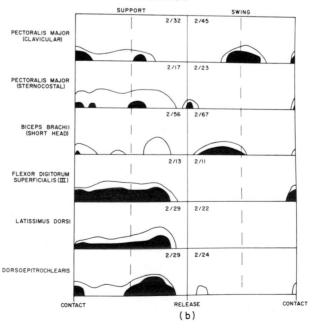

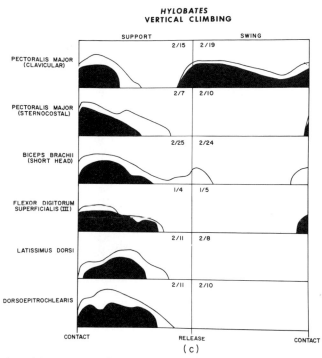

FIG. 2. Phasic activity patterns of six shoulder and forelimb muscles in gibbons (*Hylobates* sp.) during (a) ricochetal brachiation, (b) slow brachiation, and (c) vertical climbing. Data representation is the same as in Fig. 1. Note consistently very high levels of muscle activity in vertical climbing.

limb muscles in hominoids and large ceboids. If climbing is the loco-motor activity in which these muscles are most powerfully used, and climbing is known to be a major part of the naturalistic locomotor behaviour of these species (Fleagle, 1976), it seems most likely that climbing rather than brachiation is the activity for which the shoulder and forelimb muscles are being maintained; indeed it is quite likely to be the single activity that has most strongly influenced their evol-ution.

In another set of experiments, we have examined the activity of three arm muscles and the bone strain on the dorsal surface of the ulna in two spider monkeys (one *Ateles geoffroyi* and one *Ateles fusciceps*) during quadrupedalism, brachiation and climbing. Elec-trodes were placed in biceps brachii, brachialis and the lateral head of triceps brachii. In order to measure bone strain, we placed a single element semiconductor strain gauge on the convex dorsal surface of the ulna below the coronoid process, distal to the insertion of the tendon of triceps brachii (Fig. 3). This particular placement was chosen because it provided a relatively flat subcutaneous area for

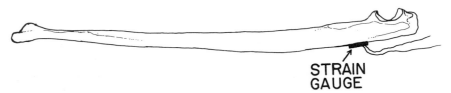

STRAIN
GAUGE

FIG. 3. Location of strain gauge placement on spider monkey (*Ateles* sp.) ulna.

attachment of the gauge, and because a gauge placed in this position on a dry bone seemed particularly responsive to anterior and posterior bending of the ulnar shaft. The results of these experiments are illustrated (Fig. 4) by representative oscilloscope displays of muscle activity and bone strain during three types of locomotion.

During the swing phase of quadrupedal walking along a branch, triceps brachii is inactive and there is little or no strain on the inferior surface of the ulna. The elbow flexors show low levels of activity as the limb is flexed and brought forward. During the support phase of quadrupedal walking, triceps brachii (the main antigravity muscle) is very active, while the two flexors (biceps brachii and brachialis) are silent. The ulna is subjected to considerable anterior bending as it becomes more curved during this phase, with the maximum strain occurring in the middle of support.

The elbow flexors are active early in the swing phase of brachiation, and the strain gauge indicates that the ulna is being slightly

QUADRUPEDAL   **BRACHIATE**   **CLIMB**

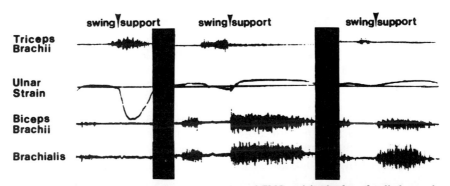

FIG. 4. Oscilloscope picture of ulnar bone strain and EMG activity in three forelimb muscles during quadrupedal branch walking, brachiation, and vertical climbing in *Ateles* sp. On the ulnar strain channel, a downward deflection indicates that the dorsal surface of the ulna is under tension and (presumably) the shaft is becoming more curved. An upward deflection indicates that the dorsal surface is under compression and (presumably) the curved shaft is being straightened. Note the similarity in both ulnar strain and EMG activity during brachiation and climbing.

straightened (probably as a result of both gravity and the action of brachialis). As the monkey reaches forward for a new grasp in the second half of the swing phase, triceps brachii is active and the flexors become silent. During the support phase of brachiation, both of the flexors are usually active and the ulnar shaft shows a strain pattern indicating that the bone is being straightened. Triceps brachii is silent.

When a spider monkey climbs a vertical pole, the patterns of muscle activity and bone strain are very similar to those observed during brachiation. During the early part of swing phase, there is activity of the elbow flexors and a slight straightening of the bone. Triceps brachii again fires near the end of swing phase as the animal reaches for a new grasp. During support phases, as the monkey pulls itself up, there is considerable activity of both biceps brachii and brachialis, and the strain pattern indicates that the ulnar shaft is being straightened.

These results concur with the electromyographic results described above in demonstrating the remarkable similarity between certain aspects of the biomechanics of brachiation and climbing. The gravitational forces acting to extend the elbow and straighten the ulnar shaft during forelimb suspension are the same forces acting on a monkey climbing up a vertical pole, and they elicit the same patterns of muscle activity and ulnar bone strain. Thus selection for such forelimb features as large elbow flexors, small extensors, and forelimb bones suited for sustaining "tensile forces" could be equally well attributed to climbing or brachiation.

HINDLIMB FUNCTION IN PRIMATES

While the musculo-skeletal similarities linking the human forelimb with apes have been well documented, such is not the case for the structure of the human hindlimb. Although the human foot shows clear affinities with that of hominoids (Morton, 1922, 1924, 1926), the musculo-skeletal anatomy of the human pelvis, thigh, and leg are largely unique and show few obvious similarities among extant non-human primates (Zuckerman *et al.*, 1973). However, Stern (1971) found that of all living primates, the extant howling monkey was the animal whose hip and thigh musculature could most easily be transformed into a human-like morphology. Since howling monkeys are not known to engage in bipedal locomotion in either naturalistic or laboratory situations, he argued that slow quadrupedal climbing must pre-adapt the hip and thigh for human-like hip and thigh musculature.

The main functional argument as to why brachiation would pre-adapt the hindlimb for bipedalism has been Morton's suggestion that during brachiation, the hindlimb is (fully) extended at the hip, an unusual position for quadrupeds. Since such extension is presumably passive, its effect would probably be most evident in the bony and ligamentous structure of the hip joint rather than in musculature.

With the increase in experimental studies, many workers have focused on the study of bipedalism in non-human primates for a comparison with the same activity in humans (e.g. Jenkins, 1972; Tuttle, Basmajian & Ishida, 1979; Ishida, Kimura & Okada, 1975; Vangor, 1979; Prost, 1967). In general, the results of these studies have been uniform in reaffirming the major differences in human and non-human bipedalism. As demonstrated in kinematic and cineradiographic studies, the bipedal walking of chimpanzees is much more like chimpanzee quadrupedalism in joint excursions than like human bipedalism (Elftman, 1944; Jenkins, 1972; Prost, 1980). Likewise bipedalism in gibbons, baboons and macaques is distinctly different from human bipedalism in utilizing far less hip and knee extension, less medial rotation of the thigh, and no real toe-off as well as other differences (Kimura, Okada & Ishida, 1979). However, Prost (1980) has recently demonstrated that vertical climbing by chimpanzees is kinematically much more comparable to human bipedalism than any other activity in their locomotor repertoire. In contrast with bipedal walking, chimpanzee climbing involves considerable extension of both hip and knee as well as medial rotation of the thigh as seen in human bipedalism (Fig. 5). Thus he argued it is vertical climbing, rather than facultative bipedalism, which is pre-adaptive for human bipedalism.

Much the same picture emerges from electromyographic studies of hindlimb muscle activity during the locomotion of non-human primates. In a long series of experiments, Japanese scientists have shown that the activities of numerous hindlimb muscles during bipedalism in five non-human species are very different in their firing patterns from those found among bipedal humans. Generally, the non-human primates use more muscles, more continuously during bipedalism and lack the phasic activities seen in humans (Ishida *et al.*, 1975; Kimura *et al.*, 1979). Tuttle, Basmajian & Ishida (1979) drew a similar conclusion from their study of bipedal walking in great apes, noting that "The EMG patterns exhibited by thigh and hip muscles in man and apes are generally strikingly dissimilar during most bipedal behavior and especially during bipedal progression" (Tuttle, Basmajian & Ishida, 1979). Vangor (1977, 1979) also noted the general dissimilarity between firing patterns of hip and thigh muscles during bipedal

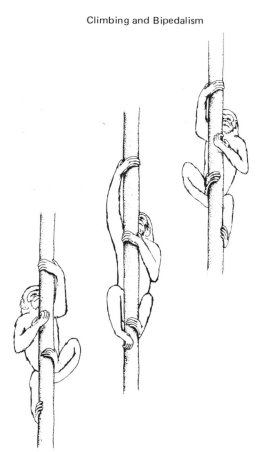

FIG. 5. Drawing of a very young chimpanzee climbing a vertical pole, traced from a video-tape. Note the medial rotation of the thigh during stance phase.

walking of four genera of non-human primates (*Ateles* sp., *Lagothrix lagotricha, Alouatta seniculus, Erythrocebus patas*) and that seen in humans. In addition, she found no indication that animals which are reported to use more bipedal behaviour in naturalistic conditions (*Erythrocebus patas*) were any more human-like in their muscle firing patterns than animals which are not known to be bipedal under naturalistic conditions. Vangor looked at the activity of 13 hip and thigh muscles during quadrupedal walking, bipedal walking and verti-cal climbing. She found that, in seven muscles, the firing patterns used during bipedalism are similar to those used during quad-rupedalism and in vertical climbing. In *Ateles* and *Lagothrix*, how-ever, there were six muscles for which the activity patterns used dur-ing bipedalism were different from those seen in quadrupedalism but similar to those used during vertical climbing. In five of these muscles (Fig. 6) the bipedal/vertical climbing pattern was more similar to that

EMG of Hip and Thigh Muscles in *Ateles*

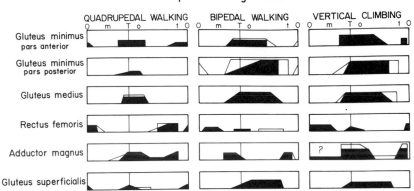

FIG. 6. Phasic activity of six hip and thigh muscles in spider monkeys (*Ateles* sp.) during quadrupedal walking, bipedal walking, and vertical climbing. O, toe-off; T, touch-down; m, midstance for limb being tested; o, toe-off for contralateral hindlimb; t, touchdown for contralateral limb. Black areas indicate very consistent presence of EMG activity. White areas indicate occasional but not consistent activity. The area beneath the "?" is probably artifact from an overlying muscle. (For details see Vangor, 1979.) Note the differences between quadrupedal walking and bipedal walking in the activity of these muscles and the similarity between bipedal walking and vertical climbing.

seen in human bipedalism than the quadrupedal pattern, suggesting that vertical climbing is pre-adaptive for the type of hip and thigh muscle activity found in bipedalism.

Kimura *et al.* (1979) have examined the locomotor behavior of six non-human primate species during bipedal (and quadrupedal) locomotion using force plate analysis. They found that the bipedal walking of chimpanzees, spider monkeys and orangutans was most similar to that of humans, with the gibbon less and Japanese macaques and baboons least like humans. Both chimpanzees and spider monkeys showed evidence of a distinct braking at the beginning of stance and a toe-off at the end. This latter feature was evident in both the force analysis and in the activity of the gastrocnemius muscle. As they noted, these are the most adept climbers of the species they studied.

In summary, both kinematic and electromyographic studies of primate hindlimb function indicate that the facultative bipedalism of non-human primates is not very similar to human bipedalism, whereas vertical climbing involves patterns of both limb excursion (Prost, 1980) and muscle activity (Vangor, 1977, 1979) that are more similar to those found in human bipedalism. In addition, there is some evidence from both EMG and force plate studies (Kimura *et al.*, 1979) that it is adept climbers such as chimpanzees, spider monkeys and woolly monkeys rather than more naturally bipedal animals such as gibbons or patas monkeys which show the greatest similarities to humans in their bipedal gaits. Thus vertical climbing, more than any other locomotor activity, would appear to be functionally pre-adaptive for the type of hindlimb function seen in human bipedalism.

DISCUSSION

In a recent review of experimental studies of primate positional behaviour, Tuttle (1979) lamented that "although during the 1970's remarkable technological advancement occurred in nonhuman kinesiology, anthropologically relevant results have not burgeoned". This feeling was echoed by Tuttle, Cortright & Buxhoeveden (1979), who noted that "Primate experimental kinesiology is such a new field that sufficient data have not been gathered to allow a probabilistic ordering of the several evolutionary models of transitional hominid positional behavior which have been proposed." By contrast, it seems to us that the available experimental studies provide a remarkably consistent picture of the locomotor adaptations of our prebipedal ancestor. They suggest that a human ancestor primarily adapted for climbing would show a forelimb morphology comparable to that previously associated with brachiation and a hindlimb morphology that is both morphologically (Stern, 1971) and functionally (Vangor, 1977) pre-adaptive for bipedalism.

The experimental finding that climbing and brachiation entail similar patterns of muscle use and bone strain in the forelimb largely obviates the need to invoke brachiation *sensu stricto* (i.e. bimanual suspension) in human ancestry. As such it accords well with the behavioural observations suggesting that bimanual suspension plays a minor role in the daily activities of most extant apes, and also accords with the comparative studies indicating the presence of "brachiating" adaptations in the limbs of many mammalian species who never brachiate. Thus a climbing theory of human ancestry (e.g. Cartmill & Milton, 1977) is consistent with all of the evidence adduced in support of brachiation theory, and also overcomes most of the difficulties suggested by naturalistic and comparative studies.

The experimental finding that vertical climbing involves patterns of limb excursion and muscle activity similar to human bipedalism, goes a long way towards solving the other major problem with brachiation theory — why should bimanual suspension give rise to bipedalism? The need to look at hindlimb function to understand the origin of bipedalism was emphasized long ago by Klaatsch (1913), but his argument was more concerned with the activities of aboriginal peoples than with comparative anatomy.

Despite all the apparent concordance with a diverse array of information, there are still numerous difficulties with a "climbing theory of human evolution". One difficulty is the heterogeneity of the behaviours that are included under the term "climbing". In the experimental studies, climbing is *vertical* climbing up a single continuous support. Climbing has much broader implications as used in naturalistic studies. Kortlandt (1975), in describing chimpanzees as basically climbing animals, refers mainly to vertical climbing up large treetrunks. However, Fleagle (1976, 1979) in describing apes as climbers, or Mittermeier (1978) and Fleagle & Mittermeier (1980) in descriptions of *Ateles* and *Alouatta* include such behaviours as three-or fourlimbed suspension, hoisting and bridging, as well as vertical climbing, under the same category. Likewise, Cartmill & Milton (1977) use climbing in this broad sense. If vertical climbing seems to be the mechanically important behaviour in human ancestry, what is the significance of these other behaviours which play a much greater role in the daily activities of living "climbers" and brachiators? As largely irregular, rather than patterned activities, they are much more difficult to analyse or describe under an experimental situation.

Just as it is difficult to see clearly what is meant specifically when one describes a behaviour as climbing, it is not at all clear what extant primate, if any, most closely resembles our climbing ancestor. As Tuttle (1974) has repeatedly emphasized, theories of human evolution have been generally modelled after specific living primates (e.g. hylobatid, troglodytian, orangutanian). Identifying climbing as the important behaviour in human evolution is of little help in choosing among Tuttle's models, since all apes, as well as several atelines (and even lorises) have been described as climbers. What these results explain is why any suspensory, climbing hominoid could become a human biped. The important problem which remains unresolved is why human became bipeds and no other climbers have followed what would appear to be a natural option. Indeed, to the extent that chimpanzees and especially gibbons have become bipedal, they have not become human-like bipeds. A likely possibility, long favoured by non-brachiationists and consistent with the fossil record, is that our suspensory ancestor was not particularly like any of the extant apes. One can only hope that when we find such an ancestor we will be able to recognize it and interpret its fossils.

ACKNOWLEDGEMENTS

We thank Lucille Betti for producing the illustrations and Joan Kelly who typed the manuscript. Much of the work reported herein was supported in part by research grants RRO 5736 from the USPHS and BNS 7683114, BNS 7724921, and BNS 7924070 from the National Science Foundation.

REFERENCES

Ashton, E. H. & Oxnard, C. E. (1963). The musculature of the primate shoulder. *Trans. zool. Soc. Lond.* 29: 553—650.

Cartmill, M. & Milton, K. (1977). The lorisiform wrist joint and the evolution of "brachiating" adaptations in the Hominoidea. *Am. J. Phys. Anthrop.* 47: 249—272.

Elftman, H. (1944). The bipedal walking of the chimpanzee. *J. Mammal.* 25: 67—71.

Fleagle, John G. (1976). Locomotion and posture of the Malayan siamang and implications for hominoid evolution. *Folia primatol.* 26: 245—269.

Fleagle, John, G. (1979). Primate positional behavior and anatomy: Naturalistic and experimental approaches. In *Environment, behavior and morphology: Dynamic interactions in primates*: 313—325. Morbeck, M. E., Preuschoft, H. & Gomberg, N. (Eds). New York: Gustav Fischer.

Fleagle, John G. & Mittermeier, R. A. (1980). Locomotor behavior, body size, and comparative ecology of seven Surinam monkeys. *Am. J. Phys. Anthrop.* 52: 301—314.

Inman, V. T., Dec, J. B., Saunders, M. & Abbott, L. C. (1944). Observations on the function of the shoulder joint. *J. Bone Jt Surg.* 26: 1—30.

Ishida, H., Kimura, T. & Okada, M. (1975). Patterns of bipedal walking in anthropoid primates. In *Symp. 5th Congr. Int. Primate Soc.* 3: 459—462. Kondo, S., Kawai, M., Ehara, A. & Kawamura, S. (Eds). Japan Science Press.

Jenkins, F. A. Jr. (1972). Chimpanzee bipedalism: cineradiographic analysis and implications for the evolution of gait. *Science, N.Y.* 178: 877—879.

Jungers, W. L. & Stern, J. T. Jr. (1980). Telemetered electromyography of forelimb muscle chains in gibbons (*Hylobates lar*). *Science, Wash.* 208: 617—619.

Keith, A. (1923). Man's posture: its evolution and disorders. *Br. Med. J.* I: 451—454, 499—502, 545—548, 587—590, 624—626, 669—672.

Kimura, T., Okada, M. & Ishida, H. (1979). Kinesiological characteristics of primate walking: Its significance in human walking. In *Environment, behavior and morphology: Dynamic interactions in primates*: 297—311. Morbeck, M. E., Preuschoft, H. & Gomberg, N. (Eds). New York: Gustav Fischer.

Klaatsch, H. (1913). Die Erwerbung der aufrechten Haltung und ihre Folgen. *Verh. anat. Ges. Jena* 27: 161—186.

Kortlandt, A. (1975). Ecology and paleoecology of ape locomotion. In *Symp. 5th Congr. Int. Primate Soc.* 3: 361—364. Kondo, S., Kawai, M., Ehara, A. & Kawamura, S. (Eds). Japan Science Press.

Mendel, F. (1976). Postural and locomotor behavior of *Alouatta palliata* on various substrates. *Folia primatol.* 26: 36—53.

Miller, R. A. (1932). Evolution of the pectoral girdle and forelimb in primates. *Am. J. Phys. Anthrop.* 17: 1–56.

Mittermeier, R. A. (1978). Locomotion and posture in *Ateles geoffroyi* and *Ateles paniscus*. *Folia primatol.* 30: 161–193.

Morton, D. J. (1922). Evolution of the human foot. *Am. J. Phys. Anthrop.* 5: 305–336.

Morton, D. J. (1924). Evolution of the human foot II. *Am. J. Phys. Anthrop.* 7: 1–52.

Morton, D. J. (1926). Evolution of man's erect posture. Preliminary report. *J. Morph. Physiol.* 43: 147–149.

Prost, J. H. (1967). Bipedalism of man and gibbon compared using estimates of joint motion. *Am. J. Phys. Anthrop.* 36: 135–148.

Prost, J. H. (1980). Origin of bipedalism. *Am. J. Phys. Anthrop.* 52: 175–190.

Stern, J. T. Jr. (1971). Functional myology of the hip and thigh of cebid monkeys and its implications for the evolution of erect posture. *Bibliothecia Primatol.* 14: 1–318.

Stern, J. T. Jr. (1976). Before bipedality. *Yb. Phys. Anthrop.* 19: 59–68.

Stern, J. T. Jr., Wells, J. P., Jungers, W. L. & Vangor, A. K. (1980). An electromyographic study of serratus anterior in atelines and *Alouatta*: Implications for hominoid evolution. *Am. J. Phys. Anthrop.* 52: 323–334.

Stern, J. T. Jr., Wells, J. P., Jungers, W. L., Vangor, A. K. & Fleagle, J. G. (1980). An electromyographic study of the pectoralis major in atelines and *Hylobates*, with special reference to the evolution of a pars clavicularis. *Am. J. Phys. Anthrop.* 52: 13–26.

Stern, J. T. Jr., Wells, J. P., Vangor, A. K. & Fleagle, J. G. (1977). Electromyography of some muscles of the upper limb in *Ateles* and *Lagothrix*. *Yb. Phys. Anthrop.* 20: 498–507.

Susman, R. L., Badrian, N. & Badrian, A. (1980). Locomotor behavior of *Pan paniscus* in Zaire. *Am. J. Phys. Anthrop.* 53: 69–80.

Tuttle, R. H. (1974). Darwin's apes, dental apes, and the descent of man: Normal science in evolutionary anthropology. *Curr. Anthrop.* 15: 359–426.

Tuttle, R. H. (1975). Parallelism, brachiation and hominoid phylogeny. In *Phylogeny of the primates: A multidisciplinary approach*: 447–480. Luckett, W. P. & Szalay, F. S. (Eds). New York: Plenum Press.

Tuttle, R. H. (1977). Naturalistic positional behavior of apes and models of hominoid evolution 1029-1976. In *Progress in ape research*: 277–296. Bourne, G. H. (Ed.). New York & London: Academic Press.

Tuttle, R. H. (1979). The problem of hominid bipedalism: what do we need in order to proceed. *Am. J. Phys. Anthrop.* 50: 457–488.

Tuttle, R. H., Basmajian, J. V. & Ishida, H. I. (1979). Activities of pongid thigh muscles during bipedal behavior. *Am. J. Phys. Anthrop.* 50: 123–136.

Tuttle, R. H., Cortright, G. W. & Buxhoeveden, D. P. (1979). Anthropology on the move: Progress in experimental studies of nonhuman primate positional behavior. *Yb. Phys. Anthrop.* 22: 187–214.

Vangor, A. K. (1977). Functional pre-adaptation to bipedality in non-human primates. *Am. J. Phys. Anthrop.* 47: 164–165.

Vangor, A. K. (1979). *Electromyography of gait in nonhuman primates and its significance for the evolution of bipedality*. Ph.D. Thesis: State University of New York at Stony Brook.

Washburn, S. L. (1973). Primate field studies. In *Nonhuman primates and medical research*: 467–485. Bourne, G. (Ed.). London & New York: Academic Press.

Zuckerman, S., Ashton, E. H., Flinn, R. M., Oxnard, C. E. & Spence, T. F. (1973). Some locomotor features of the pelvic girdle in primates. *Symp. zool. Soc. Lond*. No. 33: 77–165.

Symp. zool. Soc. Lond. (1981) No. 48, 377—427

Comparative Aspects of Primate Locomotion, with Special Reference to Arboreal Cercopithecines

J. ROLLINSON and R. D. MARTIN

Department of Anthropology, University College London, Gower Street, London WC1, England

SYNOPSIS

Most previous studies involving detailed quantification of mammalian locomotor behaviour have been confined to terrestrial, cursorial forms. This chapter deals with various quantitative aspects of locomotor behaviour in eight cercopithecine monkey species, including both guenons (*Cercopithecus* spp.) and mangabeys (*Cercocebus* spp.), maintained in captivity. The results throw light on generic distinctions between mangabeys and guenons, on the effects of terrestrial adaptation, and hence on the evolution of locomotion in Old World monkeys generally. Mangabeys (including the arboreal species *Cercocebus albigena*) are characterized by relatively slow "deliberate" locomotion which usually gives way to a gallop at fast speeds, while guenons (including the semi-terrestrial *Cercopithecus neglectus*) more commonly exhibit quite fast trotting gaits. There are also characteristic differences in tail-carriage between mangabeys and guenons, though within each genus tail length is found to be reduced in species adapted for terrestrial life. In several respects, the talapoin (*Cercopithecus (Miopithecus) talapoin*) is closer to the mangabeys than to the typical guenons, as in the greater emphasis on walking and galloping, the occurrence of digitigrady, and greater use of three-leg support patterns (rather than just diagonal couplets) in walking. This suggests that the talapoin, like the arboreal mangabey *Cercocebus albigena*, has only recently become predominantly arboreal. With respect to the general evolution of primate locomotion, confirmation of the typical use of a *diagonal support sequence* in walking gaits of the cercopithecine monkeys and other primates established this pattern as typical for the primates (excluding tree-shrews), as distinct from the *lateral support sequence* used by most other mammals (including tree-shrews). Use of the diagonal sequence is connected with the relatively posterior location of the centre of gravity in the primate body, and this in turn is linked with the consistent presence of hindlimb domination and the possession of a grasping hallux in all non-human primates (excluding tree-shrews). Analysis of certain established limb indices (intermembral, brachial, crural) in both living and fossil primates shows that they do not "eliminate" effects of body size and that in fact their variation with body size provides additional insights into the evolution of primate locomotion. The course of Old World monkey evolution can be traced from a hindlimb dominated condition in a small-bodied ancestral primate adapted for occupation of the "fine branch niche", through an ancestral simian stage in which increased body size was accompanied by a shift towards greater emphasis on quadrupedal branch-running, as in the medium-sized New World monkeys found today. Further increase in body size in various simian species in both the New and Old Worlds led to various forms of below-branch suspensory locomotor adaptations (a development paralleled by large-bodied subfossil lemurs). Among the Old World monkeys, the Colobinae probably remained predominantly arboreal after their divergence from the common ancestral stock, while the Cercopithecinae apparently became semi-terrestrial at an earlier stage. Subsequently, various cercopithecine monkey groups (*Cercopithecus* spp.; *Miopithecus*; some *Cercocebus* spp.) became re-adapted for a predominantly arboreal existence, but among the Old World monkeys generally there has been considerable lability in arboreal vs. semi-terrestrial habits.

INTRODUCTION

Background to the Study

The study of patterns of movement and speed in mammals has been of long-standing interest, but there are still relatively few comprehensive studies of non-domesticated species reported in the literature. Hitherto, investigations of locomotor behaviour have been concentrated on domesticated, terrestrial mammals because of the relative ease with which they can be studied. With arboreal mammals, particularly in their natural habitat, difficulties in obtaining standardized data for comparison are compounded by problems of reduced visibility.

Many early locomotor studies were carried out on the domestic horse, with which discrete categories of movement can be elicited on command from the rider, thus enabling each category to be defined and analysed. The earliest study recording movements of wild mammals was that of Muybridge (1899), who used still cameras to record the locomotion of a wide range of mammals. Later studies (e.g. Howell, 1944; Gray, 1944) defined differences across a broad spectrum of locomotor categories in terms of gaits and skeletal structure. More recently, a number of different approaches have been developed, such as the establishment of methods for comparing patterns of motion (e.g. Hildebrand, 1962, 1963, 1965, 1967, 1968; Prost, 1965a, b, 1969, 1970; Sukhanov, 1974), the use of mathematical techniques to analyse components of movement (e.g. Gray, 1961, 1968; McGhee & Frank, 1968; Alexander & Jayes, 1978a, b, c), and comparison within a range of species to relate differences in movement to body size, structure and habitat (e.g. Dagg, 1973, 1974, 1977; Dagg & de Vos, 1968a, b; Alexander, Langman & Jayes, 1977; Rollinson, 1975).

Turning to the primates, there have been many behavioural and ecological studies involving small numbers of individuals of single species. Field studies specifically devoted to locomotor behaviour — including assessment of postural behaviour (see later) — are becoming increasingly important (e.g. Grand, 1976; Ripley, 1967; Mittermeier & Fleagle, 1976; Morbeck, 1977; Rose, 1974, 1977; Mittermeier, 1978). Some of these investigations have been specifically motivated by the paucity of reliable locomotor information available from previous eco-ethological studies. It has emerged that primate locomotion generally appears to be a highly environment-dependent behaviour and studies of the same species in different habitats provide evidence for great variability not only in the frequency of types of movement used, but also in the types themselves (Ripley, 1967). In any study of primate locomotor behaviour, therefore, it is important to gather information on the basis of a defined methodology which allows a species profile to be established, regardless of variation due to habitat or age, for subsequent use in interspecific comparisons. However, in developing such a profile it is equally important to concentrate on aspects that contribute to interspecific differentiation. Walker (1979) followed his detailed frequency analysis of locomotion in *Galago demidovi* with a review of published descriptions of locomotion in prosimians and concluded that both greater detail and more extensive quantification would be needed to produce anything more than the present coarse locomotor classifications within the group (see later).

Most comparative locomotor studies in primates have focused either on direct morphological assessment, especially with respect to the specialized categories of brachiation, knuckle-walking and/or vertical-clinging-and-leaping (e.g. Tuttle, 1967, 1969, 1970; Lewis, 1969; Avis, 1962; Jouffroy & Lessertisseur, 1960; Erikson, 1963; Napier & Walker, 1967; Grand & Lorenz, 1968; Grand, 1976), or on the application of multivariate statistical techniques to osteological measurements to differentiate locomotor categories (Ashton, Healey, Oxnard & Spence, 1965; Oxnard, 1973a; Ashton, Flinn, Oxnard & Spence, 1976; Manaster, 1979). Relatively little attention has been directed to interspecific comparisons involving detailed quantification of actual locomotor behaviour.

Research Strategy

The Old World monkeys as a group (family Cercopithecidae) show relatively little differentiation in postcranial morphology and locomotor behaviour (see also Schultz, 1970). All lie in the restricted body weight range of 1—30 kg (Schultz, 1970) and are classified as "quadrupedal" by Napier & Napier (1967), while members of the subfamily Cercopithecinae in particular differ primarily in the degree to which they are adapted for terrestrial life (see also Rose, 1973). It is therefore an interesting exercise to compare different cercopithecine monkey species in order to identify both morphological and behavioural features of locomotion which distinguish them. The major part of this chapter is based on such a comparative study of eight species belonging to the two genera *Cercopithecus* (including the subgenus *Miopithecus*) and *Cercocebus* (Rollinson, 1975), both of which are included in Rose's (1973) arboreal branch sitting and walking sub-category. All the animals studied (total: 45 individuals) originated from Gabon in West Africa and were maintained in captivity in small groups (see Table I). With one exception (*Cercocebus torquatus*), all the species studied are broadly sympatric; indeed, most are found to occur in polyspecific associations which persist for varying periods of time (Gautier & Gautier-Hion, 1969; Gautier-Hion & Gautier, 1972; Gautier-Hion, 1978). Study of the locomotor behaviour of these species in captivity benefits from the availability of detailed information on natural ecology and behaviour (Gautier-Hion, 1971, 1973, 1978; Gautier-Hion & Gautier, 1978, 1979; Quris, 1975). In particular, it is known that within each genus (*Cercopithecus*; *Cercocebus*) there is differentiation between arboreal and semi-arboreal species, all inhabiting a predominantly forest environment. Thus, comparison of the eight species (Table I) can potentially permit identification of features which distinguish arboreal from semi-terrestrial species as well as those which differentiate between the genera (mangabeys vs. guenons).

Detailed data for comparison of patterns of locomotion between the species were obtained from film analysis, and various methods of analysis were used in pursuit of the aim of identifying species-specific locomotor characteristics. In addition to yielding some information in line with that aim, the analysis also revealed some important characteristics of primate locomotion in general. This being the case, presentation of the detailed results relating to the eight cercopithecine species studied is followed by a general discussion of the evolutionary background to primate locomotion. One important factor which must be included in any such discussion is that of *body size* (see also Aiello, this volume p. 331), since locomotion exhibits scaling effects which must be identified and taken into account. Body size is, in fact, a complicating feature of the comparison of the cercopithecine species, as the mangabeys are generally larger than the guenons and terrestrial or semi-terrestrial cercopithecines are on average heavier than arboreal species (see Table I).

POSITIONAL BEHAVIOUR

One significant aspect of locomotion, however it is defined (e.g. as "organic activity": Prost, 1965a; or as "controlled falling": Howell, 1944; Lyon, 1971), is that it occurs throughout each individual's lifetime in a continuum with postural behaviour. The combination of locomotor and postural behaviour has been called *positional behaviour* (Prost, 1965a), since it is this overall category which positions an animal within its environmental matrix. Ripley (1967) uses the concept of *total locomotor pattern* in the same sense. In a field study of six East African monkey species, Rose (1974) showed that most of the observed positional behaviour was postural, with actual locomotor activity representing only 10% of the daily total for the arboreal species and close to 30% of the total for the more ground-living species. Ideally, for any given comparison such as that involved in the present study, attention should therefore be given to the entire repertoire of positional behaviour of each species in establishing the profile. Nevertheless, taking the available data on locomotor patterns alone, an attempt can be made here to use the intrageneric arboreal/semi-terrestrial distinctions to relate locomotor adaptation to the level of the forest environment typically occupied by each species.

LOCOMOTOR DYNAMICS: SUPPORT SEQUENCE

Quadrupedal monkeys, including the eight cercopithecine species listed in Table I, are peculiar compared to non-primate mammals with respect to the sequence in which the limbs are regularly employed to propel the body during walking (see also Hildebrand, 1967; Dagg, 1977). The difference involved can be illustrated by taking the right hindlimb as an arbitrary starting-point in the walking sequence. Figure 1a illustrates the typical monkey walking sequence in a baboon: starting with the right hindlimb, the left fore is the next to come into contact with the ground, then the left hind followed by the right fore, and so on. The typical sequence for other mammals is illustrated in Fig. 1b with an atypical gait in a spider monkey. (This typical non-primate mammal sequence does occur rarely in quadrupedal primates — see later.) Here, the right hindlimb is followed by the right fore, then the left hind followed by the left fore. This sequence, although very unusual for quadrupedal monkeys, is apparently somewhat more common in species like the spider monkey which possess elongated forelimbs (Rollinson, personal observation; see also Hildebrand, 1967) and it was observed in

TABLE IA

Cercopithecine species studied: Taxonomy and ecology

Tribe	Genus	Species	Body weight[a] (kg)	Ecological aspects[a]
(1) Papionini (mangabeys, macaques, baboons, drills)	*Cercocebus*	*albigena* (Grey-cheeked mangabey)	7.8	Arboreal; prefer middle and upper canopy levels of primary forest. May occur in secondary forest.
		galeritus[b] (Agile mangabey)	7.8	Semi-terrestrial in riverine forest with variable conditions. When in trees, prefer lower canopy levels
		torquatus[b] (White-collared mangabey)	7.5[c]	
(2) Cercopithecini (guenons)	*Cercopithecus*	*nictitans* (Putty-nosed monkey)	5.4	Arboreal; prefer middle and upper canopy levels of primary forest. May occur in secondary forest.
		pogonias (Crowned guenon)	3.8	
		cephus (Moustached monkey)	3.5	Arboreal; occurring in both primary and secondary forest. Prefer middle and lower canopy levels
		neglectus (De Brazza's monkey)	5.5	Semi-terrestrial (ground and low canopy level) in riverine forest; arboreal in primary forest
	(Subgenus *Miopithecus*)	*talapoin* (Talapoin)	1.2	Arboreal, but prefer low canopy level in riverine forest.

[a] Data taken from Gautier-Hion (1978) unless otherwise stated.
[b] These two species are allopatric in Gabon.
[c] Body weight data from B. Rudder (personal communication).

TABLE IB

Cercopithecine species studied: Animals studied (N = 45)

Species		Composition						Total	No. of groups
		Adult males[d]	Adult females	Subadult males	Subadult females	Juvenile males	Juvenile females		
Cercocebus	*albigena*	2	4	1	—	1	1	9	two
	galeritus	1	1	—	—	—	—	2	one
	torquatus	—	2	—	—	—	—	2	
Cercopithecus	*nictitans*	1	2	2	—	2	—	7	one
	cephus	—	1	1	1	1	1	5	
	pogonias	2	—	1	—	—	—	3	several
	neglectus	4	1	—	—	—	2	7	several
Miopithecus	*talapoin*	2	3	—	1	3	1	10	one

[d] No more than one adult male was kept in each group.

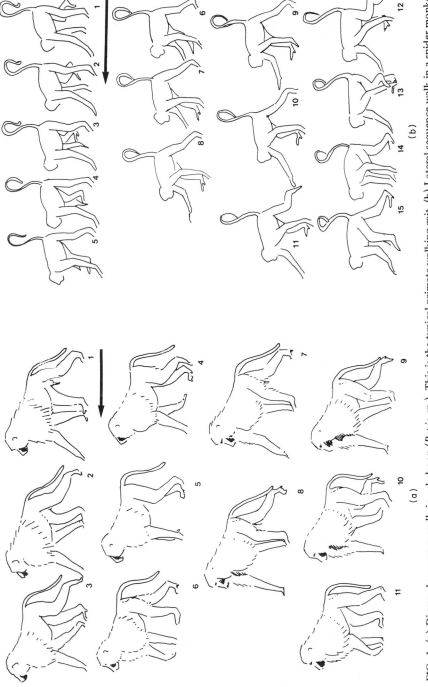

FIG. 1. (a) Diagonal sequence walk in a baboon (*Papio* sp.). This is the typical primate walking gait. (b) Lateral sequence walk in a spider monkey (*Ateles* sp.). This is the typical non-primate mammal walking gait, but also occurs as an atypical pattern in quadrupedal primates. (Line drawings from film sequences taken by J. Rollinson.)

regular use by a mangabey in the study colony which had a damaged vertebral column. In this chapter (following Hildebrand, 1967 — but contrary to Howell, 1944; Gambaryan, 1974; and Sukhanov, 1974), the typical monkey sequence of limb use in walking will be termed *diagonal*, since the footfall of either hindlimb is followed by that of the fore foot on the opposite side of the body. Conversely, the typical non-primate mammal sequence will be referred to as *lateral*, because the footfall of either hindlimb is followed by that of the forefoot on the same side of the body. The monkey diagonal sequence was first correctly recognized and figured by Muybridge (1899), who referred to it as the "pithecoid walk". Iwamoto & Tomita (1966) independently re-discovered the distinction between the diagonal and lateral sequences and referred to the different patterns as the "forward cross type" and the "backward cross type", respectively (see also Kimura, Okada & Ishida, 1979). The terminology used in this chapter is illustrated in Fig. 2.

SYMMETRICAL GAITS

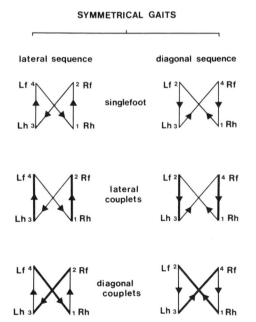

FIG. 2. Diagram illustrating the terminology for support sequences in symmetrical gaits (in agreement with Hildebrand, 1967). In this, and subsequent figures, the limbs are indicated as follows: Lf, left fore; Rf, right fore; Lh, left hind; Rh, right hind. Numbers 1—4 and arrows indicate sequences of limb movement, starting with Rh. Heavy lines indicate predominance of limb pairs (couplets) as support combinations during each cycle. For the cercopithecine monkey species studied, the usual walking sequence was diagonal with most support provided by diagonal couplets (bottom right), whereas in the typical lateral sequence walk of non-primate mammals most support can be provided by either lateral couplets (middle left) or diagonal couplets (bottom left).

J. Rollinson and R. D. Martin

Most locomotion in mammals is — at least in theoretical terms — *symmetrical*, with each couplet of feet playing an equal part in support of the body during movement. In rarer cases where the relative importance of the feet in body support is unequal, locomotion is said to be *asymmetrical* (the gallop is a typical example). In a very simple model of mammalian locomotion, each foot would be taken to act for the same amount of time, and equally spaced in time from the actions of the other feet, to give a singlefoot gait (Hildebrand, 1967; see also Fig. 2, top row). In the natural situation, however, it is more realistic to regard the feet as typically acting in couplets, which may be either diagonal or lateral. Now, if one considers the diagonal and lateral gaits in terms of body support (Fig. 2), the end results of the two different basic sequences would appear to be very similar, and several workers have accordingly dismissed the difference between the sequences as immaterial (e.g. Howell, 1944; Sukhanov, 1974). However, a fundamental difference does emerge when the forward motion of the body is taken into account in addition to the support sequence. Indeed, on the basis of mechanical models of forward motion in idealized terrestrial quadrupeds, it has been predicted that the monkey diagonal support sequence would provide an extremely unstable pattern of support (Gray, 1944) or even be completely impracticable (McGhee & Frank, 1968 — though these authors do admittedly assume a lack of grasping properties of the feet). These predictions, however, are derived from symmetrical models with fore- and hindlimbs of equal length and with the shoulder and hip joints approximately equidistant from the centre of gravity, as was recognized by Gray (1968). Thus, the explanation for the existence of the monkey diagonal gait, despite its apparent instability, must be sought in some departure from the conditions postulated in this idealized model.

The occurrence of the diagonal sequence as a norm in arboreal monkeys such as the eight cercopithecine species studied in captivity may be linked to a combination of factors. The first, and most important, of these would appear to be the position of the centre of gravity. Gray (1944) established that the position of the centre of gravity determines which foot will be put down next in sequence and that the centre of gravity will always fall within the area encompassed by supporting feet. Thus in monkeys, which spend most time supported by diagonal couplets (Hildebrand, 1967; Rollinson, 1975), the centre of gravity should most often fall somewhere along an imaginary line drawn between the feet of the supporting couplet. During the relatively long phase of support by any given couplet, the centre of gravity is therefore propelled forwards along this line (Fig. 3). In the diagonal support sequence, the next foot down will be the

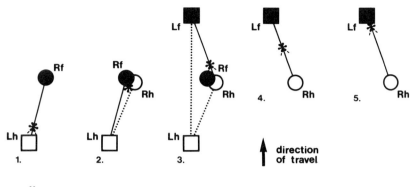

* – centre of gravity

FIG. 3. Diagram illustrating theoretical movement of centre of gravity with respect to support couplets (unbroken lines) during forward movement in a quadrupedal monkey. Left feet indicated by squares; right feet by circles; fore vs. hind indicated by black vs. white.

free hindfoot, at which stage the centre of gravity must fall within the triangle formed by the three supporting feet. When the free forefoot on the opposite side contacts the ground, the propulsive phase of the trailing hindfoot on that side is completed and a new diagonal support couplet is produced. At this stage, as the trailing hindfoot is raised from the ground the centre of gravity falls to the rear of a new support triangle formed by the other three feet. Subsequently, as the trailing forefoot of this triangle is raised in turn, support of the body is once again assured by a diagonal couplet and the centre of gravity must then move forward along the line between the supporting feet of this couplet. It follows from Gray's concept (above) that if the centre of gravity should fall too far forwards of the diagonal line of support at any stage in this process, the next foot to be placed on the ground would be the free forefoot (to impede further forward movement of the centre of gravity) rather than the free hindfoot. In other words, the typical monkey diagonal sequence would be converted into the lateral sequence typical of non-primate mammals. In the present study of captive cercopithecine monkeys, it was observed (see later) that such conversion from a diagonal to a lateral sequence occurred when the animals were descending steep inclines, the effect of which is to tilt the centre of gravity forwards relative to the limbs. Thus, since monkeys usually maintain the diagonal sequence it can be concluded that the centre of gravity is probably located more posteriorly in the body than in mammals generally. Lateral sequence walking was also commonly seen during braking and during very slow browsing locomotion in the present study of cercopithecine monkeys.

Unfortunately, although a great deal is known about the location of the centre of gravity in man (e.g. see Morton & Fuller, 1952; Steindler, 1955), there is little comparative information on other mammals, partly because of the practical difficulties involved in collecting such data. Nevertheless, some evidence is reported by Krüger (1943; cited by Lessertisseur & Saban, 1967) and by Kimura, Okada & Ishida (1979), on the basis of direct measurements. As is shown in Table II, available data indicate that in terrestrial cursorial mammals the centre of gravity is located to the fore (60:40 ratio of fore- to hindlimb body support as suggested for the cat by Gray, 1968), while in primates the situation is reversed (40:60 ratio of fore- to hindlimb support). A similar ratio was determined for the macaque by Tomita (1967), who showed that a diagonal (monkey-like) walking sequence could be induced in the domestic dog by adding weights to the animal's rear, hence shifting the position of gravity backwards. It is therefore reasonable to conclude that the typical diagonal walk sequence found in monkeys is a reflection of the fact that the centre of gravity is located further back in the body than in non-primate mammals. This interpretation is supported by the observation that the kinkajou (*Potos flavus*) has its centre of gravity closer to the hindlimbs (Table II) and is also exceptional among non-primate mammals in exhibiting a typical diagonal-sequence, diagonal couplets

TABLE II

Differential body weight distribution on fore- and hindlimbs in mammals

		Weight on forelimbs (%)	Weight on hindlimbs (%)
(1)	Terrestrial runners:		
	Dog[a,b] (various breeds)	63	37
	Cheetah[a]	52	48
	Horse[a,b]	55	45
	Ox[a]	56	44
	Dromedary[a]	62	38
	Camel[b]	66	34
	Llama[b]	55	45
	Elephant[b]	55	45
(2)	Primates:		
	Spider monkey[b]	29	71
	Hamadryas baboon[b]	44	56
	Japanese macaque[b]	46	54
	White-collared mangabey[a]	41	59
(3)	Atypical non-primates:		
	Cuban hutia[a]	45	55
	Kinkajou[a]	43	57

[a] After Krüger, 1943 (cited by Lessertisseur & Saban, 1967).
[b] After Kimura, Okada & Ishida (1979).
(Figures from two sources have been averaged).

gait (Hildebrand, 1967). In this light, it is easy to understand why McGhee & Frank (1968) concluded from their theoretical mathematical analysis that the monkey diagonal sequence would be unstable and therefore impracticable. In their model, the centre of gravity was (by implication) placed in the symmetrical mid-point of the body, and the diagonal sequence can indeed only become stable if the centre of gravity is located further back. A more recent theoretical study by Yamazaki (1976) has now shown that the diagonal sequence is a stable gait with a posterior centre of gravity. Lessertisseur & Saban (1967) concluded that the centre of gravity would typically be located forwards of the centre of the body in cursorial quadrupeds, since this would favour speed in locomotion. Although such forward location of the centre of gravity increases the possibility of forwards imbalance, it will also increase the momentum of the animal, which is important when speed is at a premium. (Hence the notion of locomotion as "controlled falling" — Howell, 1944; Lyon, 1971). On the other hand, in climbing mammals attainment of speed in horizontal locomotion may not be at such a premium, particularly in species such as the monkeys where both hands and feet have grasping powers. Where grasping extremities are present, it may be advantageous for the centre of gravity to be sited more to the rear so that it is closer to the contacts of the propulsive hindlimbs with the climbing surface. It is possible that in species possessing a grasping foot with a divergent hallux, which exerts moments about the point of contact rather than directing forces to the surface (e.g. through clawed digits), greater functional efficiency is attained in climbing if the centre of gravity is located closer to the hindlimbs. It is also important to remember that propulsion is not the only functionally significant feature of locomotion. Manter (1938) showed with the cat that the retarding (negative) effects of the limbs are as important as the propulsive (positive) ones, and it is likely that as a rule in mammals the forelimbs have a greater retarding action than the hind (see Gray, 1968). Kimura, Okada & Ishida (1979) have recently shown that the forelimb of monkeys, as in other mammals, is more retarding in its action than the hind. With non-primate mammals, however, the forefoot of a diagonal couplet lands first and provides a retardive "stop-jolt" prior to the contact of the hindfoot. In monkeys, by contrast, it is usually the hindfoot that lands first in any diagonal couplet. By the time the forefoot lands to complete the diagonal couplet in the monkey diagonal sequence, the hindlimb of the couplet has already initiated its propulsive phase, thus minimizing the retarding effect of the forelimb. Kimura, Okada & Ishida (1979) have shown that in six different simian primate species the maximum braking and propelling forces are provided by the hindlimbs, contrary to the situation in the dog and the cat (see Gray, 1968) and they distinguish primates from other mamals as possessing "front steering/ rear driving" locomotion, rather than the "front steering/front driving" type. There is also the consideration that in an arboreal environment the greater complexity and lesser rigidity of supports may favour increased emphasis on the hindlimbs in arboreal quadrupeds moving over relativley fine branches and using grasping extremities.

Overall, it is clear that the habitual diagonal walk sequence of quadrupedal monkeys is explained by the posterior location of the centre of gravity. This explanation allows for observed unusual cases of the lateral walk sequence in monkeys such as *Ateles* (Fig. 1b), where the elongation of the forelimbs (possibly combined with the forward carriage of the coiled tail) influences the gait, or in monkeys descending steep inclines, when the centre of gravity is transferred forwards with respect to the limbs. It also accounts for the occurrence of the lateral sequence gait in young monkeys, prior to acquisition of the adult body form (see later). Similarly, the typical lateral sequence of non-primate mammals may be changed during trotting, when forwards imbalance is temporarily maximized to augment momentum prior to transition to a canter or gallop.

Having established that the diagonal-sequence, diagonal couplets gait was by far the most common walking pattern in the eight cercopithecine monkey species studied, a survey was conducted to determine the frequency of occurrence of this sequence as a habitual feature in primates generally (Table III). It emerged that the diagonal sequence is typical of primates, as might be expected from the universal possession of grasping feet in non-human primates and the typical posterior location of the centre of gravity. The evolutionary implications of this observation will be examined in detail later (p. 418). Significantly, it has been found that tree shrews, which lack the grasping power of the hallux characteristic of primates, exhibit a lateral sequence with diagonal couplets during walking (Hildebrand, 1967; Jenkins, 1974) and also tend to exhibit asymmetrical, rather than symmetrical, patterns.

DETAILED ANALYSIS OF CERCOPITHECINE MONKEY LOCOMOTION

Quantification of Basic Categories

Throughout the study of the eight cercopithecine monkey species, qualitative observations were recorded in order to ensure familiariz-

TABLE III

Comparative observations of customary diagonal sequence gait in primates[a]

(1)	Lemuridae	*Lemur, Hapalemur, Lepilemur*	Personal observations[b]; Hildebrand (1967); Walker (1974).
(2)	Lorisidae	*Perodicticus, Loris, Nycticebus, Galago*	Personal observations[b]; Walker (1967,; Hildebrand (1967) and Tomita (1967) both record lateral sequence gaits for the lorisines.
(3)	Callitrichidae	*Saguinus, Callithrix*	Personal observations[b]. Hildebrand (1967) records mixture of lateral and diagonal sequence gaits for *Saguinus* spp. (limited data).
(4)	Cebidae	*Cebus, Saimiri, Cacajao, Lagothrix, Ateles*	Personal observations[b]; Prost (1965b); Hildebrand (1967); Rose (1973). *Ateles* may use lateral sequence on occasions, as may *Lagothrix*.
(5)	Cercopithecidae	*Cercopithecus, Allenopithecus, Cercocebus, Erythrocebus, Macaca, Papio, Mandrillus, Theropithecus, Colobus, Presbytis*	Personal observations[b]; Hildebrand (1967); Rose (1973); Prost (1965b) records unusual sequence at high speeds, but predominant diagonal at low speeds, for *Macaca*. Hildebrand reports one lateral sequence record for *Cercopithecus diana*.
(6)	Pongidae	*Pongo, Pan, Gorilla*	Personal observations[b]; Hildebrand (1967); Rose (1973); Kimura *et al.* (1979) report 'simultaneous contact of diagonals' for *Pongo*.

[a] Recorded either on ground or on branches.
[b] Conducted in a number of European zoos.

ation with the different species, with variations exhibited by individuals, and with general use (and changes in use) of the cage environment. Quantification of basic locomotor categories was then carried out in particular cases in order to verify specific and generic differences suggested by the qualitative observations. The effects of a cage environment on the locomotor patterns of wild-caught animals are not fully understood (see Rollinson (1975) for a brief discussion thereof). However, it is highly probable that different species are affected in different ways. For example, individuals of some species exhibit a greater tendency to put on weight in captivity, and stereotyped locomotor patterns are more likely to occur in some species than others. To some extent, the cage environment has a beneficial effect in standardizing conditions of observation, but the lack of a particular form of behaviour in a socially integrated group does not necessarily mean that it does not occur under natural conditions, nor that individuals are incapable of performing that behaviour. However, absence or low frequency of any given behaviour pattern in captivity can be taken as an indication that it occurs at a relatively low frequency in the natural species profile.

Detailed analysis of locomotion in the cercopithecine monkey species studied was achieved with the aid of film material collected under standardized conditions. The camera normally used was a 16 mm Paillard-Bolex, equipped with Som-Berthiot f1·4 and f2·5 50 mm lenses, though some sequences were filmed with a Beaulieu R16 equipped with a zoom lens. Kodak Plus-X and Tri-X 16 mm films were used and most films used in analysis were taken at a speed of 44 f.p.s. Frame-by-frame analysis of films was carried out with a spectro-analysing projector throwing the image onto a ground glass screen.

All film taken of the monkeys for use in comparisons was taken in a standardized cage (3·5 m × 5·5 m) from standardized distances (3 m, 4 m, or 5 m). A standard array of branches of fixed disposition and inclination (giving slopes of 0°, 6°, 18° and 36°) was employed and each branch bore markings at 25 cm intervals to allow for calculations of speed of movement. The major part of the quantified locomotor data was collected for movement in the branches, with the exception of *Cercocebus galeritus*, which spent more time at ground level than off the ground. The monkeys were not induced to move, nor stimulated in any way when in the cage, and as a result different individuals were filmed to different extents. Some species proved to be generally less active than others and within each species subordinate animals tended to be inhibited in their movements.

Speed of Movement

Despite the limitations of the captive environment, data on speed of movement over marked branches do reveal some differences between the cercopithecine species studied (Table IV). In the first place, the two mangabey species (genus *Cercocebus*) exhibited far more slow branch locomotion than any of the guenons (genus *Cercopithecus*, including *Miopithecus*). Whereas the latter never exhibited more than 10% of their branch locomotion at speeds of 60 cm s^{-1} or less, the former performed between 30% and 40% of their branch locomotion at such speeds. The fact that this difference cannot be attributed solely to the greater body size of the mangabeys is indicated by the data on the talapoin, the smallest species studied (cf. Table I), showing it to be intermediate between the other guenons and the mangabeys.

This information is supplemented by an analysis of the relative utilization of different locomotor categories by the different cercopithecine species (walk, trot-walk, trot and gallop; see later for a discussion of these terms). From this, it can be seen (Table V) that — as expected — the mangabeys exhibit more walking than the other species. However, this difference is not as pronounced as the data on speed (Table IV) might suggest. It is obvious that the mangabeys tend to walk relatively slowly, whereas the guenons have a higher average walking speed (despite their smaller body size). Further, when the mangabeys do move rapidly, they tend to use the gallop rather than the intermediate locomotor categories (trot-walk; trot), whereas the guenons exhibit more locomotion in these intervening categories. In fact, the mangabeys have been described as "deliberate" in their locomotion under natural conditions, because of their characteristically high frequency of slow walking (e.g. see Napier & Napier, 1967), and it is clear that this distinguishing feature persists in captivity. There may be a case for suggesting that the more ground-living species have specialized in slow movement for foraging combined with high speed movement (gallop) for alarm and escape conditions. In support of this interpretation, it is found that within the genus *Cercopithecus* (excluding *Miopithecus*) the most terrestrial species, *Cercopithecus neglectus*, exhibits the lowest frequency of trotting and the highest frequency of walking and galloping (Table V), though there is no evidence for reduced speed of locomotion (Table IV). In the field, *C. neglectus* is reported as adopting the "freeze and hide" strategy in alarm situations, rather than making a rapid escape (A. Gautier-Hion, personal communication), and this may explain why the frequency of galloping locomotion is not as high as in the mangabeys. However, if it is true that terrestrial adaptation pre-

TABLE IV

Speed of movement on branches calculated from film sequences[a]

Species	Speed of movement[b]				
	up to 60 cm s^{-1}	61–80 cm s^{-1}	81–100 cm s^{-1}	101–120 cm s^{-1}	121+ cm s^{-1}
Cercocebus albigena	38 (35)	33 (31)	23 (21)	0 (0)	6 (6)
Cercocebus torquatus	33 (32)	62 (61)	0 (0)	0 (0)	5 (5)
Cercopithecus nictitans	1 (1)	21 (20)	21 (20)	21 (20)	36 (35)
Cercopithecus neglectus	8 (8)	18 (17)	31 (30)	8 (8)	35 (34)
Cercopithecus pogonias	9 (9)	22 (21)	40 (39)	17 (17)	12 (12)
Cercopithecus cephus	1 (1)	20 (19)	64 (61)	0 (0)	15 (14)
Miopithecus talapoin	10 (10)	40 (40)	10 (10)	20 (20)	20 (20)

[a] Insufficient data were collected for the third mangabey species, *Cercocebus galeritus*, to permit calculation of reliable figures.
[b] For each species the figures indicate % of sequences at any given speed and the actual numbers of sequences used for evaluation are given in brackets.

TABLE V

Frequency of locomotor categories as a percentage of all movements[a]

Locomotor category	Species						
	C. galeritus	C. albigena	C. neglectus	C. nicritans	C. pogonias	M. talapoin
Walk	70.8	67.3	62.5	61.7	37.8	52.8
Trot-walk	2.0	10.3	14.4	13.6	3.7	6.6
Trot	3.3	1.8	9.6	21.0	45.1	14.2
Gallop	23.9	20.6	13.5	3.7	13.4	26.4
Total	100.0	100.0	100.0	100.0	100.0	100.0

[a] No adult animals were available to give figures for *Cercocebus torquatus* or *Cercopithecus cephus*.

disposes to a high frequency of walking combined with a high frequency of galloping, it would appear that both *Cercocebus albigena* and *Miopithecus talapoin* (see Table V) have developed adaptations for terrestrial life. It is possible that both these species have become secondarily arboreal relatively recently and that they have retained locomotor characteristics originally developed for terrestrial conditions (an example of "phylogenetic inertia"). In the case of the talapoin, this interpretation is supported by the observation that 50% of its branch locomotion in captivity is performed at speeds of 80 cm s^{-1} or less, whereas for other guenons the corresponding figure is less than 30% (Table IV). Similarly, although *Cercopithecus neglectus* may exhibit certain adaptations for terrestrial life in terms of its relative use of different locomotor categories, its retention of relatively high speeds of movement on branches (Table IV) could be explained as a reflection of relatively recent divergence from an arboreal common ancestor of *Cercopithecus* species (excluding *Miopithecus*). Gautier-Hion (1978) reports that in the wild *Cercopithecus neglectus* walks slowly and silently and tends to avoid leaping, but such slow locomotion is obviously not so intimately integrated in the species repertoire as to be retained under standardized conditions in captivity.

Gait Analysis

Definition of gaits involves the frame-by-frame analysis of slow motion film to record sequential contacts of the feet with the ground. This approach has been used by Howell (1944) and by several more recent authorities (e.g. Sukhanov, 1974) for a variety of animal species. Both Hildebrand (1965, 1966, 1967, 1968) and Prost (1965b, 1969, 1970) were concerned with the elimination of traditional gait labels (walk, trot, canter, etc.) and attempted to develop formulae which could be used to describe gaits without the subjective influence of traditional terminology.

The recording technique used in the present study, following the precedents established by these earlier investigations, is illustrated in Fig. 4. From the sequences of foot contacts with the branch, a *support formula* can be established, showing the actual feet in contact at various stages of the sequence. This can be contracted to give a *condensed support formula*, which merely gives the total number of feet in contact with the branch at each stage. Finally, the length of time that each foot, or combination of feet, is in contact with the support can be determined. (Time may be expressed either in terms of the numbers of frames of film, or in absolute terms by incorporating the speed at which the film was taken.)

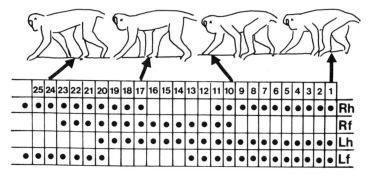

25	24	23	22	21	20	19	18	17	16	15	14	13	12	11	10	9	8	7	6	5	4	3	2	1	
•	•	•	•	•	•	•	•	•	•							•	•	•	•	•	•	•	•	•	Rh
		•	•	•	•	•	•	•	•	•	•	•	•	•											Rf
							•	•	•	•	•	•	•	•	•	•	•	•	•	•	•	•	•	•	Lh
•	•	•	•	•	•	•							•	•	•	•	•	•	•	•	•	•	•	•	Lf

• contact of foot with branch

Support formula															
Rf	Rh	Rh	Rh	Rh	Rh	Rh	Rf	Rf	Rh	Rh	Rh	Rh	Rh	Rh	Rf
Lh	Rf	Rf	Rf	Lf	Lh	Rf	Lh	Lh	Rf	Rf	Rf	Lf	Lh	Rf	Lh
	Lh	Lh	Lf		Lf	Lh	Lf		Lh	Lh	Lf		Lf	Lh	Lf
		Lf				Lf				Lf				Lf	

Condensed support formula															
2	3	4	3	2	3	4	3	2	3	4	3	2	3	4	3

Duration (no. of frames)															
9	5	1	4	12	4	2	5	9	5	1	5	8	6	2	3

Rh = right hind
Rf = right fore
Lh = left hind
Lf = left fore

FIG. 4. Illustration of the technique of recording from film sequences (top) in gait analysis. The recorded sequence of limb contacts with the substrate gives the support formula, which may be condensed into a numerical sequence. Duration is given by the number of film frames.

The condensed support formula may be used as a key to the traditional nomenclature applied to mammal gaits, in combination with the distinction between lateral-sequence and diagonal-sequence patterns. Hildebrand (1967) has given a simple definition of a gait as "an accustomed way of moving the legs in running", and this definition covers both the sequence in which the legs are moved and the pattern of support provided by the legs. It is important to bear this in mind, as in the past confusion has arisen over distinctions between the various gaits. For present purposes, discussion of terminology can be limited to the distinction between the walk, the trot and the gallop (Fig. 5). These three terms refer to different patterns which are used as increasing speed is required. As the speed of locomotion increases, there is a reduction in the length of time each limb is in contact with the ground, relative to a single cycle of limb use (a "stride", as defined by Dagg, 1977), and this is reflected in the condensed support formula.

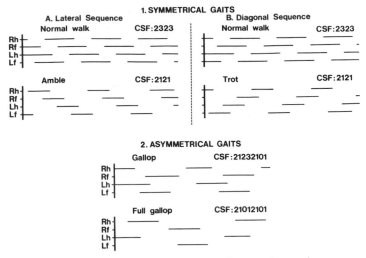

FIG. 5. Diagrammatic representation of selected different gait patterns, corresponding to traditional nomenclature (after Sukhanov, 1974), but showing the distinction between lateral and diagonal sequence gaits. Solid lines represent duration of contact of individual limbs with the substrate. Note that the monkey trot recognized here is a diagonal sequence equivalent of the lateral sequence "amble" found in non-primate mammals.

The walk

The walk in mammals is a slow, regular, symmetrical gait in which two, three or four legs may be supporting the body at any one time. In the cercopithecine monkeys studied, characteristic support formulae for walking patterns were 2.3.4.3 and 2.3.2.3 (see Fig. 5). As demonstrated above, the typical primate walking pattern has a diagonal sequence and diagonal couplets are predominant in supporting the body. In non-primate mammals, the characteristic lateral sequence walk can be associated with either lateral couplets or diagonal couplets.

The trot

The trot is a symmetrical gait of intermediate speed in which the body is predominantly supported by diagonal couplets. In theory (Sukhanov, 1974), the limbs in the diagonal couplets are swung in complete synchrony, such that the condensed support formula must be either 2.4.2.4 (slow trot) or 2.0.2.0 (fast trot). However, Hildebrand (1965) indicates that in trotting horses the diagonal couplets swing only "more or less together" and that it is typically the hind foot which is put down first in each couplet, giving a condensed support formula of 2.1.0.1. In other words, when horses adopt the trotting gait, they usually switch from the habitual lateral sequence

to a diagonal sequence. (The horses studied by Hildeb
had riders, so this adds to the artificiality introduced
ing.) It has been stated in the literature that primates
responding to the trot of other mammals (Hildebranc
1977), but in this study it was possible to recognize a
mediate speed, accompanied by bobbing of the head and body, with
a characteristic condensed support formula of 2.1.2.1 (Fig. 5). It
seems perfectly reasonable to refer to this as a "trot". With other
mammals a distinctive change in limb co-ordination must take place
so that in the trot diagonal couplets swing more or less together,
often involving a switch from the lateral sequence to a diagonal
sequence. With primates, by contrast, no such marked transition is
required to produce an intermediate speed gait of 2.1.2.1 condensed
support formula.

The gallop

The fastest gait used by primates, and by most other mammals, is the
gallop. This differs from the two previous categories in being a
typically asymmetrical gait and it is characterized in the condensed
support formula (e.g. 2.1.2.3.2.1.0.1; Fig. 5) by the presence of a
free-flight phase where no limb is in contact with the ground. In the
cercopithecine monkey species studied, there was a characteristic
bobbing movement during the gallop and in the full gallop (condensed support formula: 2.1.0.1.2.1.0.1) two free-flight phases were
recognizable during each cycle (Fig. 5).

The support relationships involved in different symmetrical patterns of locomotion can be presented in a convenient visual form
using the *gait formula* established by Hildebrand (1965). In this
formula, the first component, an indicator of locomotor speed (plotted on the abscissa), is given by the percentage of the stride interval
that each hind foot is on the ground, while the second (plotted on
the ordinate) is the percentage of the stride interval by which the
footfall of the forelimb follows that of the hindlimb on either side of
the body. (The stride interval is the time taken for completion of a
single cycle of limb movement.) In Fig. 6, the results from the present
study of eight cercopithecine monkey species are integrated with
those reported by Hildebrand for the horse (1965), for a sample of
158 genera of amphibians, reptiles and mammals (1966), and for a
sample of 26 primate genera (1967). Figure 6a emphasizes the
unusual position of the primates among terrestrial vertebrates generally in typically exhibiting diagonal sequence gaits. Non-primate
mammals which fall into this same category (according to Hildebrand,
1967) are the muntjac and the duiker, the aardvark, the giant

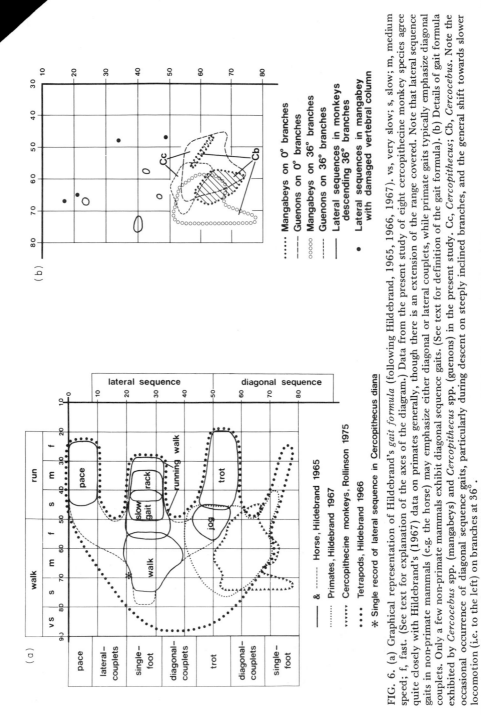

FIG. 6. (a) Graphical representation of Hildebrand's *gait formula* (following Hildebrand, 1965, 1966, 1967). *vs*, very slow; *s*, slow; m, medium speed; f, fast. (See text for explanation of the axes of the diagram.) Data from the present study of eight cercopithecine monkey species agree quite closely with Hildebrand's (1967) data on primates generally, though there is an extension of the range covered. Note that lateral sequence gaits in non-primate mammals (e.g. the horse) may emphasize either diagonal or lateral couplets, while primate gaits typically emphasize diagonal couplets. Only a few non-primate mammals exhibit diagonal sequence gaits. (See text for definition of the gait formula). (b) Details of gait formula exhibited by *Cercocebus* spp. (mangabeys) and *Cercopithecus* spp. (guenons) in the present study. Cc, *Cercopithecus*; Cb, *Cercocebus*. Note the occasional occurrence of diagonal sequence gaits, particularly during descent on steeply inclined branches, and the general shift towards slower locomotion (i.e. to the left) on branches at 36°.

armadillo, the opossum, and (as noted above) the kinkajou. Data from the present study broadly coincide with those reported by Hildebrand for the primates generally, but they also extend the range covered by diagonal sequence gaits in the diagram. It should also be noted that this diagram brings out a distinguishing characteristic of the walk identified by Hildebrand (1967): each foot is on the ground for more than half the duration of each locomotor cycle (stride) and as a result all walking gaits are recorded to the left of the vertical 50% line in Fig. 6a.

When the data for mangabeys and guenons are considered separately, and when the inclination of branches is taken into account, a number of important distinctions emerge. Comparing the data for locomotion on horizontal branches (angle: 0°) with those for locomotion on the most steeply inclined branches (angle: 36°), it is found that with the sharply sloping branches there is a greater spread of data overall, a shift towards movement at slower speeds and a marked increase in the occurrence of unusual lateral sequence gaits. (As noted above, the lateral sequence typically emerged when the monkeys were *descending* the sharply inclined branches.) At any given position of the branches (0°, 36°), the mangabeys tended to exhibit a slower pattern of movement, though on horizontal branches there was a clear separation between slow and fast locomotor patterns which was not evident in the guenons (see also Table V). Detailed comparison of the mangabeys and guenons also establishes a difference which cannot be represented in Hildebrand's gait formula diagram, and therefore highlights a limitation thereof. Although the abscissa indicates relative differences in speed of locomotion within a species, there is no provision for comparing actual speeds between species. Taking the data on gait formulae for horizontal branches, the mangabeys are found to move more slowly than the guenons at any given value of the formula. In the central zone of overlap between the mangabey and guenon data for 0° branches (see Fig. 6b), the travel speed for the mangabeys is only 66 cm s^{-1}, whereas for the guenons it is 80—96 cm s^{-1}. At the upper extreme of the mangabey range, travel speed is 82 cm s^{-1}, while guenons plotted at the same point have travel speeds of 130 cm s^{-1}. Thus, for any given pattern of limb movement, as expressed by the gait formula, mangabeys tend to travel more slowly than guenons. As Hildebrand (1965) has himself noted, identical gait formulae may be attached to "conspicuously unlike performances".

Additional information on patterns of limb-use during locomotion can be obtained by applying the methods used by Dagg & de Vos (1968a) and by Prost (1965b). Dagg & de Vos, in their studies of ruminant locomotion, listed the use of different combinations of legs in walking (expressed as a percentage of time occupied during a stride) as a means of characterizing the movements of each species. Dagg (1977) has shown that differential use of leg combinations distinguishes locomotion at the family level in ruminants. For example, Camelidae and Bovidae preferentially use lateral couplets when only two legs are in contact with the ground, while Cervidae preferentially use diagonal couplets. (Incidentally, this illustrates the fact implicit in Fig. 6a that in non-primates either lateral or diagonal couplets may be favoured, whereas in primates diagonal couplets are usually favoured.) When data on walking gaits of the eight cercopithecine monkey species are analysed in this way (Table VI), it is seen that the mangabeys and the talapoin exhibit a greater emphasis on three-leg combinations, while *Cercopithecus* species show a greater incidence of diagonal couplets. Thus, the more extensive use of trotting gaits (intimately connected with emphasis on diagonal couplets) in the *Cercopithecus* species (cf. Table V) is foreshadowed in the organization of walking gaits. Secondly, if the slower walking gait (condensed support formula: 2.3.4.3) is compared with the more rapid form (condensed support formula: 2.3.2.3), it is found that in both guenons and mangabeys the use of lateral couplets increases from zero, effectively replacing the level of support previously provided by all four limbs (see Table VI).

Following Prost's methodology (1965b), certain parameters may be extracted through gait analysis to describe locomotion in terms of limb contact periods, limb non-contact periods and speed of travel (as expressed by the "body transit period"). In the present study, apart from confirming Prost's finding for *Lagothrix* and *Macaca* of a relationship between travel speed and limb contact or non-contact periods (Rollinson, 1975), little additional insight was gained into the organization of cercopithecine locomotion. However, some clarification can be obtained by considering the *delays* separating the footfalls of individual pairs of feet. Since the monkeys studied predominantly rely on diagonal couplets in locomotion (Fig. 6; Table VI), it is to be expected that delays within the diagonal supporting couplets (RhLf, LhRf) would be shorter than those within lateral couplets (LfLh; RfRh). Further, the mangabeys generally move more slowly than the guenons and longer delays are therefore characteristically found with the former. However, even for horizontal locomotion sequences in which the speed was approximately the same (77 cm s^{-1} vs. 80 cm s^{-1}), it was found that within diagonal couplets the *Cercocebus* species exhibited longer delays than the *Cercopithecus* species (excluding the talapoin). This doubtless contributes to the "deliberate" impression given by mangabey walking.

TABLE VI

Average percentage time spent on supporting combinations of legs in walking

Species	Condensed support formula: 2.3.4.3				Condensed support formula: 2.3.2.3.			
	LC	DC	3L	4L	LC	DC	3L	4L
Cercocebus albigena	—	44	46	10	8	49	43	—
Cercocebus galeritus + *Cercocebus torquatus*	—	52	39	9	8	50	42	—
Cercopithecus neglectus	—	54	33	13	8	71	21	—
Cercopithecus nictitans	—	60	31	9	20	60	20	—
Cercopithecus pogonias	—	55	36	9	11	57	32	—
Cercopithecus cephus	—	57	29	14	N.D.	N.D.	N.D.	—
Miopithecus talapoin	—	37	54	9	9	47	44	—

LC, lateral couplet on ground; DC, diagonal couplet on ground; 3L, three legs on ground; 4L, all four legs on ground; N.D., no data for this category.

The delays within foot pairs can also be plotted against the average contact periods of the feet delimiting the delay, again following Prost (1965b). Analysis of the data in this way indicates distinctions not specifically identified by Prost. For diagonal couplets, the delays are always less than half of the average delimiting period, but there is no significant correlation between the two parameters. By contrast, with lateral couplets the delays are usually more than half of the average delimiting contact period and there is a significant correlation. In other words, the delays within the diagonal supporting couplets are consistently low, irrespective of the length of time the delimiting feet are in contact with the ground, while delays within lateral couplets decrease as locomotor speed increases (i.e. with decrease in the length of time the delimiting feet are in contact with the ground). Once again this emphasizes the consistency and importance of diagonal couplet support in the locomotion of these cercopithecine monkeys. No differences between the species were apparent in this respect, so it seems that a basic feature of cercopithecine locomotion is involved.

Overall, one of the greatest problems encountered in the analysis of monkey gaits is the variability in locomotor patterns exhibited even within a very short space of time. Such variability seems to be greater than that usually found with terrestrial cursorial mammals and as a result quantification of pure support formulae may be hindered by variation from one locomotor cycle (stride) to the next. The *Cercopithecus* species, in particular, exhibited a high incidence of incomplete support formulae. This was most evident in *Cercopithecus nictitans*, for which at least 16 variants of truncated support formulae were recorded. Qualitative observations of such locomotor patterns, which cannot be conveyed by the use of ratios and support formulae alone, reinforced the conclusion that the guenons are "trotters" in comparison to the more deliberate mangabeys.

Although the gait pattern identified above as a "trot" (condensed support formula: 2.1.2.1) occurs for relatively short periods in the cercopithecine monkeys studied (Table V), it is an important component of their locomotor behaviour. It is involved in play in animals of all ages, though particularly in juveniles, and in dominance, since it may be used in tree-shaking displays with both visible and audible components. The relatively low incidence of the trot in the locomotor repertoire of cercopithecine monkeys may indicate that it is a somewhat fatiguing gait, but when it does occur it is a vital part of the behavioural repertoire.

Relative Co-ordination

Some preliminary work was conducted on relative co-ordination of the body parts during locomotion in the eight cercopithecine monkey species, since it is to be expected that this would reflect differences in various parameters (e.g. speed of movement, inclination of substrate, age of the animal). A fairly simple approach is to take a film record of a locomotor sequence and portray movement with a "stick animal" in which the approximate sites of articulation are taken as reference points. Curves can then be derived which display the lines of travel for each articulation site in relation to the substrate. As expected such curves show that there is comparatively little vertical movement at the shoulder and hip, as opposed to the elbow and the knee. The contact and non-contact phases of the limbs may be superimposed upon the curves to associate the movement of each articulation site with its function in protraction, retraction and propulsion. For one side of the body, the propulsive phases of the fore- and hindlimbs take place at almost the same point above the substrate (though separately in time), since the extremities are closely approximated when in contact with the ground. The propulsive phases are spaced in time so that the hindlimb phase follows some time after that of the forelimb and persists for a longer period. In symmetrical movement, the propulsive phase of any limb on one side of the body is accompanied by a free (non-contact) phase of its counterpart on the other side. As has already been emphasized above, in the cercopithecine monkeys studied (and in primates generally) the combined action of diagonal limbs is of particular importance in propulsion.

In order to take account of the time lapses in the action of each limb and its component segments, it is necessary to carry analysis a step further. This was done by measuring the angles subtended by each articulation site with respect to a standard horizontal baseline (a line drawn from the base of the tail to the approximate position of the third cervical vertebra) and plotting them for successive frames of film. However, while this method is especially useful in understanding the general pattern of locomotion under different conditions, it is extremely labour-intensive and no consistent differences between the cercopithecine monkey species were found. Indeed, in one analysis of a film sequence showing a monkey moving tripedally while carrying a piece of food, no marked differences in limb co-ordination were found despite the complete exclusion of one limb from body support. This example highlights the multiplicity of additional factors (e.g. head movement, spinal flexion, spacing between feet) which are omitted in such analysis of relative co-ordination and which may compensate for variations in limb use. It did emerge that with increasing inclination of the support the hip and shoulder joints were increasingly approximated to the locomotor surface and the protractive phase was progressively shortened with respect to the retractive phase. Otherwise, the cercopithecine monkeys seemed to show a basic pattern of relative co-ordination of the limbs in locomotion which was "buffered" against variation.

Some interesting observations were collected on the *ontogeny* of locomotor co-ordination in the cercopithecine monkeys. As a rule, the initial gaits of infant monkeys are difficult to record, since they spend much of their early life clinging to their mothers. However, it proved possible to follow locomotor development in two infants (one *Miopithecus talapoin* and one *Cercocebus albigena*), since they were isolated from their mothers for other reasons at about one month of age. At this age, the infants were able to walk without difficulty, but it was found that the support sequence varied between lateral and diagonal sequences. This corresponds well with Hildebrand's (1967) observations of an infant *Macaca mulatta* from 17 hours after birth to the age of 196 days. Over this period, it was found that the infant underwent a transition from predominantly lateral sequence gaits to predominantly diagonal sequence, with the changeover taking place between 52 and 73 days of age. By day 196, the macaque was primarily using diagonal sequence gaits, but the locomotor repertoire had by that stage experienced a sharp dichotomy to retain a relatively low frequency of clearly defined lateral sequence gaits as well. Thus, it would seem that infant cercopithecine monkeys generally do not exhibit the characteristic adult concentration on diagonal sequence gaits.

With the one-month-old *Miopithecus talapoin* and *Cercocebus albigena*, it was found that acceleration of a varying lateral/diagonal sequence walk led to emergence of the characteristic "bob-walk" used predominantly by young during play (Gautier-Hion, 1971). Film analysis of the "bob" (Fig. 7) demonstrates that it occurs in mixed successions of lateral and diagonal sequences, with the bobbing motion coinciding with periods of differential coordination of the limbs. Hence, it would appear that the "bob" is symptomatic of the transition to adult diagonal sequence locomotion. Young monkeys have a relatively large head at birth, in comparison to body size, and it seems likely that the resulting forward location of the centre of gravity influences the support sequence in the expected manner. It is of interest that in older, juvenile monkeys in which locomotion is well co-ordinated according to the adult pattern, the "bob-walk" is employed in play behaviour. Apparently, this movement has

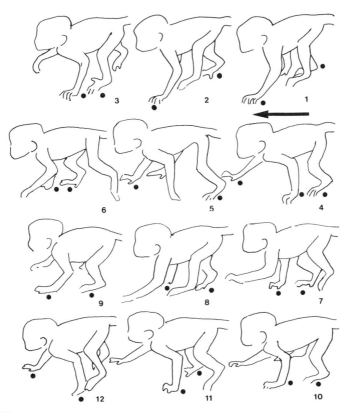

FIG. 7. Illustration of the relative co-ordination of the limbs during walking in a one-month-old *Cercocebus albigena*. A switch from lateral sequence to diagonal sequence locomotion (images 6–10) results in the characteristic "bob" (see text). Black circles indicate feet in contact with the substrate. (Line drawings from film sequences taken by J. Rollinson.)

been ritualized from the earlier transitional stage of incomplete co-ordination in infants to produce a signal for play among juveniles.

Use of the Tail

There are numerous contributory features of locomotor behaviour which are not taken into account in the established methods of gait analysis, and the most prominent among these for primates generally is the use of the tail. The tail plays an important part in overall maintenance of bodily equilibrium in the cercopithecine monkey species studied, and it also serves accessory functions in intraspecific displays. The combination of these two functions produces characteristic patterns which distinguish the genera *Cercopithecus* and *Cercocebus* and also differentiate between individual species in their finer details.

The length of the tail, relative to that of the trunk, is broadly correlated with the degree of arboreality. The more arboreal species typically have longer tails, implying an increased emphasis on the balancing function of the tail. For instance, among the mangabeys the highest tail length: trunk length ratio is found in *Cercocebus albigena* (2.8) as compared to the more terrestrial *Cercocebus torquatus* (2.5) and *Cercocebus galeritus* (2.0). Similarly, among the guenons the three essentially arboreal *Cercopithecus* species (*C. cephus; C. nictitans; C. pogonias*) have tail length: trunk length ratios ranging from 2.4 to 2.8, whereas the more terrestrial *Cercopithecus neglectus* has a ratio of only 2.0. Interestingly, the talapoin (*Miopithecus talapoin*) has a tail length: trunk length ratio of only 2.1, which allies it more to the terrestrially adapted species, despite its largely arboreal way of life.

In the standing posture, there is a generic difference in tail carriage in that the guenons typically carry the tail in a backward arc, whereas in the mangabeys it is habitually arched forwards over the back with the tip pointing to the front. This latter pattern is particularly obvious during the *callosity display*, which is performed by the mangabeys (and also by some macaques), but not by the guenons. In both the mangabeys and the guenons, during continuous motion along fairly substantial branches the tail is arched backwards, with the tip falling to one side of the branch. From time to time, the tail is raised in a looping action and may fall back to the opposite side of the branch. This action of the tail has been referred to as the "equilibrium wave" (Rollinson, 1975) and it differs in detail from one species to another. The "equilibrium wave" does not seem to be regularly associated with the pattern of limb movement. When the monkeys are moving along fine, relatively unstable supports, however, the tail is held above the back. Further, in all species except the talapoin the tail is arched over the back in presentation displays (as in the callosity display of the mangabeys). In all the cercopithecine monkey species studied, the tail was used as accessory support in various contexts (both during movement and at rest) and, in particular, it was twined round branches during descent of steeply inclined supports.

GENERAL ASPECTS OF PRIMATE ARBOREAL LOCOMOTION

Relative Proportions and Body Size

As has been succinctly demonstrated by Aiello (this volume, p. 331), limb sequence lengths in primates are influenced by an intricate combination of body size and locomotor specialization. It is therefore vital to include consideration of body size in any analysis of primate locomotor evolution. This is particularly important when the results of research on locomotion in the cercopithecine monkeys (which cover a relatively small body size range) are integrated with the general picture on primate locomotion, covering a size range from 50 g to some 150 kg.

One simple approach which has been used with the aim of accounting for differences in body size is, as noted by Aiello (this volume, p. 331), the calculation of *limb indices*. Jouffroy & Lessertisseur (1979) express a commonly-held view in stating that indices are designed to eliminate the body size factor and they draw heavily on such indices in their analysis of prosimian locomotion, though they do throw some doubt on the wisdom and the efficacy of attempting to account for body size effects in this way. In the literature on primate locomotor evolution one index that has been particularly prominent is the *intermembral index* (humerus + radius length ÷ femur + tibia length x 100). This index played a central role in the locomotor classifications used by Napier & Walker (1967) and by Napier & Napier (1967), since it distinguished reasonably well between the locomotor categories recognized. Other indices, such as the *brachial index* (radius length ÷ humerus length x 100) and the *crural index* (tibia length ÷ femur length) have been used in a subsidiary role, since their value as direct indicators of locomotor categories is generally less obvious.

It is not widely realized that any given limb index can only be taken as "eliminating" the effects of body size if direct proportionality (isometry) exists between the limb segments involved across the range of body sizes covered. Whenever the relationship between the limb segments is *allometric* rather than *isometric* (as is more usually the case), any derived limb index will vary systematically with body size. Hence, the use of complex mathematical procedures to combine a large number of limb indices (Oxnard, 1973b; Stern & Oxnard, 1973) will probably not "eliminate" the effects of body size. It is therefore instructive to take some of the standard limb indices cited in the literature (e.g. intermembral index, brachial index, crural index) and examine their relationship to body size across the Order Primates. This will reveal any residual effects of body size and also provide new insights into the meaning of such indices.

In order to interpret the patterns observed in the relationship between limb indices and body size, it is very useful to have some kind of classification which reflects actual locomotor (or, better still, positional) behaviour. A suitable basis for discussion is provided by the

TABLE VII

Locomotor categories among living primates[a]

Major category	Category	Sub-category		Representatives
(1) Vertical-clinging-and-leaping	(i) Small size		(A)	*Euoticus, Galago, Tarsius*
	(ii) Medium size		(A)	*Avahi, Hapalemur, Indri, Lepilemur, Propithecus*
(2) Arboreal quadrupedalism	(i) Small size	(a) Clawed	(B)	*Callimico, Callithrix, Cebuella, Leontopithecus, Saguinus*
		(b) Non-clawed, Agile	(C)	*Cheirogaleus, Microcebus, Phaner*
		(c) Non-clawed, slow-climbing	(D)	*Artocebus, Loris, Nycticebus, Perodicticus*
	(ii) Medium size		(E₁)	*Daubentonia, Lemur, Varecia.*
			(E₂)	*Aotus, Callicebus, Cebus, Chiropotes, Pithecia, Saimiri*
	(iii) Large size	(a) Branch sitting & walking	(F)	*Cercocebus,*[b] *Cercopithecus,*[b] *Macaca*[c] *Mandrillus*[c]
		(b) Old World semibrachiation	(G)	*Colobus, Nasalis, Presbytis, Pygathrix, Rhinopithecus*
		(c) New World semibrachiation	(H)	*Alouatta, Ateles, Brachyteles, Lagothrix*

(3) Terrestrial quadrupedalism	(i) Medium	(a) Ground standing & walking (with digitigrady)	(J) *Cynopithecus,*[c] *Erythrocebus, Macaca,*[c] *Mandrillus,*[c] *Papio, Theropithecus*
	(ii) Large	(b) Knuckle-walking	*Gorilla,*[c] *Pan*[c]
(4) Arboreal arm-swinging	(i) Medium size	(a) True brachiation	(K) *Hylobates, Symphalangus*
	(ii) Large size	(b) Modified brachiation	(L) *Gorilla,*[c] *Pan,*[c] *Pongo*
(5) Terrestrial striding bipedalism	Large size	—	(M) *Homo*

[a] Capital letters in parentheses provide the key to graphical plots in Fig. 8.

[b] Rose (1973) does not include this genus in the terrestrial quadruped category, although some species are semi-terrestrial. *Cercocebus* is digitigrade.

[c] These species are listed in two sub-categories (see text). In Fig. 8 *Cynopithecus, Macaca* and *Mandrillus* are plotted as "ground standing and walking", while *Gorilla* and *Pan* are plotted as "modified brachiators".

locomotor classification of Napier & Walker (1967), incorporating the subsequent revisions suggested by Rose (1973), as summarized in Table VII. Rose specifically noted the importance of body size in his locomotor categories. As with any classification, arbitrary dividing lines must be employed and there are, of course, numerous intermediate cases. For instance, among the bushbabies *Galago crassicaudatus* is distinctive in that it rarely leaps and typically moves around by quadrupedal running (Bearder, 1974), despite the fact that in skeletal morphology it is exceedingly similar to vertical-clinging-and-leaping *Galago* species. Further, among the simian primates (monkeys and apes) there is some overlap between arboreal and terrestrial forms. Rose (1973) included certain Old World monkey genera (viz. *Cynopithecus, Macaca, Mandrillus*) in two locomotor sub-categories (branch sitting and walking; ground standing and walking). In addition, Napier & Walker (1967) classified chimpanzees and gorillas as "modified brachiators", emphasizing their arboreal behaviour, whereas Rose (1973) classified them as terrestrial, knuckle-walking quadrupeds. As a result of such uncertainties, some authors (e.g. Cartmill, 1972; Jouffroy & Lessertisseur, 1979; Stern & Oxnard, 1973) have questioned the value of such classifications. Indeed, Stern & Oxnard (1973) went so far as to claim that they "no longer serve a useful purpose". As Walker (1979) has recognized, *accurate* locomotor characterization of primates requires detailed quantitative behavioural studies. However, if such a rigorous approach is taken to its limits in classification, there would have to be a separate locomotor category for each species, whereas the main purpose of a classification is to group species into meaningful and manageable categories for reference and analysis. When body size is taken into account along with standard indices, it is seen that the Napier/Walker/ Rose classification (Table VII) *is* meaningful in the sense that most of the different locomotor categories emerge as identifiable clusters. Obviously, one must not lose sight of the exceptions and borderline cases, but the classification does yield genuine insights into primate locomotion as a first step in analysis.

As noted by Napier & Walker (1967), the intermembral index covers a spectrum of increasing values ranging from vertical-clingers-and-leapers through quadrupeds and semibrachiators to brachiators. However, when body size is taken into account (Fig. 8, IMI) it is seen that there is an additional factor in that the intermembral index also tends to increase with increasing body size. The existence of a definite trend in this direction is suggested by the fact that among the sub-fossil lemurs of Madagascar, which evolved quite independently of the monkeys and apes, similarly high values of the intermembral index are found at large body sizes (*Megaladapis; Palaeopropithecus*). As a general rule, this trend can be related to the interaction between primate body size and branch size in an arboreal habitat. Napier (1967) has discussed this interaction and shown how small-bodied quadrupedal primates will tend to be above-branch runners, while larger-bodied forms will tend to exhibit below-branch suspensory locomotion. Apparently, suspensory locomotion in large-bodied primates is associated with increasing emphasis on the forelimbs. Nevertheless, it must also be recorded (see Rollinson, 1975) that among cercopithecine monkeys there is a tendency for increased emphasis on the forelimbs in terrestrial forms, and this development must obviously have a different functional basis. Since intermembral indices generally increase with increasing body size, a plot of index values against body size (Fig. 8) more effectively separates different locomotor categories. For example, it can be seen that the hylobatids (gibbons and siamang), which have intermembral indices overlapping the range found with the pongids (great apes), are in fact highly unusual in having such high values at a medium body size. This disparity between hylobatids and pongids is further underlined by the observation (see Aiello, this volume, p. 331) that in hylobatids high intermembral indices have been achieved by elongation of the forelimbs relative to body size, while in the large-bodied pongids the same result has been achieved by shortening of the hindlimbs relative to body size. Thus, hylobatid "brachiation" and pongid "modified brachiation" represent two quite distinct locomotor categories. (Incidentally, Jungers (1978) has shown that in the large-bodied *Megaladapis* species shortening of the hindlimbs has taken place, as in the pongids. Once again this indicates a consistent relationship between body size and intermembral index.) The plot of intermembral index against body weight also underlines the unusual nature of human locomotion, in that a low intermembral index is associated with a relatively large body weight. The sharp divergence between pongids and man indicates, as Aiello (in press) has stated, that evolution of human bipedalism from any ancestral form with definite brachiating characteristics is highly unlikely.

Overall, the fossil primates follow the living forms in the distribution of intermembral indices with respect to body weight, though the extreme specializations of man (genus *Homo*) and the hylobatids have no fossil counterparts. At the lower end of the body size range, the Eocene adapids *Notharctus* and *Smilodectes* are seen to fall within the range of the modern vertical-clingers-and-leapers (as noted by Napier & Walker, 1967), thus suggesting that these early prosimians fell within this locomotor category. However, it must be

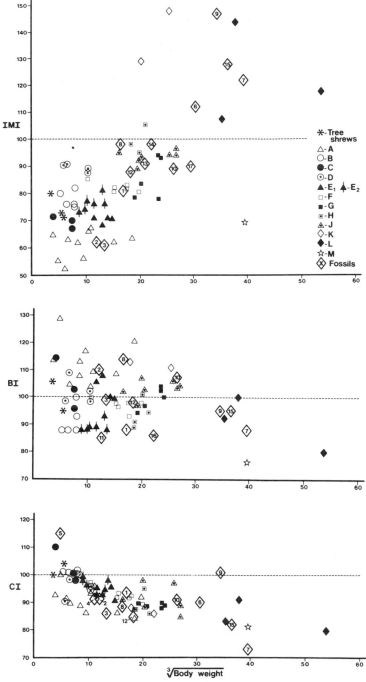

emphasized that their inclusion in such a general category does not imply that these adapids were similar in every detail. For instance, Jerison (1979) has shown that *Smilodectes* and *Notharctus* had much shorter limbs relative to trunk length than modern indriids (their nearest neighbours on the intermembral index: body size plot), while Martin (1980) has produced evidence that the neural control of the limbs was much less advanced in Eocene adapids.

Among the prosimians, both the agile quadrupedal forms and the vertical-clingers-and-leapers are characterized by low intermembral indices and only the specialized slow-climbing lorisines exhibit relatively high indices. Between these two extremes, there is a rather heterogeneous collection of generalized quadrupeds including the tree-shrews, the marmosets and tamarins and the smaller cebid monkeys. *Plesiadapis*, a supposed Palaeocene relative of the primates, also falls in this general area of the graph, though it is unusual in possessing claws at such a large body size.

Among the fossil simians, it is noteworthy that those of medium body size generally overlap with the Old World monkeys and the larger-bodied (semibrachiating) New World monkeys, while the large-bodied *Oreopithecus* overlaps with the modern pongids. Old World monkeys (as noted by Schultz, 1970) show the smallest range of values for any primate group (see also Aiello, this volume, p. 331). Finally, the trend towards increasing intermembral index with increasing body size is confirmed by the fact that the monkey-like subfossil lemur *Archaeolemur edwardsi* overlaps with the modern Old World monkeys in both intermembral index and body size.

There is also relatively clear separation between the various locomotor categories in terms of the brachial index, once body size is taken into account as well (Fig. 8, BI; Aiello, in press). In this case, although the trend is less clear-cut, there is a tendency for the index to decrease with increasing body weight (see also Aiello, this volume, p. 331 with respect to simians). Prosimians as a group typically have

FIG. 8. Plots of index values (IMI, intermembral index; BI, brachial index; CI, crural index) for living and fossil primates. For living primates, symbols correspond to the locomotor classification given in Table VII for living species. Fossils are indicated by large numbered diamonds as follows: 1. *Plesiadapis tricuspidens*; 2. *Smilodectes gracilis*; 3. *Notharctus tenebrosus*; 4. unidentified adapid (Messel); 5. *Nannopithex raabi*; 6. *Megaladapis madagascariensis*; 7. *Megaladapis edwardsi*; 8. *Varecia jullyi*; 9. *Palaeopropithecus maximus*; 10. *Archaeolemur edwardsi*; 11. *Cebupithecia sarmientoi*; 12. *Mesopithecus pentelici*; 13. *Pliopithecus vindeboniensis*; 14. *Dendropithecus macinnesi*; 15. *Oreopithecus bambolii*; 16. *Proconsul africanus*; 17. *Australopithecus "afarensis"* ("Lucy" skeleton). Sources of limb indices and body weight estimations: Aiello (personal communication); Gregory (1920); Jerison (1979); Jungers (1978); Martin (1980); Napier & Davis (1959); Napier & Napier (1967); B. Rudder (pers. comm.); von Koenigswald (1979); Walker (1974).

a brachial index in excess of 100, and high brachial indices are particularly characteristic of vertical-clingers-and-leapers. The Eocene adapids *Notharctus* and *Smilodectes* fall within the prosimian category and may be regarded as approximating the modern vertical-clingers-and-leapers of comparable body size, though the association is by no means as convincing as with the intermembral index. Interestingly, the smaller-bodied New World monkeys (marmoset, tamarins and small cebids) have strikingly low brachial indices of less than 100, as does the small-bodied Miocene fossil *Cebupithecia*, and this feature virtually separates them out as a distinct category (see also Aiello, this volume, p. 331). The hylobatids are, once again, quite distinct from other simian primates of comparable body size in having high brachial indices, but the New World "semi-brachiators" are not intermediate, as they are with the intermembral index; instead, they overlap with the Old World monkeys of comparable body size. As with the intermembral index, it can be seen that the fossil forms generally overlap the distribution for the living species, confirming the existence of broad relationships between the index and body size. The sub-fossil lemurs of Madagascar follow the overall trend in that *Varecia* overlaps the modern prosimian range, while *Archaeolemur* coincides with the intermediate-sized Old World monkeys and both *Megaladapis* and *Palaeopropitheus* fall within the range of modern pongids. *Plesiadapis*, as before, is seen to fall into a general quadrupedal category and resembles the tree-shrews in having a brachial index quite close to 100.

When the crural index is plotted against body size (Fig. 8, CI), the separation between locomotor categories is found to be far less marked than with either the intermembral index or the brachial index. Nevertheless, there is a recognizable trend in the crural index with increasing body size (reflecting the fact that the femur is positively allometric with respect to the tibia — Aiello, this volume, p. 331) and at any given body size living prosimians tend to have lower crural indices than living simians. This latter distinction is shared by the North American Eocene adapids *Notharctus* and *Smilodectes* and by the contemporary European adapid from Messel (see later), but not by the Palaeocene *Plesiadapis*, which (with a crural index of 94) more closely resembles quadrupedal Old World monkeys. The downward trend of the crural index with increasing body size in primates is, as with the brachial index, confirmed by fossil forms. *Archaeolemur* parallels the Old World monkeys, while *Megaladapis* and *Palaeopropithecus* parallel the great apes. Tree-shrews, with crural index values of 100 or more, are relatively unusual compared to most primates and their nearest neighbours are the clawed, branch-running marmosets and tamarins. Overall, however, there is relatively little diversity in the crural index (the hylobatids, for example, overlapping with both Old and New World monkeys), and this indicates that the hindlimbs have generally remained rather conservative in primate locomotor evolution, with most change occurring in the forelimbs and consequently affecting both the brachial and intermembral indices.

The Grasping Foot and Hindlimb Domination

It has already been argued (Gregory, 1920; Martin, 1972, 1979) that the common ancestor of the living primates (excluding tree-shrews) must have possessed a well-developed grasping hallux as one of its most important features. This inference is supported not only by the virtually universal possession of this feature among living primates (man is the only exception), but also by the characteristic occurrence of low calcaneal indices in both living and fossil primates (Martin, 1972, 1979). Calcaneal indices generally tend to increase with increasing body size in primates, but at body weights of up to 10 kg virtually all living and fossil primates have values of less than 100, while other living placental mammals (including tree-shrews) have values exceeding 100. The only mammals so far found to overlap with the primate distribution (Martin, 1979) are (as might be expected) certain arboreal marsupials with a grasping hallux. There are, however, two apparent exceptions to this characteristic primate pattern. First, the Palaeocene form *Plesiadapis* lacked a grasping hallux and its calcaneal index was well in excess of 100. Given the fact that *Plesiadapis* shows no identifiable similarities to generalized quadrupedal prosimians in intermembral, brachial or crural indices (Fig. 8), one must question the wisdom of including this fossil genus in the Order Primates and of using it as a basis for deriving early primate locomotor developments (see Szalay & Decker, 1974). Although the postcranial skeleton of *Plesiadapis* has been discussed in some detail (Szalay & Decker, 1974; Szalay, Tattersall & Decker, 1975), it has yet to be shown without circular argument that this early form shares anything other than undifferentiated placental mammal characters of the postcranial skeleton with primates. The second apparent exception is provided by European Eocene calcaneal specimens assigned to *Adapis parisiensis* and *A. magnus* on size grounds (see Decker & Szalay, 1974). These specimens have calcaneal indices of approximately 125 (Martin, 1979), quite unlike North American Eocene adapids (*Pelycodus, Notharctus, Smilodectes*). It has already been suggested (Martin, 1979) that the calcanei attributed to *Adapis* were misidentified, and this view has now received independent corroboration by the description of a partial adapid skeleton from the Eocene Messel deposits of Germany (von Koenigswald, 1979). In this, the only known articulated postcranial specimen from the European Adapidae, there is a well-developed grasping hallux, apparent modification of the second pedal digit for carriage of a grooming claw (as in modern lemurs and lorises), and a calcaneus comparable in proportions to that of *Notharctus* and *Smilodectes* (but quite distinct from the calcanei previously attributed to *Adapis*). Although it is not possible to give a reliable species identification to the new postcranial material from Messel, as there are no associated cranial or dental specimens from that site, it is now clear that there were medium-sized primates in the European Eocene (and adapids are the only candidates in that size range) with a grasping hallux and calcaneal proportions comparable to those of North American adapids. (Unfortunately, the specimen is still embedded in matrix such that the calcaneal index itself cannot be directly measured — W. von Koenigswald, personal communication). The new fossil material thus confirms the prediction made by Martin (1979).

Napier & Walker (1967) and Walker (1974) suggested that vertical-clinging-and-leaping could have been the ancestral locomotor pattern for primates of modern aspect, and Napier (1967) postulated an evolutionary trend from this condition through quadrupedalism and on to brachiation and bipedalism. Martin (1972) proposed a modification of the original hypothesis, using the concept of "hindlimb dominated arboreal locomotion" originally developed by Walker to characterize the ancestral primate condition, and interpreting the vertical-clinging-and-leaping of modern bushbabies, tarsiers and indriids as a more specialized derivation from such a condition. The grasping hallux, a low calcaneal index and greater dependence on the hindlimbs in locomotion are integral features of hindlimb domination, and relative conservatism in hindlimb evolution (Fig. 8, CI) is to be expected. It has now emerged from the present study that primates are highly unusual among mammals in exhibiting a characteristic diagonal walking gait which is intimately associated with a posterior location of the centre of gravity and predominance of the hindlimbs in weight-bearing and in locomotor propulsion. The concept of hindlimb domination as a central feature of primate locomotion has thus received further confirmation from a somewhat unexpected source, namely gait analysis. There is also evidence from a broad comparison of allometric variation in long bone length in modern mammals (Alexander, Jayes, Maloiy & Wathuta, 1979) that primates are relatively hindlimb dominated. All the long bones of the six primate species in their comparison (humerus, radius, ulna, femur, tibia, fibula) tended to be longer, at any given body size, than in the other placental mammals, but this distinction is most noticeable in the hindlimb, notably in the femur. This confirms the suggestion made by Napier & Walker (1967) that primates are characterized by the possession of long legs relative to body size. There is therefore a very sound basis for characterizing the primates (excluding treeshrews and plesiadapoids) as possessing a unique combination of locomotor features which can be traced to an original hindlimb dominated arboreal ancestral condition.

DISCUSSION

Although most living primates can be generally described as "quadrupedal" (up to 41 out of 52 genera, including 30 New and Old World monkeys — Rose, 1973), there has been relatively little discussion of the origins and diversification of this form of locomotion in primates. Rose (1973) correctly states that primate quadrupedalism has been largely defined by exclusion. If the concept of a hindlimb dominated ancestry for primates is accepted, some special explanation must be produced for the emergence of quadrupedalism (where such dominance is reduced) associated with intermembral indices of between 80 and 100 among New World and Old World monkeys (see Fig. 8 IMI). Further, it is necessary to trace the effects of terrestrial adaptation on locomotor functions generally, and this was one of the main aims behind the detailed behavioural study of the eight cercopithecine monkey species.

There is increasing evidence, particularly from Conroy's (1976) detailed study of simian primate postcranial material from the Oligocene Fayum deposits of Egypt, and from Aiello's allometric studies (in press), that locomotor adaptations in the early ancestors of the monkeys and apes were quite close to those in modern New World monkeys (all of which are essentially arboreal in habits). Conroy even recognizes two distinct locomotor types in the Fayum, one *Saimiri*-like and the other *Alouatta*-like. Accordingly one can picture the evolution of early simian locomotor adaptations from a prior prosimian stage in terms of a shift from the basic condition for modern Madagascar lemurs (see Martin, 1972) to the basic condition for modern New World monkeys. Such a shift requires an increase in the intermembral index, a decrease in the brachial index and (perhaps) a slight increase in the crural index (see also Delson & Andrews, 1975), all of which amount to approximation to the condition found (for example) in the clawed, branch-running, quadrupedal tree-shrews. It is therefore likely that the emergence of the ancestral simian locomotor type involved a transition from a relatively small-bodied ancestral primate adapted for saltatory locomotion in the fine branch niche (Martin, 1979) to a medium-sized ancestral simian more adapted for above-branch quadrupedal locomotion along broad branches. Further increase in body size in arboreal descendants from this ancestral simian stock would have led to various degrees of adaptation for suspensory locomotion (see Napier, 1967), as in the New World semibrachiators, the Old World semibrachiators, the hylobatids and the great apes (in addition to a number of fossil forms, such as *Oreopithecus*). It is therefore reasonable to conclude that the Old World monkeys are descended from a common ancestor of medium body size (perhaps in the region of 5 kg — see Martin, 1979) adapted as an essentially arboreal, branch-running quadruped (see also Napier, 1967; Delson, 1975). As Napier (1967, 1970) has pointed out, the background to the subsequent divergence into cercopithecines and colobines and, in particular, the phylogenetic history of the genera *Cercopithecus* and *Cercocebus* is far from clear. However, it does seem likely that the colobines remained primarily arboreal following the initial dichotomy among the Old World monkeys, while the cercopithecines initially became adapted for a semi-terrestrial existence, during which "macacoid phase" (Napier, 1970) the characteristic cheek-pouches of the subfamily were developed (Napier & Davis, 1959; Delson, 1975). Thereafter, as suggested by Napier (1967), various cercopithecines became *secondarily* essentially arboreal in habits, including representatives of the genera *Cercopithecus* and *Cercocebus*. As Napier also postulated, it is likely that this return to the trees took place earlier among *Cercopithecus* species than in *Cercocebus* species. Evidence from the present study indicates that the talapoin became secondarily readapted for arboreal life later than the other guenons, since it shares with the mangabeys certain characteristics indicative of terrestrial adaptation (relatively low incidence of trotting gaits; greater use of three-leg support patterns in walking; digitigrady (see also Rose, 1973); relatively low tail: body length index, etc.). There are other indications of such "phylogenetic inertia" in cercopithecine locomotor evolution, in that the semi-terrestrial guenon *Cercopithecus neglectus* still shares many features with its arboreal relatives (as in the retention of palmigrady in ground locomotion), while the arboreal mangabey *Cercocebus albigena* has retained many apparently terrestrial adaptations such as the slow, "deliberate" gait characteristic of the genus. Rollinson (1975) suggested that the proposed semi-terrestrial ancestor of the cercopithecines might have been associated with a riverine environment, since the link between these two aspects is found today in semi-terrestrial mangabeys, in *Cercopithecus neglectus*, in *Allenopithecus* and in some macaques.

The distinctions in locomotor adaptation found among *Cercopithecus* and *Cercocebus* species in this study are, however, relatively fine. In most respects, these cercopithecines are striking in the overall similarity of their locomotor behaviour and morphology, in accordance with Schultz's (1970) emphasis on the relative uniformity of the Cercopithecoidea. Thus, despite the basic distinction between arboreal and semi-terrestrial species, it would seem that a generalized cercopithecine locomotor pattern is suited, with only minor modifications, to life either in the trees or on the ground. Doubtless, this conservatism is connected with the fact that even semi-terrestrial primates are not completely emancipated from the trees and therefore retain the fundamental characteristics of primate hindlimb dominated locomotion. Accordingly, the hypothetical ancestral cercopithecine would also have retained a link with arboreal life, despite some apparent adaptations for a semi-terrestrial existence. The lability of arboreal/semi-terrestrial adaptation in the cercopithecines is underlined by the fact that Rose (1973) notes that the genera *Cercopithecus* and *Cercocebus* are in some respects intermediate, though he classifies them as arboreal branch sitters and walkers. Whereas some mangabey species (e.g. *Cercocebus albigena*) have apparently become adapted for arboreal life relatively recently, some of the guenons (e.g. *Cercopithecus aethiops; Cercopithecus neglectus*) seem to have recently returned to a semi-terrestrial existence. *Cercopithecus aethiops* has even developed digitigrady (Rose, 1973).

Turning to the general context of primate locomotor evolution, it is significant that the locomotor classification given in Table VII — backed up by the analysis of limb indices in relation to body size — closely follows taxonomic divisions established through consideration of predominantly non-locomotor evidence. Hence, it would seem that locomotor adaptation was an integral part of the adaptive framework of each taxonomic grouping within the Order Primates, partly as a reflection of the constraints of body size in the common ancestral stocks. But the existence of recognizable locomotor categories (whatever their limitations) also suggests a certain degree of conservatism within groups of closely related species, and there is therefore no foundation for Delson's (1975) suggestion that postcranial morphology is of little value in tracing cercopithecoid evolutionary history because it is "more reflective of habitus than heritage".

With respect to the wider context of locomotor evolution within the primates as a whole, it is obvious that much more needs to be done to assess the relationship between behaviour and individual details of postcranial morphology. In particular, it is necessary to integrate studies of postural behaviour (e.g. following Rose, 1973, 1974, 1977) with those of actual locomotion in order to understand the interaction between behaviour *in toto* and morphology. One can, however, draw a number of general conclusions from the present study. First, primate locomotor behaviour would seem to be much more variable and fluid than in the terrestrial, cursorial mammals which have been the main subject of gait analysis to date. Secondly, it is fairly clear that all modern primates share a complex of loco-motor adaptations related to hindlimb dominated locomotion in an arboreal environment, associated with (i) the quadrumanous con-dition (Rose, 1973), (ii) the separation between prehensile functions of the forelimbs and primarily propulsive/supportive functions of the hindlimbs (Rose, 1973), and (iii) a tendency towards truncal upright-ness (Napier & Walker, 1967). All these features are lacking in tree-shrews, which are now generally excluded from the Order Primates, and they are also absent from *Plesiadapis*. The allocation of the fossil plesiadapoids to the Primates, therefore, on the grounds of certain craniodental features, deserves serious re-examination. Finally, it is clear that the separation of locomotor factors *per se* from the effects of body size requires the extremely careful analysis of numerical data which cannot be achieved by multivariate statistical treatment of (for example) limb indices which vary systematically with body size.

ACKNOWLEDGMENTS

Behavioural observations on the cercopithecine monkey species were conducted by J. Rollinson at the Biological Field Station (Paimpont) of the University of Rennes from Jan. 1972 to Oct. 1973. The re-search was supported by studentship grants from the Science Research Council and by a grant from the Central Research Fund (University of London). Special thanks go to Drs A. Gautier-Hion and J.-P. Gautier for their guidance and for making available their monkey colony, to Prof. G. Richard and Dr P. Trehen for permission to work at the station, to the late Prof. N. A. Barnicot for the use of facilities at University College, and both to the Service du Film Recherche Scientifique, Paris, and to Dr G. du Boulay and Miss M. Darling of the Nuffield Laboratories of Comparative Medicine, Zoological Society of London, for permission to use film-analysing equipment. Thanks also go to Ms A. MacLarnon for assistance in certain aspects of data analysis and preparation of some of the figures. Mr I. Curtis also assisted with preparation of figures. Finally tribute must be paid to Prof. J. R. Napier, Mrs P. H. Napier and Prof. A. C. Walker, whose pioneering work and timely discussions inspired much of the thinking behind this chapter, and to Ms L. Aiello and the late Prof. N. A. Barnicot, who provided valuable ideas and information contributing to the overall assessment of primate locomotion evolution.

REFERENCES

Aiello, L. C. (In press). Locomotion in the Miocene Hominoidea. *Symp. Soc. Study Hum. Biol.* 21.

Alexander, R. McN. & Jayes, A. S. (1978a). Vertical movements in walking and running. *J. Zool., Lond.* 185: 27—40.

Alexander, R. McN. & Jayes, A. S. (1978b). Mechanics of locomotion of dogs (*Canis familiaris*) and sheep (*Ovis aries*). *J. Zool., Lond.* 185: 289—308.

Alexander, R. McN. & Jayes, A. S. (1978c). Optimum walking techniques for idealised animals. *J. Zool., Lond.* 186: 61—81.

Alexander, R. McN., Jayes, A. S., Maloiy, G. M. O. & Wathuta, E. M. (1979). Allometry of the limb bones of mammals from shrews (*Sorex*) to elephant (*Loxodonta*). *J. Zool., Lond.* 189: 305—314.

Alexander, R. McN., Langman, V. A. & Jayes, A. S. (1977). Fast locomotion of some African ungulates. *J. Zool., Lond.* 183: 291—300.

Ashton, E. H., Flinn, R. M., Oxnard, C. E. & Spence, T. F. (1976). The adaptive and classificatory significance of certain quantitative features of the forelimb in primates. *J. Zool., Lond.* 179: 515—556.

Ashton, E. H., Healy, M. J. R., Oxnard, C. E. & Spence, T. F. (1965). The combination of locomotor features of the primate shoulder by canonical analysis. *J. Zool., Lond.* 145: 406—429.

Avis, V. (1962). Brachiation: the crucial issue for Man's ancestry. *SWest. J. Anthrop.* 18: 119—148.

Bearder, S. K. (1974). *Aspects of the ecology and behaviour of the thick-tailed bushbaby*, Galago crassicaudatus. Ph.D. Thesis: University of the Witwatersrand.

Cartmill, M. (1972). Arboreal adaptations and the origin of the order Primates. In *The functional and evolutionary biology of primates*: 97—122. Tuttle, R. H. (Ed.). Chicago: Aldine-Atherton.

Conroy, G. C. (1976). Primate postcranial remains from the Oligocene of Egypt. *Contr. Primatol.* 8: 1—134.

Dagg, A. I. (1973). Gaits in mammals. *Mammal Rev.* 3: 135—154.

Dagg, A. I. (1974). The locomotion of the camel (*Camelus dromedarius*). *J. Zool., Lond.* 174: 67—78.

Dagg, A. I. (1977). *Running, walking and jumping: the science of locomotion*. London: Wykeham Publications.

Dagg, A. I. & de Vos, A. (1968a). The walking gaits of some species of Pecora. *J. Zool., Lond.* 155: 103—110.

Dagg, A. I. & de Vos, A. (1968b). Fast gaits of pecoran species. *J. Zool., Lond.* 155: 499—506.

Decker, R. L. & Szalay, F. S. (1974). Origins and functions of the pes in the Eocene Adapidae (Lemuriformes, Primates). In *Primate locomotion*: 261–291. Jenkins, F. A. (Ed.). New York & London: Academic Press.

Delson, E. (1975). Evolutionary history of the Cercopithecidae. *Contr. Primatol.* 5: 167–217.

Delson, E. & Andrews, P. (1975). Evolution and interrelationships of the catarrhine primates. In *Phylogeny of the primates*: 405–446. Luckett, W. P. & Szalay, F. S. (Eds). New York: Plenum Publ. Co.

Erikson, G. E. (1963). Brachiation in the New World monkeys. *Symp. zool. Soc. Lond.* No. 10: 135–164.

Gambaryan, P. P. (transl. Hardin, H.) (1974). *How mammals run: anatomical adaptations*. New York: John Wiley.

Gautier, J.-P. & Gautier-Hion, A. (1969). Les associations polyspécifiques chez les Cercopithécidae du Gabon. *Terre Vie* 23: 164–201.

Gautier-Hion, A. (1971). L'écologie du talapoin du Gabon. *Terre Vie* 25: 427–490.

Gautier-Hion, A. (1973). Social and ecological features of the talapoin monkey — comparisons with sympatric cercopithecines. In *Comparative ecology and behaviour of primates*: 147–170. Michael, R. P. & Crook, J. H. (Eds). London & New York: Academic Press.

Gautier-Hion, A. (1978). Food niches and coexistence in sympatric primates in Gabon. In *Recent advances in primatology 1 – Behaviour*: 269–286. Chivers, D. J. & Herbert, J. (Eds). London & New York: Academic Press.

Gautier-Hion, A. & Gautier, J.-P. (1972). Les associations polyspécifiques de cercopithéques de Plateau de M'Passa (Gabon). *Folia primatol.* 22: 134–178.

Gautier-Hion, A. & Gautier, J.-P. (1978). De Brazza's monkey — original strategy. *Z. Tierpsychol.* 46: 84–104.

Gautier-Hion, A. & Gautier, J.-P. (1979). Niche écologique et diversité des espèces sympatriques dans le genre *Cercopithecus*. *Terre Vie* 33: 493–507.

Grand, T. I. (1976). Differences in terrestrial velocity in *Macaca* and *Presbytis*. *Am. J. phys. Anthrop.* 45: 101–108.

Grand, T. I. & Lorenz, R. (1968). Functional analysis of the hip joint in *Tarsius bancanus* (Horsfield 1821) and *Tarsius syrichta* (Linnaeus 1758). *Folia Primatol.* 9: 161–181.

Gray, J. (1944). Studies in the mechanics of the tetrapod skeleton. *J. exp. Biol.* 20: 88–116.

Gray, J. (1961). General principles of vertebrate locomotion. *Symp. zool. Soc. Lond.* No. 5: 1–11.

Gray, J. (1968). *Animal locomotion*. London: Weidenfeld & Nicolson.

Gregory, W. K. (1920). On the structure and relations of *Notharctus*, an American Eocene primate. *Mem. Am. Mus. nat. Hist* (N.s) 3: 49–243.

Hildebrand, M. (1962). Walking, running, jumping. *Am. Zool.* 2: 151–155.

Hildebrand, M. (1963). The use of motion pictures for the functional analysis of vertebrate locomotion. *Int. congr. Zool.* 16: 263–268.

Hildebrand, M. (1965). Symmetrical gaits of horses. *Science, N.Y.* 150: 701–708.

Hildebrand, M. (1966). Analysis of the symmetrical gaits of tetrapods. *Folia biotheoret.* 6: 1–22.

Hildebrand, M. (1967). Symmetrical gaits of primates. *Am. J. Phys. Anthrop.* 26: 119–130.

Hildebrand, M. (1968). Symmetrical gaits of dogs in relation to body build. *J. Morph.* 124: 353–360.

Howell, A. B. (1944). *Speed in animals*. Chicago: University Press.

Iwamoto, M. & Tomita, M. (1966). [On the movement order of four limbs while walking and the body weight distribution to fore and hind limbs with standing on all fours in monkeys.] *J. anthrop. Soc. Nippon* 74: 228–231. [In Japanese].

Jenkins, F. A. (1974). Tree shrew locomotion and primate arborealism. In *Primate locomotion*: 85–115. Jenkins, F. A. (Ed.). New York & London: Academic Press.

Jerison, H. J. (1979). Brain, body and encephalization in early primates. *J. hum. Evol.* 8: 615–635.

Jouffroy, F. K. & Lessertisseur, J. (1960). Les specialisations anatomiques de la main chez les singes à progression suspendue. *Mammalia* 24: 93–151.

Jouffroy, F. K. & Lessertisseur, J. (1979). Relationships between limb morphology and locomotor adaptations among prosimians: an osteometric study. In *Environment, behaviour and morphology: Dynamic interactions in primates*: 143–181. Morbeck, M. E., Preuschoft, H. & Gomberg, N. (Eds). New York: Gustav Fischer.

Jungers, W. L. (1978). The functional significance of skeletal allometry in *Megaladapis* in comparison to living prosimians. *Am. J. phys. Anthrop.* 49: 303–314.

Kimura, T., Okada, M. & Ishida, H. (1979). Kinesiological characteristics of primate walking: its significance in human walking. In *Environment, behaviour and morphology: Dynamic interactions in primates*: 297–311. Morbeck, M. E., Preuschoft, H. & Gomberg, N. (Eds). New York: Gustav Fischer.

Krüger, W. (1943). Über die Beziehungen zwischen Schwerpunktslage und Stärke der Substantia compacta einzelner Gliedmassenknochen bei vierfüssigen Säugetieren. *Morph. Jb.* 88: 377–396.

Lessertisseur, J. & Saban, R. (1967). Squelette appendiculaire. In *Traité de Zoologie* 16: 709–1078. Grassé, P.-P. (Ed.). Paris: Masson & Cie.

Lewis, O. J. (1969). The hominoid wrist joint. *Am. J. phys. Anthrop.* 30: 251–268.

Lyon, McD. (1971). *The dog in action*. New York: Howell Book House.

Manaster, B. J. (1979). Locomotor adaptations within the *Cercopithecus* genus: a multivariate approach. *Am. J. phys. Anthrop.* 50: 169–182.

Manter, J. T. (1938). The dynamics of quadrupedal walking. *J. exp. Biol.* 15: 522–540.

Martin, R. D. (1972). Adaptive radiation and behaviour of the Malagasy lemurs. *Phil. Trans. R. Soc.* (B) 26: 295–352.

Martin, R. D. (1979). Phylogenetic aspects of prosimian behaviour. In *The study of prosimian behaviour*: 45–77. Doyle, G. A. & Martin, R. D. (Eds). New York & London: Academic Press.

Martin, R. D. (1980). Adaptation and body size in primates. *Z. Morph. Anthrop.* 71: 115–124.

McGhee, R. B. & Frank, A. A. (1968). On the stability properties of quadrupedal creeping gaits. *Math. Biosci.* 3: 331–351.

Mittermeier, R. A. (1978). Locomotion and posture in *Ateles geoffroyi* and *Ateles paniscus. Folia primatol.* 30: 161–193.

Mittermeier, R. A. & Fleagle, J. G. (1976). The locomotor and postural repertoires of *Ateles geoffroyi* and *Colobus guereza* and a re-evaluation of the locomotor category semi-brachiation. *Am. J. phys. Anthrop.* 45: 235–256.

Morbeck, M. E. (1977). Positional behaviour, selective use of habitat substrate and associated non-positional behaviour in free-ranging *Colobus guereza* (Ruppel 1835). *Primates* 18: 35—58.

Morton, D. J. & Fuller, D. D. (1952). *Human locomotion and body form: A study of gravity and man.* Baltimore: Williams & Wilkin Co.

Muybridge, E. (1899). *Animals in motion.* London: Chapman & Hall.

Napier, J. R. (1967). Evolutionary aspects of primate locomotion *Am. J. phys. Anthrop.* 27: 333—342.

Napier, J. R. (1970). Paleoecology and catarrhine evolution. In *Old World monkeys: Evolution, systematics and behavior*: 55—95. Napier, J. R. & Napier, P. H. (Eds). New York & London: Academic Press.

Napier, J. R. & Davis, P. R. (1959). The fore-limb skeleton and associated remains of *Proconsul africanus. Foss. Mamm. Afr.* 16: 1—69.

Napier, J. R. & Napier, P. H. (1967). *A handbook of living primates.* New York & London: Academic Press.

Napier, J. R. & Walker, A. C. (1967). Vertical clinging and leaping — a newly recognized category of locomotor behaviour of primates. *Folia primatol.* 6: 204—219.

Oxnard, C. E. (1973a). Some problems in the comparative assessment of skeletal form. *Symp. Soc. Study. Hum. Biol.* 11: 103—125.

Oxnard, C. E. (1973b). Some locomotor adaptations among lower primates: implications for primate evolution. *Symp. zool. Soc. Lond.* No. 33: 255—299.

Prost, J. H. (1965a). A definitional system for the classification of primate locomotion. *Am. Anthrop.* 67: 1198—1214.

Prost, J. H. (1965b). Methodology and gait analysis of monkeys. *Am. J. phys. Anthrop.* 23: 215—240.

Prost, J. H. (1969). A replication study on monkey gaits. *Am. J. phys. Anthrop.* 30: 203—208.

Prost, J. H. (1970). Gaits of monkeys and horses: a methodological critique. *Am. J. phys. Anthrop.* 32: 121—128.

Quris, R. (1975). Ecologie et organisation sociale de *Cercocebus galeritus agilis* dans le Nord-Est du Gabon. *Terre Vie* 29: 337—398.

Ripley, S. (1967). The leaping of langurs: a problem in the study of locomotor adaptation. *Am. J. phys. Anthrop.* 26: 149—170.

Rollinson, J. (1975). *Interspecific comparisons of locomotor behaviour and prehension in eight species of African forest monkey — a functional and evolutionary study.* Ph.D. Thesis: University of London.

Rose, M. D. (1973). Quadrupedalism in primates. *Primates* 14: 337—358.

Rose, M. D. (1974). Postural adaptations in New and Old World monkeys. In *Primate locomotion*: 201—222. Jenkins, F. A. (Ed.). New York & London: Academic Press.

Rose, M. D. (1977). Positional behaviour of olive baboons (*Papio anubis*) and its relationship to maintenance and social activities. *Primates* 18: 59—116.

Schultz, A. H. (1970). The comparative uniformity of the Cercopithecoidea. In *Old World monkeys*: 39—51. Napier, J. R. & Napier, P. H. (Eds). London & New York: Academic Press.

Steindler, A. (1955). *Kinesiology of the human body under normal and patho logical conditions.* Springfield, Illinois: Charles Thomas.

Stern, J. T. & Oxnard, C. E. (1973). Primate locomotion: some links with evolution and morphology. *Primatologia* 4(11): 1—93.

Sukhanov, V. B. (1974). *General system of symmetrical locomotion of terrestrial vertebrates and some features of movement of lower tetrapods.* New Delhi: Amerind Pub. Co. (for Smithsonian Institution, Washington).

Szalay, F. S. & Decker, R. L. (1974). Origins, evolution and function of the tarsus in late Cretaceous Eutheria and Palaeocene primates. In *Primate locomotion*: 223–259. Jenkins, F. A. (Ed.). New York & London: Academic Press.

Szalay, F. S., Tattersall, I. & Decker, R. L. (1975). Phylogenetic relationships of *Plesiadapis* – postcranial evidence. *Contr. Primatol.* 5: 136–166.

Tomita, M. (1967). [A study of the movement pattern of four limbs in walking.] *J. anthrop. Soc. Nippon* 75: 120–146. [In Japanese].

Tuttle, R. H. (1967). Knuckle-walking and the evolution of hominoid hands. *Am. J. phys. Anthrop.* 26: 171–206.

Tuttle, R. H. (1969). Quantitative and functional studies on the hands of the Anthropoidea. I: The Hominoidea. *J. Morph.* 128: 309–364.

Tuttle, R. H. (1970). Postural, propulsive and prehensile capabilities in the cheiridia of chimpanzees and other great apes. In *The chimpanzee* 2: 167 253. Bourne, G. H. (Ed.). Basel: Karger.

von Koenigswald, W. (1979). Ein Lemurenrest aus dem eozänen Ölschliefer der Grube Messel bei Darmstadt. *Paläont. Z.* 53: 63–76.

Walker, A. C. (1967). *Locomotor adaptation in recent and fossil Madagascan lemurs.* Ph.D. Thesis: University of London.

Walker, A. C. (1974). Locomotion adaptations in past and present prosimian primates. In *Primate locomotion*: 349–381. Jenkins, F. A. (Ed.). New York & London: Academic Press.

Walker, A. C. (1979). Prosimian locomotor behaviour. In *The study of prosimian behaviour*: 543–565. Doyle, G. A. & Martin, R. D. (Eds). New York & London: Academic Press.

Yamazaki, N. (1976). [*A study of animal walking by means of simulation methods.*] Ph.D. Thesis: Keio Gijuku University, Yokahama. [In Japanese].

Symp. zool. Soc. Lond. (1981) No. 48, 429–451

Wrist Rotation in Primates: A Critical Adaptation for Brachiators

FARISH A. JENKINS, JR

Department of Biology and Museum of Comparative Zoology, Harvard University, Cambridge, Massachusetts, USA

SYNOPSIS

Arm-swinging and other suspensory behavior of brachiating primates are facilitated by a rotatory midcarpal joint. The primary evidence is derived from cineradiographic study of the skeletal excursions of the forelimb in brachiating spider and woolly monkeys (*Ateles geoffroyi, A. paniscus; Lagothrix lagothricha*). During a swing, the trunk ordinarily rotates approximately 90° about the fixed grip of digits II-V, a movement accommodated principally at the midcarpal joint (~70°), but also by supination of the radius (~20°). Metacarpals II-V, as well as the hamate, capitate and trapezoid, remain in the plane of the swing. The proximal row of carpals, which forms a socket, rotates about the hamate and capitate, which form a ball. In this movement (termed midcarpal supination) the scaphoid (+ centrale) displaces toward the palm, and the triquetral shifts dorsally; the flexor aspect of the forearm, which faces to the side at the beginning of a swing, is thus reoriented to face anteriorly by the end of a swing. Supination of the radius represents counter-rotation to that of the ulna, and relieves the midcarpal joint of accommodating the full 90° turn. Observations on gibbons and a siamang confirm the presence of a midcarpal mechanism essentially identical to that of spider and woolly monkeys. In all these brachiators, a wedge-shaped embrasure between the capitate and trapezoid, which is widest on the palmar side, allows rotation of the scaphoid (+ centrale) around the medial side of the capitate; in gibbons, narrowing of the embrasure at its dorsal end limits rotation in the opposite direction. In macaques (*Macaca* spp.), the shape of this embrasure is reversed, preventing the kind of movement critical to brachiators but permitting rotation in the opposite direction. These findings bear on the interpretation of the locomotor adaptations of extinct primates.

INTRODUCTION

Wrist morphology has emerged as an important component in the study of primate adaptations and evolution. Most influential has been a series of papers by Lewis (1965, 1969, 1971a, b, 1972a, b, 1974) which presents detailed anatomical data on the carpus of monkeys, apes, and

man. Many investigators have subsequently addressed such fundamental problems as primate relationships and locomotor evolution
through comparative studies of primate wrists. Corruccini, Ciochon
& McHenry (1975) and Corruccini (1978) employed multivariate
statistics to assess metric differences in bone size and shape, and
sought to derive phylogenetic implications. O'Connor (1975) evalu-
ated the cercopithecoid wrist in functional terms, and Ziemer (1978)
described osteoligamentous features in the hand of the woolly
monkey that have functional implications. Tuttle (1967, 1969a, b,
1972) and Jenkins & Fleagle (1975) undertook analyses of various
musculoskeletal features of the wrist in conjunction with observations on live primates in an effort to relate structure to locomotor
behavior. These investigators recognized the potential of such studies
for reconstructing facets of the locomotor behavior of extinct primates, as have others who have focused principally on fossils (Conroy & Fleagle, 1972; Corruccini, Ciochon & McHenry, 1976;
Morbeck, 1975, 1976, 1977; O'Connor, 1976; Schön & Ziemer,
1973; Zwell & Conroy, 1973).

The anatomical complexity of the wrist and the diversity of locomotor and other uses of the hand are, taken together, formidable
obstacles to interpreting primate carpal structure in phylogenetic or
adaptive contexts. Thus, despite many thorough studies, there is as
yet no general consensus as to how primate wrist anatomy bears on
questions of systematics, evolution, or function. The central problem
is that of relating structure and function, and this is directly
approachable only through in vivo studies.

Although only a few extant primate species habitually arm-swing,
they none the less represent a distinctive locomotor type. Many
investigations and much debate in recent years have focused on the
question of identifying, both structurally and functionally, the
specific musculoskeletal adaptations that have evolved for brachiation
and other suspensory activities. The present paper provides evidence
that one such adaptation is to be found in the structure of the midcarpal joint and its capacity to rotate. This study also explores the
functional implications that hand support above and below the body
have with respect to rotatory movements of the primate carpus.

Brachiation and other forms of suspensory behavior are employed
regularly by hylobatids (Hylobates, Symphalangus) (Carpenter,
1976; Fleagle, 1976) and by spider monkeys (Ateles) (Eisenberg &
Kuehn, 1966; Mittermeier & Fleagle, 1976; Mittermeier, 1978).
Other genera (e.g., Lagothrix, Brachyteles) are also reported to be
brachiators although their naturalistic behavior is less well docu-

mented (Erikson, 1963). Spider monkeys were selected as the principal subjects for this study because of their tractable nature and availability.

MATERIALS AND METHODS

Three spider monkeys (two female *Ateles geoffroyi*; one male *Ateles paniscus*) and one female woolly monkey (*Lagothrix lagothricha*) were conditioned to brachiate on a rope mill. The mill, consisting of an overhead rope loop driven between two pulleys by an electric motor, permitted the monkeys to brachiate in place at speeds of 3—4 km h^{-1} while their skeletal movements were recorded radiographically on 16 mm cine film at approximately 150 frames/s. For a full description of methods and a figure of this apparatus, see Jenkins, Dombrowski & Gordon (1978). Twelve hundred feet of film, recording various sequences of the hand, arm and shoulder, were studied with a Vanguard film analyser.

In order to obtain better resolution of carpal relationships, conventional 20 × 25 cm radiographic films were taken of spider monkey hands in the following phases of a swing along the rope mill: beginning of swing, mid-swing, and end of swing. The time exposures for this film required that the subject be momentarily stationary (and the rope mill stopped). Thus, radiographs of hand postures at the beginning and end of a swing were taken when the animal paused between swings. Mid-swing postures were recorded by manually halting the animal in mid-swing. The woolly monkey used in this study would not submit to this procedure; nonethless, radiographs were obtained of its hands as the monkey hung in various postures and orientations on the stationary rope mill.

Conventional radiographs were taken of the hands of a young adult female siamang (*Symphalangus syndactylus*), five adult gibbons (two male, two female *Hylobates lar*, one female *H. agilis*), and six macaques (two male, two female *Macaca mulatta*, one male, one female *M. fascicularis*). The siamang and one gibbon, under light sedation, hung in various postures from a horizontal pole. The macaques and the rest of the gibbons, under heavy sedation, were radiographed with their hands passively manipulated into various positions.

Osteological and ligamentous preparations of hands from all species mentioned above, together with preserved forelimbs of several spider monkeys, were studied by conventional anatomical methods.

RESULTS

Wrist Rotation in *Ateles*

The rotatory mechanism employed by brachiating spider monkeys is readily observed in the change of forearm orientation relative to that of the fixed hand grip. At the beginning of a swing (Fig. 1a), the extensor surface of the forearm and dorsum of the hand are seen to be facing laterally. By the end of the swing (Fig. 1b), the dorsum of the hand is still visible, but the forearm is reoriented so that the flexor surface is evident. Thus, rotation takes place between the forearm and a fixed part of the hand, i.e., at the wrist. The details of this movement will now be described with reference to radiographic observations.

Beginning of swing (Figs 2a, 3a, 4a)

At the beginning of a swing, *Ateles* secures a grip by flexing the middle and distal phalanges of digits II to V over the support; the proximal phalangeal segments are not flexed, but contact the side of the support as part of the grip. The metacarpals and distal carpals (hamate, capitate, trapezoid and trapezium) are aligned in the plane

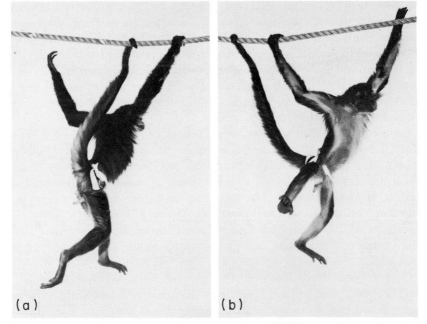

(a)   (b)

FIG. 1. A female spider monkey brachiating on a rope mill. Wrist rotation is evident in the change of forearm orientation from (a) the beginning of a swing to (b) the end of a swing.

of movement and remain so throughout the swing (with the exception of the trapezium and pollical metacarpal). The proximal carpal row (pisiform, triquetral, lunate, scaphoid and centrale) and the radius and ulna lie in a neutral position, as follows. The elongate pisiform is directed palmarward (and thus is seen in end-on projection in Figs 2a, 3a, 4a). The triquetral is positioned in the plane of the hamate along its ulnar border. Proximally, the triquetral forms with the pisiform a shallow fossa for reception of the styloid process of the ulna. The lunate articulates principally with the capitate, overlapping the hamate only slightly. The centrale makes the principal contact with the remainder of the proximal as well as the medial aspect of the capitate; it is overlain by the scaphoid proximally, which articulates with the capitate only by a narrow facet along the scaphoid tubercle. The radial styloid process lies on the medial side of the scaphoid.

Mid-swing (Figs 2b, 3b)

At mid-swing, the *Ateles* forearm is vertically overhead, and carpal rotation is under way. The digits, metacarpals, and distal carpals (except the trapezium and pollical metacarpal) follow the plane of the swing but otherwise maintain the same orientation as at the beginning of the swing. The trapezium and pollical metacarpal displace palmarward.

The principal rotation takes place at the midcarpal joint with the proximal carpal row, which forms a socket, moving about the hamate and capitate, which form a ball. As the triquetral moves onto the dorsal aspect of the hamate's articular surface, the centrale and scaphoid shift palmarward around the capitate. In this movement, the centrale passes through an embrasure between the capitate facet and an opposing facet on the trapezoid. The lunate approximates the center of the rotation, but also shifts to a more dorsal position on the "ball" formed by the hamate and capitate. Both the radius and ulna follow this movement; thus, by mid-swing, the radial and ulnar styloid processes displace palmarward and dorsally, respectively, relative to the distal carpal row.

End of swing (Figs 2c, 3c, 4b)

As the swing is completed, the wrist achieves maximum rotation. The triquetral overlaps the dorsal aspect of the hamate facet. The long axis of the pisiform comes to lie in the plane of movement (that is, with the apex of the bone pointing forward, rather than laterally as in the beginning of the swing). The centrale, together with the scaphoid with which it is ligamentously conjoined, shifts farther toward the palmar side between the capitate and trapezoid. The lunate con-

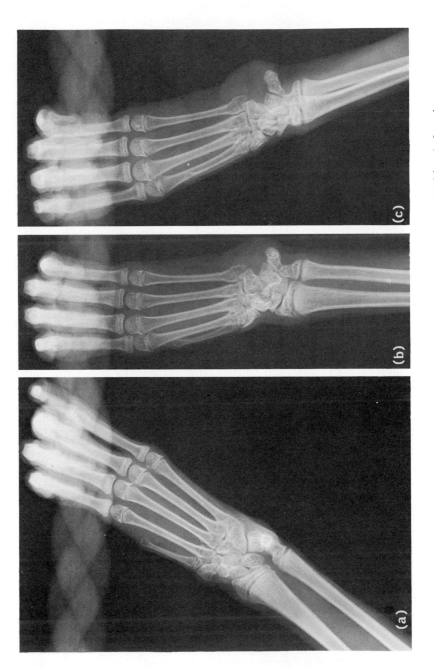

FIG. 2. Radiographs of a spider monkey wrist in postures used at (a) the beginning, (b) middle, and (c) end of a swing.

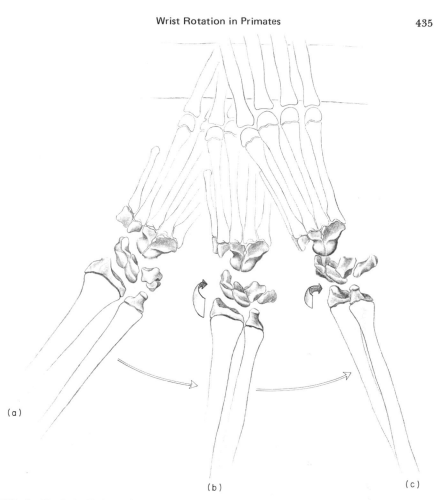

FIG. 3. "Exploded" view of carpal relationships in a spider monkey, seen from dorsal aspect in (a) the beginning, (b) middle, and (c) end of a swing. Reconstruction based on cineradiographic and other radiographic observations.

tinues to rotate approximately in place, the dorsal half of its distal articular facet sliding onto the dorsal aspect of the capitate. Both radial and ulner styloid processes maintain their relation to the proximal carpal row throughout this movement; thus, at the end of a swing they lie in a plane that is nearly normal to the direction of the swing (Fig. 2c), whereas initially they lay in the plane of the metacarpals and were aligned in the direction of movement (Fig. 2a).

The rotation thus described, which permits the forearm and proximal carpal row to achieve a supinated position relative to the fixed digits and distal carpal row, may be appropriately termed midcarpal supination. This movement occurs as a result of the spider monkey holding onto the right side of a support with the right hand, or the left side with the left hand (an ipsilateral grip). Regularly, however,

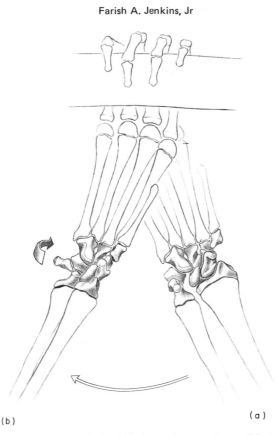

(b)   (a)

FIG. 4. Palmar view of carpal relationships in a spider monkey at (a) the beginning and (b) the end of a swing, as interpreted from cineradiographic and other radiographic observations.

spider monkeys and other brachiators use a reverse, or contralateral, grip (e.g., taking hold of the left side of the support with the right hand); in this case the hand is initially hyperpronated, probably in part assisted by medial rotation of the humerus at the glenohumeral joint, and substantial midcarpal rotation does not occur during the swing.

Wrist Rotation in *Lagothrix*

In this study, the woolly monkey used a style of brachiation that differed in certain respects from that of the spider monkeys. Most noticeable was the tendency to maintain some degree of elbow flexion throughout the swing, thus avoiding a fully pendular excursion of the trunk. Nonetheless, cineradiographic records of the woolly monkey brachiating on the rope mill, as well as conventional radiographic films taken as the subject suspended in various postures from a stationary overhead support (Fig. 5), demonstrate a midcarpal rotation essentially identical to that in *Ateles*.

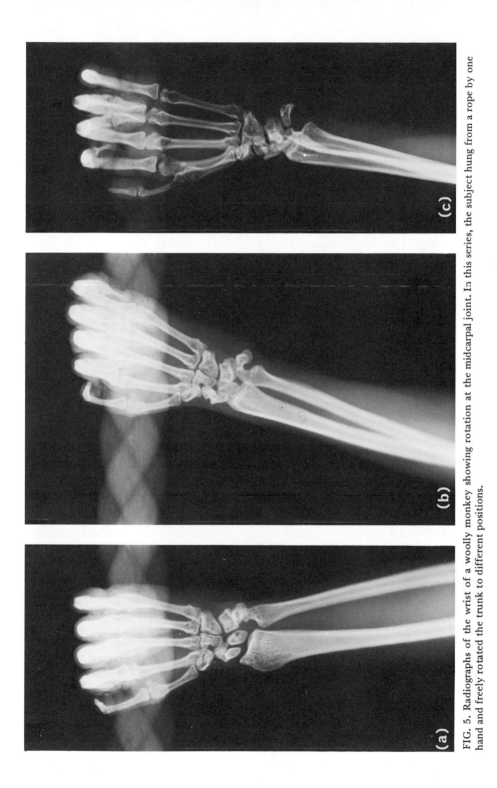

FIG. 5. Radiographs of the wrist of a woolly monkey showing rotation at the midcarpal joint. In this series, the subject hung from a rope by one hand and freely rotated the trunk to different positions.

(a)   (b)   (c)

Trunk and Forelimb Excursions in *Ateles*

During brachiation the spider monkey's trunk normally rotates approximately 90°. At the beginning of a swing, the chest faces about 45° away from the line of progression and toward the side of the free (non-suspending) limb (Fig. 1a). By the end of the swing, the trunk has turned toward the side of the suspending limb, again at about 45° to the line of progression (Fig. 1b). Greater rotation is employed (up to 150°) to reach farther, and typically occurs during rapid brachiation. When a spider monkey pauses between swings, the chest faces 90° from the line of progression. Thus, with a single swing, a spider monkey is capable of rotating the trunk no fewer than 180°.

The 90° of rotation that typically occur between the swinging trunk and the first digits could be accommodated potentially at three sites: between the scapula and thorax, at the glenohumeral joint, and at the wrist (including the proximal radiohumeral joint, as in pronation-supination).

Jenkins *et al.* (1978) documented that the three principal components of scapular excursion in brachiating spider monkeys are rotation, caudal translation, and medial translation. The 15° rotation during a swing increases the rostral orientation of the glenoid as the body passes below and ahead of the arm. Caudal translation of the scapula, most of which takes place on the upswing, may contribute to the propulsive thrust of the body (for example, by contraction of the latissimus dorsi; see Stern, Wells, Vangor & Fleagle, 1976) and, at the same time, shorten the distance between the centre of gravity and the handhold, thus reducing loss of momentum due to gravitational deceleration (see Fleagle, 1974). However, neither rotation nor caudal translation of the scapula contributes to truncal rotation. Medial translation of the scapula could possibly contribute to truncal rotation if the scapula were displaced from the side of the thorax to the back. But Jenkins *et al.* (1978) demonstrated that the spider monkey scapula is positioned on the dorsum throughout the swing and the small amount of medial translation involves no major reorientation of the plane of the scapula on the thorax. They concluded that medial translation aids in positioning the shoulder beneath the overhead support, and tends to reduce lateral oscillation of the body. Scapulo-thoracic excursion, therefore, does not appear to contribute significantly to any rotation of the trunk.

Cineradiographic evidence discounts the possibility that truncal rotation, at least the 90° that normally occur during brachiation at moderate speeds, is accommodated by rotation at the glenohumeral joint. At the beginning of a swing, the long axis of the humerus intersects the glenoid fossa, and the lesser tuberosity is aligned approximately with the caudal (axillary) border of the scapula (Fig. 6a). By the end of the swing, the long axis of the humerus intersects the plane of the scapula at approximately 90°, but the lesser tuberosity remains aligned with the caudal border of the scapula (Fig. 6b). Thus, the humerus abducts substantially, but does not rotate relative to the scapula. It is possible that glenohumeral rotation occurs during rapid brachiation, or during attempts to span unusually long distances; these conditions were not observed. As noted above, however, glenohumeral rotation may be employed to place the hand in the hyperpronated posture initially used to secure a contralateral grip (e.g., the right hand taking hold from the left side of the support), in which case the necessity of substantial midcarpal rotation during the swing is avoided.

Rotation of the trunk about an ipsilateral grip, therefore, is primarily accommodated by movements in the forearm and wrist. The trochlea ensures that the ulna conjointly rotates with the humerus through approximately 90°; thus, the dorsum of the forearm faces laterally at the beginning of a swing, and posteriorly at the end of a swing (Figs 1 and 6). The radius participates in this movement but does not turn through a full 90° because it simultaneously counter-rotates into a position of supination. This complex movement is best understood with reference to the radial tuberosity. When the grasp initially is made, the hand is pronated; in this posture the radial tuberosity is directed *toward* the ulnar shaft (Fig. 6a). During the swing, the radius supinates so that the radial tuberosity rotates *away* from the ulnar shaft (Fig. 6b). This excursion differs from conventional supination, which involves movement of the entire hand and radius, because only the proximal carpal row and radius rotate. Although the forearm as a whole swings forward and turns, radial supination provides, in effect, a retarding counter-rotation. The distal end of the radius thus lags behind the distal end of the ulna (Fig. 6b), consequently reducing the amount of rotation necessary at the wrist.

The relative contributions of radial and midcarpal supination to accommodating the 90° turn of the trunk about the fixed hand were estimated with the aid of a revolving microscope stage. An *Ateles* ulna, radius, and proximal row of carpals were mounted in the middle of the stage, and oriented to a line-of-sight in a position replicated from *in vivo* radiographic projections characteristic of the beginning of the swing (that is, at the time the hand initially grasps the overhead support, as in Fig. 2a). The stage was then turned until the orientation matched the radiographic projection typical of the terminal phase of the swing (as in Fig. 2c). This method confirmed that the ulna is reoriented approximately 90°, and established that

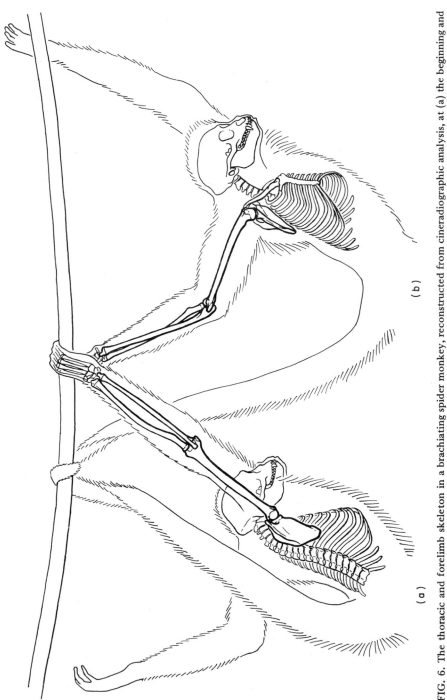

(a)

(b)

FIG. 6. The thoracic and forelimb skeleton in a brachiating spider monkey, reconstructed from cineradiographic analysis, at (a) the beginning and (b) the end of a swing.

the distal end of the radius is reoriented about 70° ± 5°. Therefore, of the full 90° of rotation that occur between the arm (humerus and ulna) and the metacarpals and distal carpal row (which are fixed in the plane of the swing), 70° ± 5° of rotation occur at the midcarpal joint and the remaining 20° ± 5° are accommodated by radial supination.

Wrist Rotation in Hylobatids and Macaques

Observation of brachiating gibbons reveals that the forearm rotates relative to the fixed hand in a manner similar to that seen in spider monkeys. Baldwin & Teleki's (1976: fig. 97) photograph of a hanging gibbon is directly comparable in terms of forearm posture to that shown by the spider monkey in Fig. 1b. Carpenter (1940) first called attention to the fact that " . . . brachiation involves an extraordinary rotation of the wrist, arm and shoulder." Noting that "the patterns of pronation and supination are complex," Carpenter did not attempt to analyse these movements although he continued to recognize their importance to the unique locomotion of hylobatids (Carpenter, 1976). Radiographs of the wrists of a sedated siamang (Fig. 7a, b) and a gibbon, hanging from a horizontal pole, show that midcarpal supination is at least partly responsible for the observed rotation.

The ability of *Hylobates* to accommodate substantial rotation at the midcarpal joint was further confirmed by manipulating the hands of sedated individuals. With the forearm held fixed and the wrist in neutral position (Fig. 7c), the hand may be readily supinated to the point where the plane of the metacarpals is rotated about 80° (or approximately at a right angle to the neutral position; Fig. 7d). The scaphoid, lunate, and triquetral essentially retain their original positions during this manoeuver; the pisiform shifts only slightly. Rotation is effected at the midcarpal joint. The nature of this movement is revealed by the fact that the soft tissue "space" between the capitate and hamate seen in Fig. 7c is obliterated as those bones are superimposed by rotation (Fig. 7d); metacarpals III, IV, and V are carried with them in this movement. The trapezoid (with metacarpal II, to which it is firmly joined) and trapezium (with the pollex) also participate in this rotation.

Attempts at counter-rotation from the neutral position to a "hyperpronated" posture yield little gross movement of the hand (Fig. 7e). The proximal carpal row remains stationary. The capitate is displaced slightly distally with respect to the hamate, but the fact that the "tissue space" between the two bones is still visible is evidence that no significant rotation has taken place as in the case of "hypersupination" (Fig. 7d). Furthermore, metacarpals III, IV, and V remain aligned in the neutral position. Metacarpal II and the

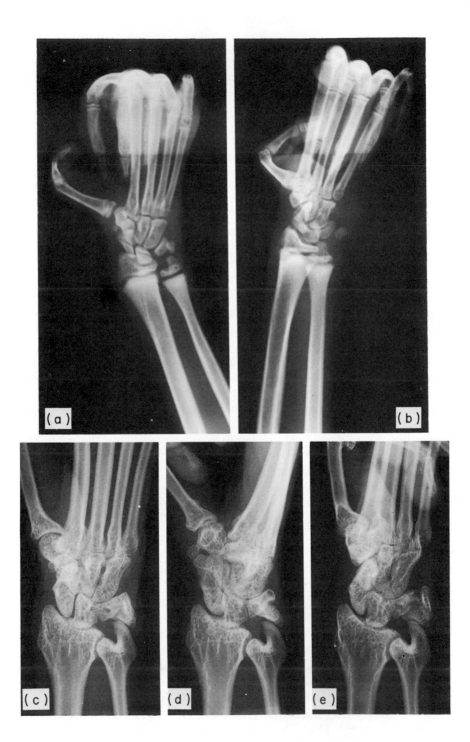

trapezoid are shifted somewhat towards the center of the palm by this manoeuver, and thus the adjacent margins of metacarpals II and III are seen to overlap. However, the hand as a whole cannot be manipulated significantly farther into pronation beyond that achieved in neutral position.

Similar manipulations carried out on sedated rhesus monkeys (*Macaca mulatta*) reveal rotatory capabilities quite different from those in gibbons. With the forearm held fixed and the wrist in neutral position (Fig. 8a), attempts to supinate the hand yield relatively little rotational movement (Fig. 8b). The proximal carpal row remains stationary. The hamate displaces palmarward relative to the capitate, resulting in a "cupping" of the palm by virtue of the volar shift of metacarpals IV and V, which are carried with it. The capitate and trapezoid are unaffected by this manoeuver, and thus metacarpals II and III retain essentially their original positions.

Rotation of the macaque wrist in the opposite direction (i.e., "hyperpronation") is freely effected by passive manipulation, in contrast to the condition in gibbons. Figure 8c shows that, as a result of this manoeuver, the margins of the scaphoid, lunate and triquetral are seen to overlap, and thus the row as a whole has undergone slight rotation. The pisiform is completely reoriented. However, the major rotatory movement appears to occur at the midcarpal joint. The hamate, capitate and trapezoid are all superimposed to a significant degree, and thus the metacarpals II, III and IV come to lie more or less in the same plane. The freedom with which wrists of rhesus monkeys "hyperpronate" appears comparable to the amount of "hypersupination" possible in the wrists of gibbons and other brachiators.

Wrist Morphology in Relation to Rotation

Several features of the midcarpal joint differentiate the wrists of brachiators from those of other primates. The most obvious is the configuration of the proximal articular facets of the capitate and hamate. In spider monkeys and gibbons (Fig. 9), these bulbous surfaces have a relatively uniform radius of curvature; together they appear to represent half of a ball-and-socket joint. In comparison, the

FIG. 7. (a and b) Radiographs of the wrist in a sedated siamang showing rotation comparable to that in Figs 2 and 5. (c-e) Radiographs in postero-anterior view demonstrating wrist rotation by passive manipulation in a sedated male gibbon, *Hylobates lar* (LEMSIP No. 1). Note that the forearm has been fixed in the same position in all three views. (c) right hand in a prone (neutral) position; (d) the hand has been manipulated into supination; (e) an attempt has been made to rotate the hand from the neutral (prone) position in the opposite direction to that shown in (d), i.e., to "hyperpronation."

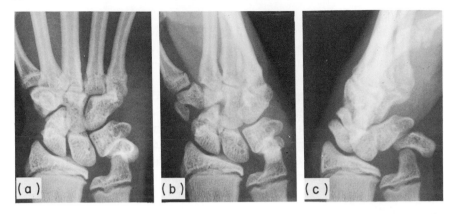

FIG. 8. Radiographs in a postero-anterior view demonstrating wrist rotation by passive manipulation in a sedated male rhesus monkey, *Macaca mulatta* (LEMSIP No. 117). (a) right hand in a prone (neutral) position. (b) an attempt has been made to rotate the hand from the neutral position to "hypersupination"; little rotatory movement occurs. (c) the hand is readily manipulated into "hyperpronation."

same facets in macaques are less symmetrical in outline; furthermore, the curvature of the surfaces is less uniform.

Features of the midcarpal joint that appear to determine the difference in rotation exhibited by brachiators and *Macaca* are the shape of the capitate facet, the orientation of the proximal facet of the trapezoid, and the configuration of the space between them. In spider monkeys and gibbons, the centrale and part of the scaphoid pass through an embrasure between the capitate and trapezoid (Fig. 9). The wedge-shaped embrasure widens transversely toward the palm, and is curved by virtue of the hemispheroidal shape of the capitate facet and the opposing (but less pronounced) concavity of the trapezoid facet. Thus, the embrasure guides the rotatory movement of the scaphoid (+ centrale) toward the palm. In gibbons, counter-rotation (i.e., midcarpal pronation) is limited by impaction of the wider, palmar half of the centrale in the narrower, dorsal end of the capitate-trapezoid embrasure. In spider monkeys, the dorsal end of the capitate-trapezoid embrasure is relatively unconstricted and the corresponding wedge-shaped articular facets of the centrale are of a uniform shape; it is therefore possible that *Ateles* is capable also of significant midcarpal pronation, although this was not specifically tested during the study.

In *Macaca*, the medial aspect of the capitate facet is oriented dorsally as well as medially, and overall is much flatter than the hemispheroidal form of brachiators; the joint surface is thus not structured for substantial rotation. Furthermore, the embrasure between the capitate and trapezoid facets narrows toward the palm.

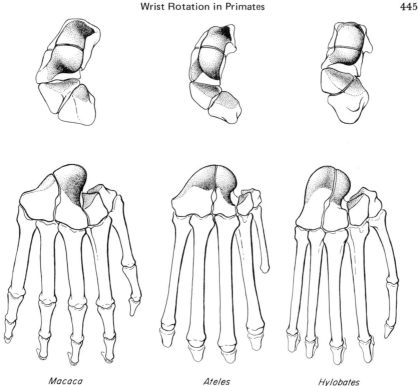

Macaca   Ateles   Hylobates

FIG. 9. The articular surfaces of the distal carpal row in a macaque, spider monkey and gibbon seen in proximomedial (top row) and proximodorsal (bottom row) views, showing the shapes of the capitate-trapezoid embrasure and the proximal capitate facet. The extension of the triquetral facet onto the dorsal surface of the hamate, a characteristic of spider monkeys and gibbons which differs from that of macaques, is also shown.

The scaphoid (+ centrale), if moved on the capitate in the direction of the palm (as occurs in midcarpal supination), becomes impacted in the narrowed, palmar end of the capitate-trapezoid embrasure. Counter-rotation (midcarpal pronation) occurs freely without any bony impediment.

Another feature related to midcarpal supination is the configuration of the hamate facet for the triquetral. In spider monkeys and gibbons, this facet extends onto the dorsal aspect of the hamate (Fig. 9), and thus accommodates the rotatory, dorsal displacement of the triquetral during midcarpal supination (Figs 2c, 3c, 4b). In contrast, the hamate facet for the triquetral in macaques does not extend dorsally, but is confined to the proximolateral (i.e., ulnar) aspect of the bone (Fig. 9); the triquetral therefore cannot be rotated onto the dorsum of the hamate as occurs in midcarpal supination in brachiators.

DISCUSSION

Midcarpal and radial supination permit brachiators to maintain a stable, hook-like grip while the body rotates to position the free limb for the next handhold. Rotation of the trunk has two effects. First, the distance spanned by successive handholds is increased by moving one shoulder and limb ahead of the other (see Avis, 1962); in those species that employ a pendulum-like swing, truncal rotation may also accelerate the center of gravity on the downswing, lengthening the arc of the swing, and thus extend the reach to the next handhold. Second, in feeding and other suspensory activities, the repertoire of postures and especially the range of the free limb are substantially increased by the rotational freedom of the wrist. The parallel evolution of this mechanism in New World and Old World primates is testimony to the adaptive significance of midcarpal supination for brachiators.

The identification of a rotational mechanism in the wrists of brachiators raises several questions. First, to what degree is midcarpal supination possible in other primates, especially those which are predominantly arboreal and therefore may employ suspensory behavior? The larger members of the family Cebidae are of interest from this perspective because of the many structural similarities they share with their ateline relatives. Osteologically, however, the midcarpal joints of *Alouatta belzebul*, *Cebus apella*, *Pithecia chiropotes* and *Cacajao rubicundus* do not clearly possess those features that facilitate midcarpal supination in *Ateles*. Typically, the proximal capitate facet is flatter, and lacks a uniform, hemispheroidal curvature. The triquetral facet of the hamate does not extend onto the dorsal surface of the bone. In *Alouatta*, the capitate-trapezoid embrasure is narrower on the palmar than on the dorsal side, as in macaques. Field studies by Richard (1970) of *Alouatta villosa* and Mendel (1976) of *Alouatta palliata* confirm that these howlers do not brachiate; however, both studies emphasize suspensory behaviors during feeding ("swing and grasp" (Richard, 1970), and hanging (Mendel, 1976)), in which at least some rotational ability might be employed. In both *Cebus* and *Pithecia*, the structure of the midcarpal joint is sufficiently different that the embrasure analogy is inappropriate; however, a tongue-and-groove-like interlocking of the centrale between the capitate and trapezoid would appear to restrict rotation. Only in *Cacajao* are the relationships of the centrale in the capitate-trapezoid embrasure suggestive of the possibility of limited midcarpal supination. Inferences such as these will remain tenuous until the wrists of living representatives of these genera can be studied.

Among pongids, the orang-utan possesses carpal features that appear to be consistent with a rotational ability appropriate for its suspensory habits. In contrast, the wrist of chimpanzees is not readily classified in either of the "brachiator" or "pronograde" categories by the structural criteria employed here (see Jenkins & Fleagle, 1975: fig. 1); furthermore, radiographs of the wrist manipulated in two sedated individuals (as in Figs 7c-e, 8) reveal that both midcarpal pronation and supination are possible. The observations by Avis (1962) on the suspensory behavior of gibbons and great apes provide evidence that, if not conclusive, is at least suggestive. She called attention to some basic similarities in arm-swinging by apes, among which is *supination* of the forearm. However, she interpreted the glenohumeral joint as the site of rotation of the trunk about the grasping arm. The one figure in her paper that illustrates brachiation (an orang-utan on a single and a multiple overhead support) is not sufficiently detailed or complete to assess the relative contributions of glenohumeral rotation and forearm supination (and the concomitant midcarpal supination) to truncal rotation.

The possible significance of midcarpal pronation in macaques is also interesting. Whether such rotation, observed in this study on sedated monkeys, is actually employed by cercopithecoids remains to be demonstrated. It appears possible that midcarpal pronation accommodates a rotatory component between the body and a fixed hand on a branch, and is a consequence of gripping a substrate below the body rather than an overhead support. For each of these cases — cebids, pongids, and cercopithecoids — one point is worth emphasizing: accurate assessment of the rotational capability of the wrist cannot be based on morphology alone, but most involve *in vivo* techniques such as radiography. Preferably the wrist should be studied in free-ranging animals; passive manipulation of sedated or dead specimens may not exactly duplicate normal movements.

Another question warranting investigation is the degree to which rotational mechanisms involve related modifications of the carpal ligaments and the muscles of the forearm and hand.

The identification of major bony features related to carpal rotation in primates offers a new basis for interpreting the locomotor adaptations of extinct forms. For example, the Miocene hominoid *Dryopithecus africanus* has been cast variously as a knuckle-walker, brachiator, or palmigrade quadruped (for a review see O'Connor, 1976). In one specimen from Rusinga (Kenya National Museum RU-2036), a nearly complete carpus is preserved with the forelimb skeleton, and associated skull fragments ensure positive taxonomic identification (Napier & Davis, 1959). Unfortunately, the trapezoid is lacking and therefore an assessment of the shape of the capitate-trapezoid embrasure is impossible. However, a cast of the capitate reveals that the proximal facet for the scaphoid (+ centrale) does not have the uniform radius of curvature characteristic of brachiators, although the facet as a whole *appears* to have been oriented somewhat more medially (i.e. radially) than in macaques. The possibility that in this subadult specimen the carpals may not have been fully ossified or, additionally, may have been plastically deformed after burial, warrants caution. Another capitate, an isolated bone from Songhor tentatively referred to *Dryopithecus* (Kenya National Museum SO-1002), represents a more robust and probably older individual. A cast of this specimen unequivocally displays a proximal facet of the pronograde, rather than the brachiator, type. Other isolated capitates available as casts (Kenya National Museum RU-1907, SO-1000, SO-1001) are approximately three-quarters the size of the two larger specimens, and present further variation of the basic facet form observed (some of which may be postmortem artifacts). If these isolated capitates actually represent *Dryopithecus*, the evidence corroborates the interpretation of the KNM-RU-2036 as a primate that probably did not possess the abilities of a brachiator.

The results of the present study not only invite re-examination of carpal structure in fossil primates, but also point to the need for futher *in vivo* investigations of the wrists in living monkeys and apes, not for the purpose of assigning their structure and habits to man-made classifications, but because only through detailed investigation of functional anatomy and behavior can we begin to appreciate the significance and richness of primate diversity.

ACKNOWLEDGEMENTS

For assistance in various phases of this study, I express sincere thanks to Dr John G. Fleagle, Mr J. Q. Landers, Mr Philip J. Dombrowski and Ms Elizabeth P. Gordon. I am grateful to Dr Daniel R. Snyder (Neurobehavioral Research Unit, Yale University), Dr Wendell H. Niemann (Laboratory for Experimental Medicine and Surgery in Primates, New York University), and Dr William C. Satterfield (Boston Zoological Society) for making available the live primates used in this study. Dr John G. Fleagle (Department of Anatomical Sciences, State University of New York at Stony Brook) and Dr M. E. Morbeck (Department of Anthropology, University of Arizona) provided constructive reviews of the manuscript. I thank Mr A. H. Coleman and Ms P. Chandoha for photographic work, Mr Laszlo Meszoly for rendering the figures, and Mrs Lillian L. W. Maloney and Mrs Alice Blatchley for secretarial assistance. This study was supported in part by NSF Grant PCM 73-00702.

REFERENCES

Avis, V. (1962). Brachiation: The crucial issue for man's ancestry. *SWest. J. Anthrop.* **18**: 119–148.

Baldwin, L. A. & Teleki, G. (1976). Patterns of gibbon behavior on Hall's Island, Bermuda. In *Gibbon and Siamang* **4**: 21–105. Rumbaugh, D. M. (Ed.). Basel: S. Karger.

Carpenter, C. R. (1940). A field study in Siam of the behavior and social relations of the gibbon (*Hylobates lar*). *Comp. Psychol. Monogr.* **16**(5): 1–212.

Carpenter, C. R. (1976). Suspensory behavior of gibbons *Hylobates lar*, a photoessay. In *Gibbon and Siamang* **4**: 1–20. Rumbaugh, D. M. (Ed.). Basel: S. Karger.

Conroy, G. C. & Fleagle, J. G. (1972). Locomotor behaviour in living and fossil pongids. *Nature, Lond.* **237**: 103–104.

Corruccini, R. S. (1978). Comparative osteometrics of the hominoid wrist joint, with special reference to knuckle-walking. *J. hum. Evol.* **7**: 307–321.

Corruccini, R. S., Ciochon, R. L. & McHenry, H. M. (1975). Osteometric shape relationships in the wrist joint of some anthropoids. *Folia primatol.* **24**: 250–274.

Corruccini, R. S., Ciochon, R. L. & McHenry, H. M. (1976). The postcranium of Miocene hominoids: Were dryopithecines merely "dental apes"? *Primates* **17**: 205–223.

Eisenberg, J. F. & Kuehn, R. E. (1966). The behavior of *Ateles geoffroyi* and related species. *Smithson. misc. Collns* **151**(8): 1–63.

Erikson, G. E. (1963). Brachiation in New World monkeys and in anthropoid apes. *Symp. zool. Soc. Lond.* No. 10: 135–164.

Fleagle, J. G. (1974). Dynamics of a brachiating siamang [*Hylobates (Symphalangus) syndactylus*]. *Nature, Lond.* **248**: 259–260.

Fleagle, J. G. (1976). Locomotion and posture of the Malayan siamang and implications for hominoid evolution. *Folia primatol.* **26**: 245–269.

Jenkins, F. A., Jr., Dombrowski, P. J. & Gordon, E. P. (1978). Analysis of the shoulder in brachiating spider monkeys. *Am. J. phys. Anthrop.* **48**: 65–76.

Jenkins, F. A., Jr. & Fleagle, J. G. (1975). Knuckle-walking and the functional anatomy of the wrists in living apes. In *Primate functional morphology and evolution*: 213–227. Tuttle, R. H. (Ed.). The Hague: Mouton Publishers.

Lewis, O. J. (1965). Evolutionary change in the primate wrist and inferior radioulnar joints. *Anat. Rec.* **151**: 275–285.

Lewis, O. J. (1969). The hominoid wrist joint. *Am. J. phys. Anthrop.* **30**: 251–267.

Lewis, O. J. (1971a). Brachiation and the early evolution of the Hominoidea. *Nature, Lond.* **230**: 577–578.

Lewis, O. J. (1971b). The contrasting morphology found in the wrist joints of semibrachiating monkeys and brachiating apes. *Folia primatol.* **16**: 248–256.

Lewis, O. J. (1972a). Osteological features characterizing the wrists of monkeys and apes, with a reconsideration of this region in *Dryopithecus (Proconsul) africanus. Am. J. phys. Anthrop.* **36**: 45–58.

Lewis, O. J. (1972b). Evolution of the hominoid wrist. In *The functional and evolutionary biology of primates*: 207–222. Tuttle, R. (Ed.) Chicago: Aldine-Atherton, Inc.

Lewis, O. J. (1974). The wrist articulations of the Anthropoidea. In *Primate locomotion*: 143–169. Jenkins, F. A., Jr. (Ed.). New York and London: Academic Press.

Mendel, F. (1976). Postural and locomotor behavior of *Alouatta palliata* on various substrates. *Folia primatol.* **26**: 36–53.

Mittermeier, R. A. (1978). Locomotion and posture in *Ateles geoffroyi* and *Ateles paniscus. Folia primatol.* **30**: 161–193.

Mittermeier, R. A. & Fleagle, J. G. (1976). The locomotor and postural repertoires of *Ateles geoffroyi* and *Colobus guereza*, and a reevaluation of the locomotor category semibrachiation. *Am. J. phys. Anthrop.* **45**: 235–255.

Morbeck, M. E. (1975). *Dryopithecus africanus* forelimb. *J. hum. Evol.* **4**: 39–46.

Morbeck, M. E. (1976). Problems in reconstruction of fossil anatomy and locomotor behaviour: The *Dryopithecus* elbow complex. *J. hum. Evol.* **5**: 223–233.

Morbeck, M. E. (1977). The use of casts and other problems in reconstructing the *Dryopithecus (Proconsul) africanus* wrist complex. *J. hum. Evol.* **6**: 65–78.

Napier, J. R. & Davis, P. R. (1959). The fore-limb skeleton and associated remains of *Proconsul africanus. Fossil Mammals Afr.* **16**: 1–69.

O'Connor, B. L. (1975). The functional morphology of the cercopithecoid wrist and inferior radioulnar joints, and their bearing on some problems in the evolution of the Hominoidea. *Am. J. phys. Anthrop.* **43**: 113–121.

O'Connor, B. L. (1976). *Dryopithecus (Proconsul) africanus*: Quadruped or non-quadruped? *J. hum. Evol.* **5**: 279–283.

Richard, A. (1970). A comparative study of the activity patterns and behavior of *Alouatta villosa* and *Ateles geoffroyi. Folia primatol.* **12**: 241–263.

Schön, M. A. & Ziemer, L. K. (1973). Wrist mechanism and locomotor behavior of *Dryopithecus (Proconsul) africanus. Folia primatol.* **20**: 1–11.

Stern, J. T., Jr., Wells, J. P., Vangor, A. K. & Fleagle, J. G. (1976). Electromyography of some muscles of the upper limb in *Ateles* and *Lagothrix. Ybk phys. Anthrop.* **20**: 498–507.

Tuttle, R. H. (1967). Knuckle-walking and the evolution of hominoid hands. *Am. J. phys. Anthrop.* **26**: 171–206.

Tuttle, R. H. (1969a). Quantitative and functional studies on the hands of the Anthropoidea. I. The Hominoidea. *J. Morph.* **128**: 309–363.

Tuttle, R. H. (1969b). Knuckle-walking and the problem of human origins. *Science, N.Y.* **166**: 953–961.

Tuttle, R. H. (1972). Functional and evolutionary biology of hylobatid hands and feet. In *Gibbon and Siamang* **1**: 136–206. Rumbaugh, D. M. (Ed.). Basel: S. Karger.

Ziemer, L. K. (1978). Functional morphology of forelimb joints in the woolly monkey *Lagothrix lagothricha. Contr. Primatol.* **14**: 1–130.

Zwell, M. & Conroy, G. C. (1973). Multivariate analysis of the *Dryopithecus africanus* forelimb. *Nature, Lond.* **244:** 373—375.

Author Index

Numbers in italics refer to pages in the References at the end of each article

A

Abbott, B. C., 233, *237*
Abbott, L. C., 362, *373*
Able, K. P., 215, *217*
Aiello, L. C., 331, 333, 345, 347, 348, 354, *357*, 380, 408, 409, 413, 415, 416, 419, *423*
Alaba, J. O., 97, *108*
Alexander, R. McN., 73, 88, *106*, 185, *195*, *196*, 230, *236*, 243, 245, 248, 250, *263*, 269, 271, 272, 273, 274, 275, 276, 277, 278, 280, 281, 282, 283, 285, *286*, *287*, 313, *328*, 352, *357*, 378, 418, *423*
Aleyev, Yu. G., 48, *51*, 150, 151 *169*
Altenbach, J. S., 166, *169*
Amelink-Koutstall, J. M., 103, *112*
Andersson, M., 182, *196*
Andres, P., 419, *424*
Anton, L., 46, *51*
Ariano, M. A., 290, 293, 296, *303*
Armstrong. R. B., 234, 235, 289, 290, 292, 293, 294, 295, 296, 297, 298, 300, 301, 302, 303, *303*, *304*
Arnold, G. P., 53, *68*
Ashton, E. H., 362, 367, *373*, *375*, 379, *423*
Avis, V., 379, *423*, 446, 447, *449*
Awan, M. Z., 235, *236*

B

Badrian, A., 360, *374*
Badrian, N., 360, *374*
Baggott, D. G., 310, 315, 316, 321, *328*, *329*
Baharar, D., 230, 231, *238*
Bainbridge, R., 10, *26*, 66, *68*, 120, *134*
Baird, D., 262, *263*
Bakker, R. T., 240, 242, *263*
Baldwin, K., 290, 295, *303*
Baldwin, L. A., 441, *449*

Balsley, B. B., 215, *217*
Bando, T., 121, *134*
Bannister, E. W., 234, *236*
Bárány, M., 94, *106*
Barclay, O. R., 253, *263*
Barets, A., 84, 85, 87, *106*
Barnard, R. J., 289, 290, 291, 295, *303*, *304*
Barnett, C. H., 246, 255, *263*
Bartel, D. L., 278, *287*
Bartelemez, G. W., 129, *134*
Basmajian, J. V., 242, 246, 251, 255, *265*, 362, 368, *374*
Batchelor, G. K., 40, *51*
Batty, R. S., 53, 105, *106*
Baudinette, R. V., 221, 222, *236*
Bayers, J. H., 306, *328*
Bearder, S. K., 412, *423*
Beis, I., 95, *106*
Belterman, Th., 247, 249, 253, 254, *264*
Bennet-Clark, H. C., 280, *286*
Bennett, A. F., 72, 99, *106*, 240, 241, 243, *263*
Bennett, M. V. L., 130, *135*
Benzonana, G., 91, *106*
Berger, M., 158, *169*
Bergman, R. A., 75, 87, 88, *106*
Bernstein, J. J., 128, 129, *134*
Bernstein, M. H., 164, *169*, 222, *236*
Best, A. C. G., 85, *106*
Betters, B., 289, 296, 297, 299, 301, 302, *304*
Biegert, J., 336, 337, 340, 342, *357*
Biggs, N. L., 295, 296, *304*
Bilinski, E., 100, 103, *106*, *110*
Bilo, D., 158, *169*
Black, E. C., 96, 101, 103, *106*, *112*
Blackwell, F., 215, *218*
Blagosklonov, K. N., 154, *169*
Blake, P. D., 306, *329*
Blake, R. W., 29, 41, 42, 43, 44, 45, 46, 47, 48, *51*, 53, 60, *68*
Blanton, P. L., 295, 296, *304*
Blaxter, J. H. S., 200, *217*

Subject Index

Numbers in italics refer to figures